SECULARISM

The Hidden Origins of Disbelief

Mike King

James Clarke & Co.

James Clarke & Co
P.O. Box 60
Cambridge
CB1 2NT

www.jamesclarke.co.uk
publishing@jamesclarke.co.uk

First Published in 2007

ISBN: 978 0 227 17245 2

British Library Cataloguing in Publication Data
A catalogue record is available from the British Library

Contents

For Douglas Harding
1909-2007

Introduction

This book sets out a radically new exploration of the question: what are the origins of the secular mind? This question has an urgency at the start of the twenty-first century that modernist thinkers of the twentieth century could not have anticipated. According to mainstream Western ways of thinking, religion was a relic of pre-Enlightenment culture, and its disappearance was inevitable, as a fog dispersed by the sun. This 'sun', according to the view of the secular mainstream, was science and rationalism. But 9/11, the rise of militant Islam, and the rise of the religious right in America, suddenly made the secular assumption of the death of religion untenable. It became necessary, urgent, even, to attempt a new understanding of religion: everywhere the question of religion has had new debating grounds. But what is missing – and which this book attempts to address – is a new debate about ***secularism***.

Painful as it might be, this book takes apart the assumption that science and rationalism are as the sun to the fog of religion and spirituality. The uncomfortable fact needs acknowledgement: that a huge range of scientists and thinkers, from the Enlightenment onwards, find no contradiction between their outer pursuit of science and rationalism, and their inner pursuit of faith and spiritual enquiry. What makes it possible to gradually understand this so-called contradiction is a recognition that 'religion' is not the monolithic entity that either Western faith traditions like to assume, or that secularists have *inherited* as an assumption, exposed as they are mainly to defensive accounts of religion put forward by theologians. This book explores instead the idea of 'spiritual difference' – religion, or spirituality, as an expression of a wide range of spiritual impulses, some of them congenial to science and rationalism, and some of them deriving instead from the devotional, the moral, the intuitive, the poetic, and the aesthetic. When secularism applies its own well-developed language of pluralism to the phenomenon of religion and spirituality, instead of lazily assuming the monolithic nature of religion, then the first step is taken of engaging with the subject. To taxonomise a domain is to have respect for it. But, crucially, this articulation of spiritual difference also allows for a radically new understanding of the origins of the secular mind.

In exploring the history of spiritual difference the turning point in Western history – the Enlightenment – is reconsidered. Instead of understanding this moment in history as the beginnings of a process that would inevitably bring about the death of religion, we can see instead that the secular mind arose out of a purely negative legacy of that period. It was not the death of religion that was made inevitable by the Enlightenment, but the gradual confounding of its secular legacy. Slowly, it becomes clear that the adventure of secularism has gone wrong by betraying its Enlightenment roots. Secularism became a caricature of Enlightenment thought, an unbalanced extremity of assumption, and, in its forced and reluctant re-engagement with religion, it now has the opportunity to repair a breach in the very fabric of society.

This book opens up a conversation between religion and secularism in a way not previously attempted. It sees their apparently opposing worldviews as not intrinsically antagonistic, but only so through accidents of history. Once those accidents are exposed and accounted for as partially wrong turnings, a new rapprochement can take place. But the conversation pursued here has a third partner: the New Age. Usually dismissed by both the other parties, it has elements within it that are important to the debate, particularly its open enthusiasm for the spiritual life. By the term 'New Age' the broad phenomenon of new religious movements, adoption of Eastern practices, and revival of pre-Christian religions is meant.

Hence we have in this book a conversation, and the development of a vocabulary for the wider dissemination of a conversation, between the secularist, those who pursue faith traditions ('old religion'), and those who pursue new religions or popularisations of Eastern religious movements. For simplicity, these three groups will be referred to as secularists, religionists, and New Agers: we can understand the new cultural landscape of the twenty-first century as a three-way contest between them. Each group has unique insights and limitations of thought, and a conversation must start with the awkward acknowledgement of those limitations. As this book begins the conversation with the idea of spiritual difference, we need to address it through that virtuous hallmark of the secular era – pluralism. Religionists, secularists and New Agers, each in their own way deny the idea of spiritual difference and spiritual pluralism: old religion through its absolutist certainties and war on 'heresy'; the secular world through indifference, and the New Age through an uncritical acceptance of all and everything – its mantra that 'all is one'.

Freud criticised monotheistic religion in particular for its 'narcissism of minor difference'. He meant by this the small differences in religious belief that separated groups, which otherwise shared vastly similar ideas, and which could lead to the most appalling violence between them. If we characterise old religion as suffering from the 'narcissism of minor difference' then we can characterise secularism as suffering from the 'narcissism of self-sufficiency', indicating the false belief that humanity had grown up and out of the need for any kind of spirituality (a key shibboleth of secularism). The New Age in turn can be said to suffer from the 'narcissism of difference denied' – its tendency to downplay spiritual difference in the interests of a superficial harmony.

Chapter One outlines the contours of the contemporary secular worldview, using some case studies in cultural production, including literature and film. It explores the role that science, philosophy, language, psychology and neurology play in constructing the shibboleths of the secular mind.

Chapter Two presents a two-fold Model of Spiritual Difference to provide the necessary detailed articulation of spirituality for the subsequent chapters. It highlights the difference between the devotional and non-devotional spiritual impulse (usefully articulated in Hinduism as *bhakti* and *jnani*), and looks at pathologies and correctives in the spiritual life.

Chapter Three starts with a detailed consideration of the alienation of the modern, secular mind, and then works backwards through Western history to identify the major factors in the abandonment of religion as a credible cultural force. It considers both the rise of science and the legacy of religious cruelty as factors, and concludes that these were insufficient to account for the 'death of God'.

Chapter Four considers the rise of Christianity as a devotional religion with Hebraic roots, as juxtaposed to the Hellenic modality of the spirit: a contrast between the profoundly different spiritualities identified through the terms *bhakti* and *jnani*.

Chapter Five reconsiders the Enlightenment, not as the triumph of rationalism, but as the re-assertion of a non-devotional world-curious spirituality: the atheist Enlightenment thinker being the exception, not the rule. It is shown that the 'death of God' was not intended by almost any serious Enlightenment thinker, but that they defended a conception of 'God' that was more Hellenic (or *jnani*) than Christendom could bear.

One:
Issues in Secularism and Culture

In this chapter we begin our exploration of the secular mind and how it has placed itself beyond any discourse of the spirit in secularism's overarching cultural dominance. The word 'spiritual' is broadly rejected by the secular mind, and in secular Western culture, permitted only to poets whose aesthetic vocabulary requires the use of old-fashioned and vaguely provocative terms. However, at least a place-holder definition of spirituality is required here, so we start by saying that it is simply to do with a *profound connectedness*. Crucial to the understanding of spirituality is the idea that it is multiple, not one; that there exists spiritual difference, spiritual pluralism, a variety of spiritual impulse. Its detailed articulation is kept for Chapter Two, and in Chapter Three we will show how it also delineated through the manifestation of its opposite: a debilitating sense of alienation.

We are particularly interested to show here how contemporary Western culture reflects the assumptions, the shibboleths, of secularism. A shibboleth is a culturally received but unexamined assumption that is both a requirement of membership and outmoded: the group clings to it for identity but forgets its original meaning. Professor Mike Tucker of Brighton University puts it well:

> ours is the first culture to proclaim with hubristic certainty that history and politics together constitute the sole ground of our being, and that any sense of the 'vertical' or 'cosmic' dimension in life is but a reactionary remnant from the irrationalities of pre-Enlightenment thought.[1]

Tucker is right about politics: it is secularism's first cultural reality, and we will show that political thinkers, in particular Marx, are amongst the prime architects of the secular mind. Marx effectively denied the spiritual or religious interiority of the mind when he said of the human essence: 'In its reality it is the ensemble of the social relations.'[2] We can call this strand of thought a *social* secularism. Tucker is also right to highlight the Enlightenment as the point in Western development that led to the secular mind. We suggest that contemporary Western culture is born out of a trauma which came to a head in the Enlightenment, one that has made words like 'spiritual' problematic. We will also explore psychology in this chapter as a key part of the secular mind, a *psychological* secularism. We also have new insights from it to bear on our problem here, one of which is the idea of 'cathexis', which the Chambers dictionary defines as 'a charge of mental energy attached to an idea or object'. Social secularism, due to Marx, and psychological secularism, due to Freud, have a third and over-riding partner: scientific secularism. The association of secularism with science goes deep: the conventional view is that the mechanistic view of the universe originating in Newton – once applied to the world of living things through Darwinism – made secularism inevitable. This is the greatest shibboleth of the secular mind.

The words 'religion' and 'spirituality' have different emotional charges for different groups. For the committed secularist the word 'religion' represents everything they reject, while 'spiritual' may be just as bad or merely cause embarrassment. For the committed religionist the word 'religion' conveys the deepest, best and most important aspects of their life and community. However the word 'spiritual' may be suspect, and is often used to dismiss narcissistic New Age self-preoccupation. For the New Ager, 'religion' is often the bathwater they reject in favour of the baby of 'spirituality'. Hence, because of these cultural complexities, we cannot simply adopt the formula 'religion bad, spiritual good', which some commentators use. Instead a critical approach is taken which stresses that both organised religion and self-directed spirituality answer deep and genuine human needs, but not avoiding criticism of either where justified.

The question in this chapter is not so much *why* secularism dominates contemporary culture – that question is pursued over the following chapters – but *how* it does so. This chapter provides an anatomy of the secular mind as it is currently portrayed through aspects of culture that include the arts, science, philosophy, language, psychology and neurophysiology. However the attack, for such it is, is only against the *shibboleths* of secularism, its negative and weak points, its unthinking dismissal of the spiritual. Its strengths – which undergo regular re-evaluation and testing in democratic societies – include the core values of pluralism and democratic freedoms forged in the Enlightenment. It is easy to be cynical about these, especially when a simplistic attempt to impose them on others devalues their currency, but the easy route of cynicism is not taken here.

1.1 Secularism, Atheism and Scientism

Secularism, atheism and scientism are related ideas that are central to the modern mind, and only to some degree mitigated within the cluster of ideas known as 'postmodernism'. Their presence within mainstream Western culture represents the barrier to the spiritual life. 'Secularism' as a term became significant in the mid-nineteenth century, with the establishment of the National Secular Society in the UK, along with many grass-roots organisations affiliated to it. These secular societies were anticlerical, anti-religious and mostly working-class (though often led by middle-class activists and thinkers). Similar forces were at work in the US at this period, though its history of non-conformist religiosity meant that the debates played out differently there.

'Atheism' is a term that became important in Europe in the seventeenth century as a term of the utmost censure, but has lost its capacity to shock in twenty-first century Western culture. 'Scientism' means a narrowly scientific world-view, one which insists on empirical science as the sole arbiter of meaning. We will explore the history of scientism in some detail, in order to show that while science and religion are not at all incompatible, *scientism* is an implacable opponent of spirituality and religion. We can understand the term 'secularism' to includes atheism, scientism and also what we understand by humanism, a term indicating the Enlightenment re-orientation away from religion towards human-centred values. For the committed secularist the secular world-view is hard to analyse from the vantage point of complete immersion: hence a considerable historical exposition is required. For the committed religionist secularism simply stands for a limiting cultural milieu that might be a mild irritant, or, in the case of traditional Jews, Christians and Moslems, a threat to an entire way of life. For the less thoughtful New Ager the question might be: what secularism? This is because New Agers are not taken seriously enough by secularist thinkers to be much criticised, and also because New Agers live in a cultural bubble of their own. The very term implies and encourages an ignorance of previous ages, including recent history, preferring instead to imagine a utopian near-future.

Secularism contains a spectrum of views, from a benign tolerance of religion to what we can call a *vituperative* atheism – that is an atheism which is unexpectedly angry. The tolerant end of the spectrum expresses itself in the democratic rights to self-expression and freedom of worship. The vituperative end of the spectrum, while upholding those rights as inalienable, avails itself of those rights to shower contempt upon old religion, indignant that such 'irrationalities' survive into the modern era.

Secularism and Statehood

Since the end of the cold war and the end of the capitalist-communist divide as the defining global schism, a new ideological battleground has emerged: between the West and Islamic fundamentalism. This is presented as the struggle between the ideologies of the secular state and the ideologies of the *presecular* state: between democracy and 'theocracy'. Communist countries used to represent the earlier ideological 'other' to secular Western democracies, but with the collapse of the old Soviet Union the spreading adoption of democracy and capitalism is eroding that contrast: it is the Islamic world that has come into sharp relief now as the 'other'. Its theocratic objections to the twin secular gods of democracy and capitalism are nowhere understood in the secular world, as the attempt by the US to impose them on Iraq demonstrated.

The status and involvement of a presecular religion with its secular government varies, and the different ways in which the US, Britain, France and Turkey have approached this is instructive. The US has constitutionally separated state and religion, deriving this approach from its history as a haven for the non-conformist of all persuasions in the seventeenth and eighteenth centuries. On the surface the UK may look like the US in all practical applications of the law, but never got round to the legal separation of church and state (known formally as disestablishment). In fact this remains one of the founding goals of the Secular Society of Great Britain, as we shall see.

The Guardian newspaper (itself a useful benchmark of contemporary secularism) revealed in a 2006 poll that 63% of the British people are not religious at all.[3] But statistics for the year 2001 suggest that 79% of Americans, 72% of Britons, and 70% of Canadians are nominally or fully Christians.[4] Clearly such statistics are unreliable indicators of the extent of secularism, relying on simplistic questions to elicit responses. In addition these figures hide very big differences in the history and role of Christianity in each nation. In the US election year of 2004, liberal US commentators grumbled that George W. Bush (the incumbent Republican contender) was running a theocracy, and that it was easier for candidates to profess 'freethinking' tendencies at the founding of America. British journalist Alexander Chancellor suggested that 'In Britain, flaunting religious faith is a political liability; in America, it is a political necessity.' His article, in *The Guardian*, was a tongue in cheek appeal to the Democrat candidate, John Kerry, to talk up his religious credentials.[5] The *Guardian* represents the contradictions of secular left-wing Britain: it wanted to see a Democrat victory in the US election, but was ambivalent on the necessity for Kerry to speak on religion.

France's revolt against religion, in the Revolution of 1789, was more violent and bitter than in most other European countries, but since then it has become host to one of largest Muslim populations in Europe. These factors combine to make the issue of the headscarf worn by Muslim women a site of bitter contention, leading to attempts at legislation to ban its wearing by school pupils that are mostly absent in the UK and the US by comparison. (The ban finally entered French law in 2004.)

Turkey's recent history represents a struggle between a 90% Muslim population and its secular government, a balancing act that has been fraught with difficulties since its inception by Atatürk in the 1920s. The wearing of the Muslim headscarf has been equally a difficult issue for Turkey, particularly within the Universities where thousands of students have been banned for wearing it. However a vital difference between Turkey and Iraq remains: Turkey is America's model for a secular Islamic state, but Turkey's secularism was not imposed by military force from outside. Rather it was the result of a visionary Turkish leader working within a Europe-oriented political climate (which culminated in 2005 in the acceptance of its application to join the European Union).

The secular world guarantees the right to private religious beliefs and spiritual practices, and to private gatherings to celebrate shared ritual. It also guarantees the right to express disbelief and even hostility to religion, for example a 2001 letter to a British academic journal was headed 'In favour of religious hatred', and turned out to be a plea from an Oxford chemist to encourage the hatred of religion itself (perhaps now less acceptable since the passing of new British legislation on religion).[6] It was a reaction to the events of 11 September. There is no doubt that Peter Atkins, the author of this letter, speaks for a small, but vocal minority in the Western world, a group that we have characterised with the term 'vituperative atheist'. A number of scientists fall into this camp, along with other liberals. One from the literary world is Gore Vidal (prolific American novelist, playwright and essayist), who says for example 'I favour an all-out war on the monotheists', and 'The great unmentionable evil at the centre of our culture is monotheism.' Such writers have a range of motives for their attacks on religion, but one is tempted to wonder at the extremity of their language, given the muted *public* face of religion prior to 11 September. (Speculation that vituperative atheism is a reaction to a collective experience well in the past is presented in section 3.3, *The Spiritual Wounds of the West*.)

Secularism in the Twentieth Century

The National Secular Society is an organisation explicitly formed to promote and protect secularism in Britain (the American equivalent is the Council for Secular Humanism). Charles Bradlaugh, a notorious atheist of the Victorian period, helped to establish the National Secular Society in 1866 as an organisation opposed to Christian dogma. Its present General Principles are as follows:

> Secularism affirms that this life is the only one of which we have any knowledge and human effort should be directed wholly towards its improvement.
>
> Affirming that morality is social in origin and application, Secularism aims at promoting the happiness and well-being of mankind. Secularism demands the complete separation of Church and State and the abolition of all privileges granted to religious organisations.
>
> Secularism affirms that progress is possible only on the basis of equal freedom of speech and publication; that the free criticism of institutions and ideas is essential to a civilised state.
>
> It asserts that supernaturalism is based upon ignorance and assails it as the historic enemy of progress.
>
> It seeks to spread education, to promote the fraternity of all peoples as a means of advancing universal peace to further common cultural interests and to develop the freedom and dignity of mankind.
>
> To remove an impediment to these objectives, we demand the complete separation of Church and State and the abolition of all privileges granted to religious organisations.[7]

In fact, the goals of the National Secular Society represent much that is laudable, and a rejection, not of the spiritual life as we will broadly paint it, but of a particular culturally embodied form of a particular religion – Christianity – in a particular country, with an alleged adherence to an afterlife and a narrow conception of the supernatural. Gore Vidal is more outspoken on the subject of religion:

> Now to the root of the matter. The great unmentionable evil at the centre of our culture is monotheism. From a barbaric Bronze Age text known as the Old Testament, three antihuman religions have evolved – Judaism, Christianity, Islam. These are sky-god religions. They are, literally, patriarchal – God is the omnipotent father – hence the loathing of women for 2,000 years in those countries afflicted by the sky-god and his earthly male delegates. The sky-god is a jealous god, of course. He requires total obedience from everyone on earth, as he is in place not for just one tribe but for all creation. Those who would reject him must be converted or killed for their own good. Ultimately, totalitarianism is the only sort of politics that can truly serve the sky-god's purpose.[8]

Vidal is explicitly attacking *monotheism*, rather than religion in general, and associating it with all that is totalitarian in Western society. Polly Toynbee, a British journalist with an attitude to religion comparable to Vidal's, writes on current affairs in the *Guardian*, and is a supporter of the National Secular Society. The following is a typical quote from her:

> The only good religion is a moribund religion: only when the faithful are weak are they tolerant and peaceful. The horrible history of Christianity shows that whenever religion grabs temporal power it turns lethal. Those who believe theirs is the only way, truth and light will kill to create their heavens on earth if they get the chance. Tolerance only thrives when religion is banished to the private sphere.[9]

Toynbee's analysis is similar to that of Marx, but in the intervening century and a half it seems that our contact with the East has not brought any counter-examples into contemporary consciousness. To say that 'whenever religion grabs temporal power it turns lethal', is to have at one's disposal only one example: monotheism. The counter-examples of Buddhism, Hinduism, Confucianism, Taoism and Shintoism as mostly non-lethal are rarely commented upon. Nor do examples of absolutism and lethal *temporal* power quite unrelated to religion seem to change the equation: it is a common proposition that far more have died in the name of Marxism than in the name of religion. The US academic R. J. Rummel estimates that 110 million citizens were murdered by their own communist regimes so far, excluding those who died in war.[10] After the publication of new biographies of Mao Tse-Tung, Rummel has revised the figure upwards to include the millions who died in the Great Famine of 1958-1961, because these now seem clearly avoidable. Having said all this it is clear that the National Secular Society, Vidal and Toynbee are all defending an essential principle, established in the Enlightenment and enshrined in the US Constitution: that religion should be separated from state, that it rightly belongs to the private sphere. Their arguments are atheist and secular, but our argument is from a spiritual perspective: given the variety of spiritual impulse that we will be arguing for the existence of, how could any single state religion provide for that? The secular framework of human rights provides exactly the guarantees of freedom necessary for spiritual pluralism and the celebration of spiritual difference; it is contemporary *culture* that promotes scepticism of it.

Scientism

Newton, effectively the founder of modern science, was a deeply religious man. So is Sir John Polkinghorne, a long-time quantum physicist at Cambridge, more recently an ordained Anglican priest and recipient in 2002 of the million-dollar Templeton prize for the advancement of religion. We shall argue that religion (when interpreted more broadly than a narrow Christian monotheism) and science are not contradictory, as these two examples (and dozens more amongst prominent scientists) show. The perceived hostility is due to nothing in science itself, but to what we are calling 'scientism' – the attempted arrogation to science of all domains of human experience. Scientism is a cluster of issues argued for very often by those who don't understand the science, though also by trained scientists who don't understand religion (we look at the examples of Daniel Dennett, Francis Crick and Richard Dawkins in section 3.1.1 *Science and the Century of Alienation*).

A key feature of scientism includes a rejection of the subjective, emotional and spiritual dimensions of experience and a reduction of values to *usefulness*. 'What Use is Religion?' is the title of an essay by biologist Richard Dawkins in the *Humanist* (a journal published by the Council for Secular Humanism in the US).[11] Dawkins considers that a medieval cathedral for example is an 'affront to Darwinism', in that it does nothing to ensure the survival of the species, the *only* criteria for action he will allow. This form of reductionism – a diminishing of criteria to the simplest and most basic – works well as a tool in science, but becomes an absurdity when applied to the breadth of human experience. The postmodernist philosopher Lyotard regrets 'performativity' as the driving force in society, meaning the continued appeal to a narrow material usefulness.[12] Dawkins's 'Darwinian performativity' must be the ultimate example of this.

Scientism as a world-view says that we are analytical beings (interested only in the *explanations* of things), and denies that we are feeling beings, emotional beings, communal beings, aesthetic beings and spiritual beings (and why should even this list be exhaustive?). We shall explore this diminishing of self; also the 'psychologising' of self (through analytical psychology) and the 'neurologising' of self (through brain science). *Science* however is not the issue: neither experimental science nor its Pandora's Box of technological fallout is what is opposed here, rather it is the curious habit within the last three hundred years of taking science as the first arbiter and sole mirror of human nature.

In the 1920s Carl Gustav Jung visited New Mexico and reported the following revelation in his autobiography *Memories, Dreams, Reflections*:

> I asked him [a Pueblo Indian] why he thought the whites were all mad.
>
> 'They say that they think with their heads', he replied.
>
> 'Why of course. What do you think with?' I asked him in surprise.
>
> 'We think here', he said, indicating his heart.
>
> I fell into a long meditation. For the first time in my life, so it seemed to me, someone had drawn for me a picture of the real white man.[13]

Jung was propelled by this into a long post-colonial reverie about the Western subjugation of native peoples by Christianity (going back to Roman times), and on the imbalance within the 'white man': the dominance of his intellectual function. It is only recently however that new ideas of 'emotional intelligence' or EQ are emerging. These have been popularised by Daniel Goleman[14] and others as an adjunct to intelligence quotient (IQ). Further still, authors Dana Zohar and Brian Marshall are suggesting that we take 'spiritual intelligence' or SQ seriously.[15] This debate, which is playing out in business management (Zohar and Marshall are business consultants when they are not Oxford academics), points to what we will identify as a 'cultural autism' of the West: a preoccupation with the intellect at the expense of feeling. (Business culture is not typical of the mainstream however.) Thomas Berry, a Roman Catholic theologian and environmental thinker, considers that 'we find ourselves in an autistic situation' that cuts Western people off from Nature.[15]

We are suggesting that Western culture is over-intellectualised, to the point of it suffering a mild Asperger's syndrome. This is a point on what is known as the autism spectrum, a range of conditions where a person cannot emotionally engage with another. Hence we will use the somewhat colourful expression 'cultural autism' to flag up the aspect of Western culture which is rationalistic, reductionistic and leads to what is known as scientism – the tendency to explain away emotion and spirituality in mechanistic, often Darwinian terms.

1.2 Secularism and Contemporary Culture

Given the large proportion of the population reckoned to be Christian in the US and the UK, and the growing realisation that the rest of the world is more religious still, why the perception that we live in an intensely secular society? The answer is culture. It is the outpourings of Western culture – film, TV, radio, newspapers, novels, theatre and so on – that portray an unrelentingly secular world, and which up to relatively recently have ignored the religiousness of the non-Western majority of the globe. Of course there are many interesting exceptions, but they prove the rule both in their scarcity and in the secular assumptions that usually frame them. Most media have a 'God slot', but these effectively comprise a media ghetto echoing the cultural ghetto that religion occupies. The same holds for 'God channels' – televangelism. Contemporary Western culture is intensely secular when compared to either seventeenth century or earlier Western culture, or, for example, to the Islamic world. Whatever the true picture of religion in Western society, its scarce and sceptical portrayal in Western culture creates the impression of an intensely secular worldview. (Muslims for example must form their impression of Western society to a large extent on this basis, so it is not surprising that the more religious are led to reject it outright.) That Western culture largely rejects religion is clear, but what makes culture worth examining in detail in this respect is its potential for articulating the contours of that rejection.

We now look at four examples from recent Western culture: a novel by Jean-Paul Sartre, *Nausea;* a film, *Monty Python's The Life of Brian;* an episode of a US TV series *The Simpsons;* and another novel, *Thinks . . .* by David Lodge. *Nausea* was first published in 1938, much earlier than the other examples, and can be considered as being amongst the first truly atheistic novels of the twentieth century, describing an alienation prefigured in more romantic writers such as Thomas Hardy, John Steinbeck and C.S. Forester. The sense of alienation appears in their novels as a vague nihilism, in fictional characters that would have been at home in the nineteenth century, whereas Sartre's anti-hero, Antoine Roquentin, is a truly twentieth-century invention. *The Life of Brian* (1979) is a satire on the life of Jesus, and attempts to deal with the issue of religious gullibility, as a growing crowd of followers gather round the unlikely figure of confused Jewish revolutionary Brian. In an episode of *The Simpsons* called 'The Joy of Sect' (1998), issues of religious groups such as the Heaven's Gate cult, whose members committed mass suicide in 1997, are explored. Finally we look at a twenty-first century novel, David Lodge's *Thinks . . .* , which explores a very contemporary topic, cognitive science, and its relation to the arts.

Sartre's *Nausea*

Nausea was Jean-Paul Sartre's first novel, started in 1930. It conveys a secularism that is perhaps typically and existentially French. It was enormously influential, and its existential legacy is seen after WW2 as far afield as in the plays of Harold Pinter, in novels like *American Psycho* (Bret Easton Ellis), or the nature writings of Annie Dillard. While 'The Joy of Sect' and *Monty Python and the Life of Brian* show a secular concern with the spiritual that is shallow and driven by a quick eye for the comic, *Nausea* is atheist to the core, and driven by quite different agendas.

Sartre's existentialism lived within his philosophy alongside an equally demanding partner: Marxism. We shall see that the cold atheism of Sartre was heralded in Marx nearly a hundred years previous, and from that perspective *Nausea* is not so radical. Nevertheless no work in fiction before *Nausea* had told the world what the future would look like, stripped of all the values of the past except one: honesty. (These values were agonised over but not yet completely abandoned by Nietzsche for example.) This was a new honesty however, deriving philosophically from the principles of phenomenology, elevated by Heidegger as *authenticity*, and in keeping with the clinical nature of scientific enquiry. This honesty was an imperative to oneself, and completely incompatible with the prescriptive morality of religion, at least as it was handed down in the West. Sartre himself grew up without a father, and, like many European intellectuals and artists of his time, was traumatised by

WW1, leading to a complete revolt against the traditions of authority and power that had started it. He saw only hypocrisy in the institutions and ideals of his father's generation, and became part of the wholesale revolt against them.

Sartre's *Nausea* is permeated with an atheism that neither engages with religion, nor even takes the time to dismiss it: the question simply isn't on the horizon. Yet, paradoxically, the novel contains within it reflections and experiences that touch directly on the mystical. In reality, the factor that led to its resonances with the spiritual life was the taking of a hallucinogenic drug. Sartre was persuaded in 1935 by a doctor friend, working in a sanatorium for the mentally ill, to inject himself with mescaline.[16] Sartre had a bad trip on the drug, out of which he completed *Nausea,* and in this experience anticipated Aldous Huxleys' famous experiments by two decades. Huxley popularised the possible link between drugs and mystical experience in his novel *The Doors of Perception,* but few have noticed this link in *Nausea,* couched as it is in the language, familiar to us now, of existentialism. What marks the novel as so significant for the spiritual life of the twentieth century is that it presents us with an almost pathological *inversion* of mysticism, an indictment both of the secular mindset and of the use of drugs for spiritual purposes. In this passage its protagonist finds himself 'stranded in the present':

> I looked anxiously around me: the present, nothing but the present. Light and solid pieces of furniture, encrusted in their present, a table, a bed, a wardrobe with a mirror and me. The true nature of the present revealed itself: it was that which exists, and all that which was not present did not exist. The past did not exist. Not at all, neither in things nor even in my thoughts. True, I had realised a long time before that my past had escaped me. But until then I had believed that it had simply gone out of my range.[17]

The quality of being in the present, though never part of mainstream religious heritage, is nevertheless a central theme in the writings of the world's mystics, and is always presented, firstly as a difficult goal to attain, and secondly as blissful. It is difficult because of the workings of the mind, which is continuously active in the sites of past and future (as memory and imagination). Sartre's experience with mescaline is typical in that it, and many other psychotropic substances, has the effect in a few moments that years of meditation are designed to bring about: the stilling of the mind and the entering into the silence of the present moment. The novel reaches a climax when its protagonist, Roquentin, sits on a park bench and is faced by the terrifying 'nowness' or 'suchness' of the trees, in particular the root of a chestnut tree. There are many parallels in mystical writing conveying mystical realisation in such settings, and many attempts to convey in writing the profundity of that moment. The term 'tathata' is used in Buddhism for example to convey this spiritual quality in moments of attention to ordinary things, moments that are sought after, and usually the product of a committed spiritual practice (meditation). Sartre gives us potentially one of the greatest descriptions of that moment in Western mystical literature, except that is presented to us as merely pathological and the existential antithesis to mysticism.

That Sartre's psychical response to mescaline was to produce a bad trip, where Huxley's generally produced good trips, can be speculated upon. What is interesting is that despite the fact that Sartre had read the mystics (according to Simone de Beauvoir)[18] he could not make the link that Huxley did between his experience on drugs and that of the mystics, perhaps solely because of its unpleasantness. Yet of the two it is Sartre that has had a profound impact on Western thought, Huxley variously dismissed as a snob or romantic.

Monty Python and the Life of Brian

The film *Monty Python and the Life of Brian* was made in 1979, directed by Terry Jones and co-written by Graham Chapman and John Cleese. It was made exactly one hundred years after Dostoyevsky wrote his classic novel *Brothers Karamazov* in which he included a story about Jesus returning in the time of the Spanish Inquisition and being crucified again. Whereas Dostoyevsky was genuinely concerned to probe the spiritual realities behind both Jesus and the Inquisition, the Monty Python film is pure satire. It starts with Brian's mother, Mrs Cohen, receiving three unexpected guests at the birthplace of her son in a stable: the Wise Men bringing their gifts. These are promptly reclaimed when they realise that they should be paying homage at the stable over the road. Brian as a young adult and his mother are then seen listening to a sermon by Jesus where he tells the crowd: 'Blessed are the meek.' Getting bored, Mrs Cohen takes her son off to a stoning, and we follow the camera to Reg, leader of the Judean People's Front muttering sarcastically: 'Blessed is just about everyone with a vested interest in the status quo.' (This conforms to a popular Marxist dismissal of Jesus as failing to rise as revolutionary leader.)

The film continues in this satirical vein, targeting some typical

British institutions such as the Trades Unions and public school education, though its principle target is of course Christianity. Brian joins the Judean People's Front and is soon in trouble with the Roman authorities. The film at this point perfectly illustrates Freud's 'narcissism of minor difference': the Judean People's Front spends more time fighting its rivals, the People's Front of Judea, than the Romans.[19] In a narrow escape Brian finds himself forced to preach to cover his identity, and mouths some trite religious platitudes. A crowd gathers, and despite his best efforts to get rid of them they follow him, with an apparently endless appetite for such instruction. The more he protests his incompetence as Messiah, the more the crowd adore him, leading up to this exchange, fondly etched into the memory of a generation that grew up with Monty Python:

> **Brian**: I'm not the Messiah! Will you please listen? I am not the Messiah, do you understand?! Honestly!
> **Girl**: Only the true Messiah denies His divinity.
> **Brian**: What?! Well, what sort of chance does that give me? All right! I am the Messiah!
> **Followers**: He is! He is the Messiah!
> **Brian**: Now, fuck off!
> [silence]
> **Arthur**: How shall we fuck off, O Lord?

Britain at the time of the film's release was ripe for this gleeful, naughty irreligiousness, with only a few lone voices, such as broadcaster and journalist Malcolm Muggeridge voicing disapproval (he called it 'This squalid little film'). Its reception in the US was less enthusiastic, and screenings were picketed by religious groups. Its dominant theme, as we shall see with the 'The Joy of Sect', is of gullibility, and there is no doubt that the Monty Python team explored this with comic genius. It focuses on the weaknesses of followers rather than on the exploitativeness of leaders, its central secular assumption being that there can be no genuine 'seeking' or genuine relationship between spiritual aspirant and teacher. The makers of the film could not understand the offence given to people of Malcolm Muggeridge's generation or outlook, having in their own eyes merely tapped a rich vein of comic potential. The atheism implicit in the film is casual, unconsidered and culturally received: we can usefully call it a 'Monty Python atheism', in contrast to the vituperative atheism of Vidal, Atkins and Toynbee.

Other modern films have been made on the subject of Jesus, including *The Last Temptation of Christ* directed by Martin Scorsese,

1988. Such films take a more serious look at the phenomenon of a cult leader involving rather different, but still quintessentially secular, assumptions: that Jesus was or could or should have been a political leader, and that sex is a cure for all this nonsense. D. H. Lawrence wrote a short story called the *Escaped Cock* where a Jesus figure, surviving crucifixion, also wanders off and abandons pretensions at religious leadership in favour of sex. Jesus, as is the case with other founders of religion, becomes either the cultural property of a stagnant Church, or is the cultural property of secular revision with its Freudian or Marxist analysis.

'The Joy of Sect'

'The Joy of Sect' episode of *The Simpsons* responds specifically to the issue of religious cults, and involves the family joining up, signing over their property to a sect, and reduced to slave labour picking lima beans. In *The Simpsons* religion is generally examined through the character of Ned Flanders, the Simpsons' neighbour and born-again Christian. It is indicative of the mainstream cultural perception of the spiritual life in the US that religion is fundamentalist Christian, or there lies the only alternative: weird religious sects. For example, the Heaven's Gate cult members committed mass suicide in March 1997, in Del Mar, California. The 39 victims apparently believed they were going to join a starship hiding behind the Hale-Bopp comet, which was to carry them to a better life, to the 'Kingdom of Heaven'. The flying saucer in which the cult leader escapes in 'The Joy of Sect' may be a reference to this, or the story might have drawn on the cult of Bhagwan Shree Rajneesh (referenced via the 'Leader' in a Rolls Royce), or the Church of Scientology (referenced by 'brainwashing'). Scientology is a cult founded by science fiction author L. Ron Hubbard and, interestingly, actress Nancy Cartwright, voice of Bart Simpson, is known to be a Scientologist, which may have constrained the satire.

'The Joy of Sect' is typical of popular culture in its approach to the spiritual. Homer Simpson is not smart enough in the first place to resist the lure of the cult's promises, but on the other hand this does not mean he accepts the cult's spiritual premise either. In fact the cult's 'spirituality' is shown to be entirely bogus, and is used only as a device for exploitation. Homer is rescued when his more resourceful wife, Marge, offers him beer, which brings him to his senses. 'Beer' can be taken here as symbolic of the life of the senses, so the implication is that the spiritual, even if it were genuine, involves renunciation of the pleasures of the world. In short the programme, while ostensibly exploring 'issues' around cult membership, was content to conform to two widely held secular shibboleths: firstly that cults peddle pseudo-spirituality for the sake of exploitation, and secondly that spirituality itself equates with self-denial.

Thinks . . .

Thinks . . . is a novel by David Lodge, Honorary Professor of Modern English Literature at Birmingham University. *Thinks . . .* is a 'two cultures' novel about consciousness, pitting male cognitive scientist Ralph Messenger against female creative writer Helen Reed. Their developing affair allows Lodge to examine the new claims of science to investigate consciousness, the field he believes to be traditionally the domain of the reflective arts such as literature. Helen resents Messenger's scientism: 'Hasn't science already appropriated enough of reality? Must it lay claim to the intangible invisible essential self as well?'[20] This is Lodge's basic secular assumption: that the 'intangible invisible essential self' is the 'province of the arts, especially literature, and most especially the novel.'[21] Why is it not also the province of religion and spirituality? Lodge's novel updates the two cultures debate between science and the arts initiated by C.P. Snow in 1959, which are represented in the novel by a cognitive scientist known throughout by his surname, and a novelist known by her first name. Messenger is director of the Centre for Cognitive Science at a fictional British University, and Helen wonders whether the mirror-glass used for the walls of the Centre represents the vanity of cognitive scientists.[22]

Of our four examples, Lodge's novel best charts the contemporary contours of the 'Guardian-reading' secular mind, including as it does a far-reaching investigation of the 'intangible invisible essential self', an investigation which cannot allow that this might be the domain of the *spiritual*. Roquentin in *Nausea* concludes that fame is his only possibility for an entry into eternity; so does Messenger in *Thinks . . .* 'Anyway, I hope to leave a permanent mark on the history of cognitive science before I go. Just as you must hope to do in literature. That's a kind of life after death. The only kind.'[23] Helen is a lapsed Roman Catholic, but religion is presented throughout the book in the most conventionally secular terms: as a massive mistake. In other words it shares with *Nausea* a now-classical casual atheism, one that was ground-breaking in the 1930s, but seemingly incontrovertible at the start of the twenty-first century, and which we can conveniently call 'Monty Python' atheism.

1.3 Philosophy and Secularism

There is no doubt that Western philosophy has contributed to shaping the secular mind. Sartre can be taken to represent the secularism of most twentieth-century philosophers, even when they belong to radically different schools of thought. But, when taking a longer view of the history of Western philosophy – on the alert to the possibility that earlier philosophers may have held far from secular views – it becomes obvious that philosophy in earlier times served deeply religious purposes. It will become a central issue in later chapters to explore how philosophy so shifted its ground within the modern period: this transition is crucial to the origins of the secular mind.

Western philosophy is considered to have three important periods: Ancient Greek, Scholastic, and Modern (which includes Postmodernism). In religious terms we could say that the Greek period is pre-Christian, the Scholastic is Christian and the Modern is post-Christian. We will develop over several chapters the idea that some Greek, Scholastic and early Modern philosophers were actually deeply spiritual thinkers, but whose spirituality was not easily accommodated within Christianity. We will term their spirituality 'non-devotional' (to be explained in detail in Chapter Two), and will suggest that it aligns itself better to Buddhism and other Eastern traditions, than to Christianity. Hence we adopt for some key Western philosophers this term: 'the lost buddhas of the West.'

The term 'Buddha' was originally a generic one in India, meaning an 'enlightened one', and only gradually became exclusively associated with *the* Buddha. We are using the term 'lost buddha' for these philosophers to suggest two things: firstly that their spirituality is similar to the Buddha's (though in no way derived from it), and secondly that they have become lost to Western history – particularly to the secular mind – as mere philosophers. The 'lost buddhas of the West' might include Pythagoras, Heraclitus, Socrates, Plotinus, Nicholas of Cusa, and, more controversially, Descartes, Spinoza and Leibniz, and some of the German idealists. Not all of them by any stretch of the imagination rank with the Buddha, but the comparisons for Socrates and Plotinus hold well, and the others may be considered to be minor lost buddhas. Once we begin to see these philosophers in this light, we also see that many other philosophers share with them at least a nascent non-devotional spirituality, though lost to the secular viewpoint.

Ancient Greek philosophy, as our list of lost buddhas suggests, was particularly strong in its non-devotional spirituality. In contrast the Scholastic period was located in a fervently devotional context, the pious mysticism of the late Middle Ages. The Modern period grew out of and rejected the devotional context, and hence it is also useful to say that Greek philosophy was not a theology (at least not in the Christian sense); during the Scholastic period philosophy *was* theology; and that subsequent to the Scholastic period theology and philosophy diverge. Theology becomes an intellectual ghetto, isolated from mainstream thought, a discourse that flies below the philosophical and cultural radar of the West. Scholastic philosophy is usually ignored by the secular mind: indeed the period from the sixth to the seventeenth centuries is dismissed as philosophically barren, for example by Bertrand Russell. For the atheist, rationalist secularist the arrival of the Enlightenment philosophers of the seventeenth and eighteenth centuries heralds the end of religion and the dawn of the age of reason. We shall attempt to show that this is a fiction: one of the shibboleths of the secular mind. Most of the philosophers of this period were deeply religious.

The fact that we will need to turn to the East to provide us with a suitable vocabulary for a non-devotional spirituality gives rise to the question: what about Eastern philosophy? Does it have a parallel development leading to the modern era? No, is the simple answer, and perhaps even simpler is the proposition that the East has no philosophy at all, not in the modern Western sense at least. This is to say that it has never been secularised, either in the Greek style or in the modern/postmodern idiom, until Western influence began to dominate. Yet the term 'Eastern philosophy' is a common one and is normally assumed to include Hindu, Buddhist, Taoist and Confucian writings. What separates by far the bulk of it from Western philosophy however is that, like the Scholastic writings, Eastern philosophy is always embedded in a religion, and its proponents have been practitioners of that religion. *Eastern 'philosophy' is nearly always grounded in spiritual practice.*

The historical antipathy between the Christian theological mind and the Greek philosophical mind is an important factor in all of Western history: this can be understood as a creative tension between the *Hebraic* (Judaeo-Christian) and the *Hellenic*, one that forged the Western mind. It is also the tension between a largely devotional monotheistic spirituality and a non-devotional, potentially esoteric and pluralistic spirituality. We can take two terrible murders as iconic of this division: the first of Hypatia in 415 CE and the second of Giordano Bruno in 1600 CE. Hypatia was the head of a Neoplatonist school in Alexandria, famous for her learning and also her beauty, but was horribly murdered by a fanatical Christian mob that associated learning and science with 'paganism'. The subsidiary daughter library of the great Library of Alexandria in the Temple of Sarapis had already been sacked by such mobs in 391 CE. (Alexandria went into decline as a centre of scholarly activity with the death of Hypatia and the loss of the Library.) Christian mobs were responsible for vandalism against art as well: countless art treasures of the ancient world were defaced or destroyed by the early Christians, including many of the sculptures of the Parthenon (this is an issue of *iconoclasm,* to be dealt with later on).

Giordano Bruno was burned at the stake in the year 1600 by a system that had formalised the early Christian mob violence: the Inquisition. His crime, like Hypatia's, was to be interested in what we will call the esoteric (in the ancient world science and the esoteric were intermingled). His philosophy would be unrecognisable to modern-day philosophers, but his free-thinking and dreadful fate has made him a hero to many secularists who reject religion. That Christianity should have developed an early and ongoing visceral distrust of the scientist/philosopher/intellectual is a crucial factor in Christianity's downfall, as we shall demonstrate. But, we will suggest that the murders of Hypatia and Bruno are symptomatic of the antipathy of Christianity not so much to philosophy and science as to *non-devotional spirituality*.

In the modern / postmodern period there emerges a consistent theme at the heart of what is otherwise a bewildering array of speculative systems. Philosophy in the Scholastic period was to the glory of 'God' and to the glory of the earthly reflection of his divinity, as shown in the divine order of Nature and society. Philosophy in the modern / postmodern period has increasingly taken on the role of subversion: to challenge the accepted order, to question the divine origin of laws, to question the traditional power structures of society and to question the very roots of such assumptions in language itself. If Kant begins the dismantling of the certainties that were carried through into the Enlightenment from the Scholastic era; if Nietzsche takes a hammer to these fractured structures (in his own words he 'philosophises with a hammer'); then twentieth-century philosophers appear to be grinding the rubble into ever smaller fragments. The progressive dismantling of the older structures of knowledge and power leaves a 'flatland' of postmodern thought (as it is often claimed). But something of vital importance has been lost in this trajectory: the spiritual.

1.4 The Role of Language

The role of language in spiritual life is complex. Three aspects are briefly touched upon here: first the universal insight of the mystics that the essence of the spiritual life is located *beyond* language; second, that philosophy has traditionally over-emphasised the intellect as a *linguistic* entity (culminating in what is known as the 'linguistic turn'); and third, the way in which words take on a life of their own.

To mention again the 'mystics' at this point is to use a word with different connotations for different groups. For the secularist it has little currency; for the religionist it may signify the core of their religion or alternatively a largely heretical group of non-conformists; while for the shallow end of the New Age it means crystals, tarot cards and past-life reading. When De Beauvoir recorded that Jean-Paul Sartre 'read the mystics' she meant Christian mystics – Teresa of Avila, John of the Cross and Meister Eckhart. Mystics can be regarded as specialists in religion, while priests can usually be regarded as functionaries. At times priests have been mystics – Eckhart is a good example – but the roles of priest in a community are usually too demanding to allow the depth of contemplation required of the mystic.

But to say as we did above that mystics have 'universal' insights is already to take sides in a continuing debate as to whether the differences in the way that mystics present their experience merely masks that universality, or whether their traditions and cultural contexts determine those experiences in the first place. These two viewpoints are usually referred to as the 'perennialist' and the 'contextualist' positions respectively. The term 'perennialist' derives from what is known as the Perennial Philosophy, which term was popularised by Aldous Huxley in his book of the same name, an annotated compilation of mystical writings. He introduces it with this statement:

> *Philosophia Perennis* – the phrase was coined by Leibniz; but the thing; the metaphysic that recognizes a divine Reality substantial to the world of things and lives and minds; the psychology that finds in the soul something similar to, or even identical with, divine Reality; the ethic that places man's final end in the knowledge of the immanent and transcendent Ground of all being – the thing is immemorial and universal. Rudiments of the Perennial Philosophy may be found among the traditionary lore of primitive peoples in every region of the world, and in its fully developed forms it has a place in each of the higher religions.[24]

Huxley notes that few of the extracts in *The Perennial Philosophy* are in fact by professional philosophers, confirming that the contemporary philosopher has an entirely secular role, in contrast to their closer affinity to the mystic in earlier times. We can as a first approximation use the term 'perennial philosophy' and the term 'Neoplatonism' interchangeably, as both approach the spiritual life in terms of spiritual *knowledge*, rather than devotion. (Neoplatonism is discussed further in section 4.2.2.) However, the point about the perennialist position is that mystics, whether of the devotional or non-devotional type, are speaking about essentially the same experience of mystical union, a *given* of the human condition across the centuries and continents. This experience is held to be beyond language, but in order to share it with others recourse has to be made to the spoken or written word, inevitably inferior to the condition described. The contextualists on the other hand argue that there is no universal mystical experience, and that the differences in mystical accounts show that they derive from cultural conditions and expectations. Hence a Christian mystic might see visions of Christ, while a Hindu mystic might see visions of Krishna, and so on. (Note also that Huxley's perennialism conflates different types of spirituality that we shall carefully disaggregate in Chapter Two.)

The contextualist position emerged in the 1970s as a direct result of postmodern thinking in philosophy about language. In its milder form it proposes merely that whatever the experience of the mystic, their reports are expressed in, or at least highly coloured by, the religious (or secular) language of the day. The more extreme position denies that mystics have experiences that are universally valid in the first place, and that all we have are *texts* that are open to whatever interpretation we choose to give them. Scholar of mysticism Steven Katz sums up the contextualist approach:

> It is my view, argued in detail elsewhere, that mystical reports do not merely indicate the postexperiential description of an unreportable experience in the language closest to hand. Rather, the experiences themselves are inescapably shaped by prior linguistic influences such that the lived experience conforms to a preexistent pattern that has been learned, then intended, and then actualised in the experiential reality of the mystic.[25]

To understand why Katz should focus on *language* as the inescapable prior influence rather than tradition or culture in a broader sense, we need to look at the 'linguistic turn', mentioned above. However we should add here that this book adopts the opposite position to Katz, though it does not support a naïve perennialism either.

The Linguistic Turn

The deconstructionists in postmodern philosophy, such as Jacques Derrida, drew on early ideas about language from Ferdinand de Saussure and Ludwig Wittgenstein. The linguistic turn can simply mean the greater emphasis or even the sole emphasis placed on language as the topic of study for contemporary philosophy. At its extreme however it suggests that our *entire* experience is constructed out of language. The origins of this go back to Saussures's ideas on words defined in terms of each other, 'a sliding chain of signifiers' terminating in no referent, or real-world object. This idea derives in turn from Kant, who famously denied that we could know the 'thing in itself' – we know only our mediated perceptions and the constructs we read into them. Another way to put it is in a denial of the 'given', a formulation pursued by philosopher Richard Rorty in his well-known *Philosophy and the Mirror of Nature*. For Rorty, the 'mirror' is the myth of a mind that can accurately represent the world, and he proposes a philosophy that does without this mirror. Accurate representation for Rorty is 'simply an automatic and empty compliment which we pay to those beliefs which are successful in helping us do what we want to do.'[26] Wittgenstein in his *Philosophical Investigations* also denied that words refer to things, rather that they should be understood in the context of a 'language game', with arbitrary rules or relationships between the symbols. Wittgenstein explicitly attacks St Augustine's model of language, the common-sense notion that words refer to things in the world, and are learned by the repetition of association between words and things.[27] The perennialist position relies on the Augustinian model, only the 'things' referred to by the mystics are universal inner realities rather than physical objects. Hence the linguistic turn largely works against a real engagement with the spiritual, because it denies inner realities other than those constructed arbitrarily through language.

The motivations for the linguistic turn are part of the general postmodern outlook, involving a rejection of traditional monolithic systems of thought. If there is a 'given' world, waiting to be explored and understood, then this (classical) view presents two problems to the postmodern mind: who 'gave' it in the first place, and whose account of it should we trust? The first problem is a legacy of the creator-'God' and the second one a rejection of the certainties of religious and scientific authority. The apparently overriding necessity to reject received wisdom has created a new system of thought that allows each one of us to be our own authority, to interpret the world and texts as we see fit. We can understand postmodernism as an expression of democratic and pluralistic impulses, yet it creates as many contradictions as it resolves. On the one hand the pluralistic impulse is central to our discussion, but on the other hand the denial that any account is closer to the 'truth' than any other would make the notion of discernment in the spiritual life redundant. More than that, it denies validity to an ancient and universal figure in human history: the seeker of truth. For Monty Python such a figure is merely delusional in its hunger for any trite religious platitude, while for *The Simpsons,* such a figure will inevitably succumb to exploitation unless rescued by secular values. Ridiculous claims of 'truth' deserve only parody, but the abandonment of the very possibility of distinguishing between truth and falsehood leads to a universal despair. The linguistic turn shares with the New Age a reluctance to make such distinctions – we call this the 'narcissism of difference denied'. While the New Age does so out of naivety, postmodernism does so out of sophistication.

In an extraordinary book called *Mysticism after Modernity* (1998) the British theologian Don Cupitt approaches mysticism entirely through the linguistic turn, coming to the absurd – and postmodernist – conclusion that mysticism is solely an act of writing. Cupitt usefully pursues the positive sides of the spiritual life through the idea of religious happiness, but his wholesale adoption of the secularist philosophy of language creates an impossible contradiction in his work. The anthropologist David Abram has suggested that European thought has traditionally sought a justification for elevating man above the animals, so, having lost 'God' and religion as this justification, it has turned to language instead.[28] Abram looks to the phenomenology of Husserl and Merleau-Ponty as a better source of understanding about language, one that is rooted in the body and Nature. But any adoption of secular philosophical principles will at some point or other undermine the foundations of spirituality. Intellectuals who are otherwise convinced of the spiritual life cannot find, it seems, an alternative intellectual framework outside of Western philosophy and its fascination for *words*.

The Power of Words

Psychiatrists are well aware that particular words can hold emotional charge for their patients, due to painful associations. The technical term in psychoanalysis for the process whereby an event or word attains this charge is *cathexis,* as mentioned earlier. (Neuroscientist Antonio Damasio uses instead the term 'somatic marker' for this charge, to suggest its location in the body.[29]) The same is true for whole societies, that words hold emotional charge, relating to their usage in periods of collective trauma. Wittgenstein, in his ideas about language, proposed that the meaning of a word was its current usage. This is fair to a point, but has resulted in a loss of interest in etymology, that is in the origins and history of a word. But in any attempt to create a discussion around the spiritual life we are forced to use many words, often ancient, that carry emotional charge, and hence their history and etymology is important. An improvement on Wittgenstein's position would be to say that the meaning of a word, *for a particular user-community*, lies in its usage in that community. If the user community for a particular English word happens to be the whole English-speaking world, then the meaning of that English word is uncontested. 'Table' would be a good example. 'God' however is not, but we will leave the discussion of that most difficult of words for section 2.5 *The Problem of 'God'*.

Six emotionally charged words

In the spiritual life we have inherited words with emotional charge, born out of the collective trauma around religion in the West – charged because they are related to discrimination, persecution, torture and death. We will look at four words which are thus highly charged: 'idolatry', 'heretic', 'pagan' and 'heathen', and two more that are also problematic: 'pantheist' and 'syncretist'.

The word 'idolatry' is used throughout the English-speaking world in an entirely negative way, though a secularist or Eastern thinker would not guess this from the dictionary definition and etymology. It means to worship idols, and comes simply from the Greek for worship and idol. (Socrates was condemned to death partly because he was found guilty of *not* being properly idolatrous.) But to Jews, Christians and Moslems it is one of the most condemnatory terms that one can apply to a religionist: in the Hebrew language the term originally just meant 'strange worship'. But the secular world has inherited its usage in a way that *ought* to strike people as bizarre. Aldous Huxley, thinking and writing far from the Christian mainstream, used the term as a disparagement for the worship of material goods and other secular valuables.[30] The great American writer on Eastern thought, Alan Watts, used the term to condemn the absolutising of the concepts of 'God'. [31] Karl Popper used it in his book *The Open Society and Its Enemies* to dismiss Hegel's theistic historicism.[32] Most shocking perhaps is that Mahatma Gandhi, probably because of his Western education, also showed no hesitation in using the term 'idol-worshipper' to disparage.[33] The theologian Phillip Blond uses the term no less than 23 times in a recent book on postmodern theology as a term of high and indignant condemnation.[34] Behind the simple etymology and neutral dictionary definition lies a history of appalling brutality and persecution, all hidden in and cathected to the word 'idolatry'. We will explore this more in section 3.3 *The Spiritual Wounds of the West*, but for now we can say that the transition to monotheism in the West was no mere evolution of spiritual thinking, but was accompanied by bloodthirsty repression of the earlier modes of religion, namely polytheism, Goddess worship and shamanism. The secular world is indifferent to the religious niceties implicit in the term 'idolatrous', but, like a child who hears a word always repeated with a certain negative feeling-tone, has simply adopted it to convey condemnation.

Spiritual development took place in India just as much as in the Middle East, but for some reason it was rarely accompanied by violence or denigration of earlier forms. The Western world in contrast inherited the savagery by which shamanism and polytheism were rejected in the Middle East, and it became the pattern for European Christianity. Hence the word 'idolatry' became for the Christian a word by which to smear and vilify those who did not follow the mainstream faith. Those tarred with the label could face excommunication, torture or death. It has been the experience of India however that idolatry *may* represent a primitive kind of faith, but that it can also be an expression of part of the *highest* and most transcendent spiritual genius, as for example in the great nineteenth century Indian mystic Ramakrishna. The *outward* form of religious expression was never a divisive issue in India (prior to the arrival of Islam) as it came to be in the West – after all, how can one judge another person's interiority? If one sees an individual, group, or whole society engaged in a religious practice that involves the creation and use of artefacts – for example a figure representing a deity or an aspect of a deity – then how can an outside observer judge their inner motivations? Or worse, call it an abomination and feel justified in violence towards such people? Hence we promote here the idea of idolatry as a *legitimate* mode of spiritual expression. To reclaim highly charged words for a neutral usage has, for example been attempted by black Afro-Caribbeans with the word 'nigger'. A less politically sensitive word undergoing rehabilitation is 'witch', through the efforts of feminist writer on Goddess religions, Starhawk. So why should someone not be proud to call themselves an idolater?

The word 'heretic' is linked with 'idolatry' in the pantheon of difficult religious words. Its etymology speaks volumes for its sorry history: it means 'to choose', from the Greek *hairesis*. 'To choose' is precisely what the European history of religion did not allow, leading to the horrors of the Inquisition (explored in depth in section 3.3.1). Like 'idolatry' the word 'heretic' has unthinkingly fallen into contemporary secular usage, meaning someone who goes against the prevailing doctrine. Its negative charge in its use today possibly lies with the unconscious and automatic association that we have in Western culture between 'heretic' and 'burn'.

The words 'pagan' and 'heathen' are sometimes used in popular speech to dismiss someone as primitive or 'unbelieving'. But in religious terms they are closer to 'idolatrous' in their negative connotation – pre-Christian or non-Christian. Both words have an etymology deriving from 'countryside', which tells us a great deal about the development of religion in the West (pagan from *pagus*, Latin for countryside, and heathen from *Heide*, German for open country). We will suggest that city-dwellers were responsible for the shift from shamanic to polytheistic forms of religion, involving a more symbolic treatment of the spirits residing in Nature. Alongside this process evolved a certain contempt for the country-dweller, or at the least a gulf of comprehension, tinged perhaps with anxiety that the country and its mysterious ways were the source of urban sustenance (as they remain). David Abram points out that as far back as the fifth century BCE Plato dismissed what lay outside the city walls of Athens.[35]

The word 'pantheist' may to the secular mind conjure up the quaint but not unreasonable idea that 'God' is in everything, but with little understanding of the negative connotations for the conventional Christian. It is only when one imaginatively immerses oneself in the Christian tradition of 'God' as a personal 'God', one who is separate from the world and can act in it, that one might form an intellectual appreciation of why 'pantheism' is unacceptable. But it is hard to feel the emotional charge associated with the word in Christian tradition, one that relies on its closeness to the terms pagan and heathen. Evelyn Underhill, author of the great work *Mysticism, The Nature and Development of Spiritual Consciousness*, and otherwise mostly presenting a balanced view of Western mysticism, has this to say:

> Unless safeguarded by limiting dogmas, the theory of Immanence, taken alone, is notoriously apt to degenerate into pantheism; and into those extravagant perversions of the doctrine of 'deification' in which the mystic holds his transfigured self to be identical with the Indwelling God.[36]

Anyone brought up in a Buddhist, Hindu or Taoist tradition (to cite just a few Eastern systems) would be baffled by Underhill's statement:

why 'degenerate' for a move towards pantheism? Why 'perversion' for the mystical union with 'God'? Keith Ward, the Oxford theologian and author of *God: A Guide for the Perplexed*, unconsciously confirms the unthinking Christian prejudice against pantheism when he says: 'It is very doubtful whether anyone has ever been a pantheist. It is one of those words that you apply to other people when you want to criticise their opinions without actually reading them.'[37] Ward doesn't consider that the use of such unexamined and culturally received terms of abuse is precisely the mechanism by which prejudice is spread, though in this case the user-community for the word and its surrounding prejudice is mostly the Christian one. We can only speculate that 'pantheist' provokes a visceral, anxiety about the shamanic or the mystical that mainstream monotheism has denied for millennia, and which must grip the mind of theologians with a negative cathexis.

Likewise, 'syncretism', which means an approach based on spiritual pluralism, might strike the modern mind as quite reasonable, but is in fact anathema to the traditional Christian. Pico della Mirandola (1463–1494) was exiled for his attempts during the Renaissance to create a spiritual pluralism: his crime was 'syncretism'. But, when Caesar conquered Gaul, one of his first tasks was to find the correspondence between the gods of Rome and the gods of the conquered people: this is an earlier form of 'syncretism', obviously alien to Christianity.

But how shall we be able to articulate spiritual difference (necessary for the analysis in this book) with such a heritage of loaded words? We need to use the old vocabularies – as well as the new and those from the East – but have to be careful about words that carry negative emotional charge. This is often first communicated in the way that others *use* a word, not in a grammatical sense, but more to do with tone of voice, gesture and facial expression. It is also possible that the emotional charge lies in the collective unconscious, rather than with the personal history of an individual, a theme explored in section 3.3.5 *The Unexamined Collective Trauma of Western Religion*.

We can give a contemporary example of how a single word can cause problems through its emotional charge: in August 2000 vigilantes daubed 'paedo' on the house of a woman doctor in Wales.[38] There had been anti-paedophile riots in Portsmouth, far from Wales, but what had provoked the attack was that the woman was a *paediatrician* – as her nameplate spelled out. No doubt the initially frightening experience was resolved for both the doctor and the vigilantes, but the incident is a reminder that human behaviour over *words* has often been historically graver. We condemn the vigilantes for their ignorance, but is Keith Ward's use of the word 'pantheist' really less worrying?

1.5 Psychology, the Brain and Secularism

It can be argued that philosophy and in particular its more arcane recent developments such as the 'linguistic turn' have a somewhat remote influence on day-to-day secularism. More obvious in our everyday secular cultural exchange there lie quite separate strands of influence: the psychological, going back a least a century, and the neurological, a more recent development. It is in the work of Sigmund Freud that we find one of the most vigorous rejections of religion, and the most pervasive mechanistic influence on the contemporary secular worldview. Freud's psychoanalytical theories provide not only a detailed dismissal of religion through a psychological perspective, but also a new language of interiority that has quite replaced the religious language of old. Another way to put it is that Freud completed the removal of the religious 'self', and initiated its replacement by the psychological 'self'.

Freud and other early psychoanalysts like Jung made strenuous efforts to persuade themselves and others that the work they were doing was 'scientific'. Within academic psychology however their work was largely ignored and a quite different approach was taken: firstly in behavioural psychology and then in cognitive psychology (as in Lodge's novel). These are scientific in the more rigorous sense of conducting empirical research, meaning laboratory testing of various kinds, the assumption being that a person's *interiority* was not directly open to scrutiny. Instead one can observe behaviour in a variety of controlled circumstances (behavioural psychology) or one can investigate perception in a variety of controlled circumstances (cognitive psychology). Neuroscience has extended this using non-invasive methods for investigating changes in the brain that accompany perceptual and other kinds of experience.

Psychology

There are four main branches of psychology: first, the cognitive-behavioural strand; second, psychoanalytical or depth psychology; third, humanistic psychology; and fourth, transpersonal psychology. Humanistic psychology breaks with analytical and cognitive-behavioural psychology because it goes beyond the pathological, taking an interest in the psychology of well-being. Transpersonal psychology goes beyond the humanistic because it places an emphasis on spirituality. Of these four strands it is the psychoanalytical that has the most direct impact on culture, and therefore on the prevailing secularism. The ideas of Freud permeate art, literature, poetry, theatre and film, while the ideas of Jung also readily enter popular culture through its productions, particularly mainstream cinema. Jung's theory of the archetypes, particularly of the hero archetype, influenced the American author Joseph Campbell, whose popular writings in turn influenced Hollywood through Christopher Vogler's classic text on screenwriting, *The Writer's Journey*. George Lucas, in the writing of the *Star Wars* films, consulted widely with Campbell for example. However, it is Freud's reworking of the Oedipus myth that appears again and again in American cultural production (some say even in its politics: as in the two Presidents Bush). And it is Freud's influence that is so deeply secular, in that it denies any interiority beyond the simplest drives of sex and death (*eros* and *thanatos*).

The great writer on philosophy, psychology and religion William James (1842-1910) had already separated out psychology and religion in his two major works; Freud went further and developed a psychological dismissal of religion as 'patently infantile'. Freud's work was anticipated in various ways in the nineteenth century, and was expanded upon enormously by his followers and fellow travellers, but his work is unique in its influence and its position as a cornerstone of the secular mind. His psychoanalytical theories are widely taught within cultural studies programmes, and underpin 'critical theory' through their reframing by French postmodernist Jacques Lacan and others. Whole generations of post-war humanities graduates have been exposed to these theories. The United States in particular exhibits a 'shrink culture' where novels, films and so on continuously draw on simple psychoanalytical staples, such as parent-child complexes, transference, and the language of ego. Fictional characters will often be in therapy, even if that is not particularly relevant to the plot: it is a simple fact of American social and cultural life.

We can comment here on one of Freud's best-known responses to the religious impulse. In the following extract Freud is writing about his friend Romain Rolland, the great French religious writer (whose work on Ramakrishna brought him to European attention), and Rolland's conception of the 'oceanic':

> I had sent him my small book that treats religion as an illusion and he answered that he entirely agreed with my judgement upon religion, but that he was sorry I had not properly appreciated the true source of religious sentiments. This, he says, consists in a particular feeling which he himself is never without, which he

> finds confirmed by many others, and which he may suppose is present in millions of people. It is a feeling which he would like to call a sensation of 'eternity', a feeling as of something limitless, unbounded – as it were, 'oceanic'. This feeling, he adds, is a purely subjective fact, not an article of faith; it brings with it no assurance of personal immortality, but it is the source of religious energy which is seized upon by various Churches and religious systems, directed by them into particular channels, and doubtless also exhausted by them. One may, he thinks, rightly call oneself religious on the grounds of this oceanic feeling alone, even if one rejects every belief and every illusion.

It is interesting to note that Rolland has taken care to use a non-theistic, non-devotional language in his response to Freud that he presumably hoped would fit with Freud's 'scientific' outlook, though it may also have come from his exposure to Indian religious thought. (We will see that Rolland's choice of terms, including 'eternity', 'unbounded', and 'oceanic', are universal to non-devotional mysticism across the world.) But Freud's response is summed up: 'I cannot discover this *oceanic* feeling in myself. It is not easy to deal scientifically with feelings.' From a Western theocentric tradition and viewpoint Freud is doubtful whether the 'oceanic' is the origin of religion, particularly as the term corresponds to nothing in his own experience. We can better articulate this by saying that 'oceanic' is a non-theistic or non-devotional conception, and hence does not fit so well with monotheism, as it does with non-theistic traditions like that of some Greeks (e.g. Pythagoras, Heraclitus, and Socrates) or Buddhism and other Eastern traditions. Freud speculates that 'it is a feeling of an indissoluble bond, of being one with the eternal world as a whole', which he then dismisses as an 'intellectual perception', perhaps because he had not come across anything like it in the Judaic tradition. We can only suggest here that Freud's spiritual instincts – such as they are – locate themselves in a rejected Judaic monotheism, and that he finds nothing spiritual in the Hellenic tradition which might use abstract words like 'eternity' or 'unbounded'.

Freud articulates the whole secular objection to *anything* spiritual, which is that the spiritual cannot be readily explained in terms of scientific, psychological, psychoanalytical, evolutionary or genetic *purpose*. At best it can be explained *away*, by the arguments made by Freud which are so familiar now, that the 'oceanic' and similar feelings are an inappropriate regression to an infantile state of unity with the mother. The following list (kindly provided by the online

Freud Museum) sums up the comprehensive onslaught against religion that Freud mounted within his writings:

- Religion is a 'universal obsessional ritual'.
- Religion is an attempt to master the Oedipus complex.
- Religion is the return of the repressed.
- Religion is a reaction to infantile helplessness.
- Religion echoes infantile states of 'bliss'.
- Religion is a mass delusion or paranoid wish-fulfilment.
- Religion is a way to hold groups together.[39]

It is hard to find a more forceful and vivid rejection of religion in all the annals of secularism, a useful reminder of the arguments ranged against any renewal of the spiritual life at a broad cultural level. There is some truth in each accusation of course, but by focussing solely on the negative no quarter is given for the counterbalancing positive. In this listing Freud provides the secular mind with an articulation of its core narcissism: that of self-sufficiency. It doesn't need religion because religion is irrational, ritualised, Oedipal, delusional, paranoid, regressive, and so on. If Marx provides the social component of secularism and Darwinism provides its scientific component, then Freud provides its psychological component, later buttressed (culturally at least) by cognitive science.

However psychology and spirituality do find common ground in some New Age thinking, epitomised perhaps in the work of popular philosopher Ken Wilber. His first book, *The Spectrum of Consciousness*, published in 1977, is one of a handful of seminal New Age works. He is known both as a pivotal thinker in transpersonal psychology and perhaps the most important philosopher of the New Age, with an extraordinary breadth of sources in both Eastern and Western spirituality. He is also the straight inheritor of the Western philosophical, scientific and Freudian tradition, adhering to principles of evolutionary thinking deriving from Darwin, and principles of developmental psychology deriving from Freud. Hence Wilber's psychology cannot ultimately escape its reductionist basis, one which inevitably leads to a psychologising of the self. The essence of the psychological self lies in a *mechanistic* understanding of the personality and its subjective experience. The psychological self, as a mechanical assemblage of internal entities such as complexes, exists only in relationship to other such selves, whereas the spiritual self is indivisible and exists in respect to spirits, 'God', or eternity. The psychological as a language of interiority has replaced the religious as a language of interiority, or, in the New Age setting, it becomes its gatekeeper.

The Neurologising of Self

The psychologising of self undertaken in the service of secularism is about a hundred years old. Much newer is what we identify as the 'neurologising' of self, a move to understand self in terms of the workings of the brain. The recent and growing status of the brain in secular culture is attested to in the vote of the US Congress to designate the last ten years of the twentieth century as 'The Decade of the Brain.'[40] We will see that biologist and Nobel Prize winner Frances Crick asserts that 'we are nothing but a pack of neurons', which sums up this trend. Neuroscience is in fact making extraordinary progress in mapping out the functions of the brain, now known to be more complex even in its *physiological* structure than the rest of the body put together.

Lodge's character Ralph Messenger in *Thinks . . .* epitomises the approach of the reductionist scientist. As conscious experience can be correlated with specific neuronal activity through non-invasive techniques, the conviction has grown that all human experience can mapped out like this. Brain scientists suggest that we can discover the 'neural correlates' of the experience of pain, of the colour red, and even of consciousness itself. Neuroscientist and gifted popular science writer Antonio Damasio has written a series of books that describe the latest research in brain science, findings that look at what Ralph Messengers' creative writing antagonist and lover, Helen Reed, regarded as the domain of the arts: *feelings*. His thesis is that feelings are the later mental correlates of prior bodily events he terms emotions, a process that relies on certain areas of the brain mapping out physical responses in the body.[41] We saw above that Freud held that 'It is not easy to deal scientifically with feelings', but science has moved on apparently. Molecular biologist Candace Pert goes a step further to show that the process is not just a matter of electrical circuitry – the neuronal system – but that the endocrinal system is involved through the production of neuropeptides.[42] 'You are nothing but your peptides', is the latest message from science it seems.

Francis Crick says we are nothing but our neurons, Richard Dawkins says that we are nothing but our genes, others offer peptides: who knows what is next. This is pure scientism talking – the science itself does nothing more than offer accounts of *mechanism*, but scientism as a worldview always reduces us to this. It is comforting of course to know that when the mechanism goes wrong there might be 'mechanics' with the knowledge and skill to fix it. But there is nothing in direct human experience that corresponds to neuron, gene or peptide; we have no *experiential correlates* for these scientific concepts, to turn the scientific parlance on its head. (Even the science tells us that there are no nerve endings in the brain, so how could we experience neuron or peptide?) But the grip on the contemporary mind of mechanistic explanations is hard to prise apart: this is why we will look in detail at how we got here from the science and philosophy of the seventeenth and eighteenth centuries.

New Age Science

The psychologising and neurologising of the self are typically secular strategies of understanding, but might psychology and neuroscience paradoxically also provide a new language of the spiritual? Certainly Ken Wilber and others in the transpersonal tradition think that psychology can take us there. Some even think that neuroscience will prove the validity of spiritual experience. One clear strategy emerging since the 1970s is to use science itself – whether of physics, biology or psychology – to prise open the claustrophobic constrictions of scientism. For example, neurologist and Zen practitioner James Austin has brought science and religion together in his book *Zen and the Brain*, where he argues that meditation and enlightenment in the Zen tradition can be properly understood through neuroscience.[43] Michael Persinger, a neuropsychologist at Canada's Laurentian University in Sudbury, Ontario, has been investigating what he calls the 'God module' in the brain, a region that becomes stimulated during religious experience. Popularised as the 'God Spot', it has brought about opposing sets of responses: from the sceptics like Polly Toynbee, who can now dismiss religion on the grounds of a brain malfunction,[44] and from the religionists, who can now claim scientific evidence for the validity of religious experience – even for the existence of 'God'. This is a good example of a neutral science provoking radically opposing cultural receptions.

The Institute of Noetic Sciences (IONS) in the United States, and the Scientific and Medical Network (SMN) in the UK are examples of organisations largely pursuing this strategy of using science to fight against reductionism. Both include scientists of high international standing and New Age thinkers of the more serious type. Not surprisingly, it turns out that James Austin is associated with both organisations. His approach in *Zen and the Brain* is fundamentally different to Damasio's, despite working out of the same discipline: he takes a 'both and' line, accepting both the validity of Western neuroscience and the profound truths of a tradition that go back to the Buddha. Despite Austin's sincere respect for Zen, and his commitment to meditation practice, his book, however, betrays his secular Western upbringing. On page one he says: 'Long ago in a distant land, a man's brain abruptly changed.. . . His transformation was so complete, enduring, and influential that he is still remembered as the Enlightened One.' In the preface he tells us that 'awakening, enlightenment, occurs only because the human brain undergoes substantial changes.' Yet his book presents no convincing evidence of this, that the Buddha and all those who undergo spiritual enlightenment have radically, or even marginally, changed brains. In this respect he is no different to Damasio, in adhering to what is rapidly becoming *the* secular shibboleth of the early twenty-first century: that we should understand who we are through understanding the brain.

But Austin's impulse to bring together religion and the brain is a departure from mainstream science, and is one we characterise as 'New Age Science'. It is New Age because it is motivated by a desire to integrate body and spirit into a single discourse, if not in actuality. Old religion, leading up to Descartes, separated the two, while secularism took only the material side of the 'Cartesian split' as valid. We can say that the New Age is characterised by the imperative to unite them. We have more to say later on about Descartes and his supposed central role in Western dualism, but here we just return to a more obvious New Age piece of science: Candace Pert's *Molecules of Emotion*. Her book of that name is in fact part autobiographical – of a woman struggling to make a mark in the male-dominated world of science – and part excellent exposition of her scientific discoveries for the lay reader. It is her appearance in the film *What the Bleep do We Know*, produced by effectively a cult organisation,[45] that has made Pert known to the wider New Age audience. Wittingly or unwittingly her discovery – that the body participates in brain states through the release of immuno-peptides – has made her an apparent keeper of the flame of that cherished New Age ideal: holism. To those desperate to find an integration between mind and body, the discovery that the brain is not after all a computer in total control, rather that 'intelligence' is equally corporeal, is manna from heaven. Of course it is welcome news to those who regret that both old religion and brain science have marginalised the life of the body, but an expanded reductionism is still a reductionism. Whether we reduce mind to brain alone, or to brain-plus-endocrine-system, Pert's discoveries do not help us understand 'mind' or whatever term we want to give to the subjective consciousness that physical science can never locate.

1.6 The Shibboleths of Secularism

We have suggested that a key characteristic of the secular mind, its key narcissism, is self-sufficiency. It is the assumption that the arts, politics and science, as conceived in reductionist, materialist terms, are sufficient to encompass all that is truly human. The Renaissance motto 'man is the measure of all things', picked up from Protagoras, seems to be the basis for this secular assumption, but the fact is that 'man' was conceived in those times primarily as a spiritual being, in relationship to an ultimate sacred entity, and reflecting of that divine nature. Man was potentially divine, and as the measure of all things he reflected what was sacred in creation. What the secular revolution of thought has achieved is the opposite: it managed firstly to posit the world as a machine, and then to conceive of man as made in its image. It is the narrowness of the secularist conception of a human being that has become the narcissism of self-sufficiency, a picture of the human being as a biological machine, bounded by its skin, disconnected from cosmos, and only made relational through society. Secular politics takes each person as an atomic unit, related through the means of economic production, conceived of through scientific materialism, and whose arts and culture must reflect that alienating vision. Scientific, social, and psychological secularisms form three pillars of the secular mind: Newton and Darwin stand for the scientific, Marx and Bentham for the social, and Freud and cognitive studies stand for the psychological. (More on Jeremy Bentham, the eighteenth-century Utilitarian philosopher, in later chapters.)

Sartre's *Nausea* is an early benchmark of modern secularism, yet it contains compelling descriptive passages reminiscent of the accounts by many mystics of their most valued experiences. How could Sartre get so close, yet completely miss this parallel? How did the Monty Python team in their scatter-gun satirising of all and everything, come to represent in the modern mind, not a schoolboy prank in respect to religion, but the essence of contemporary thinking on it? How was it possible, in a country as religious as America, for *The Simpsons* to represent religion as either fundamentalist or cultish, with no representation of the middle ground of true religion and of the religious happiness to be found there? How is it possible for a novelist like David Lodge to set out to defend the subjective realm of the arts from predation by the reductionism of cognitive science, but to fail so thoroughly? And which broader failure leads to what we are calling 'cultural autism'? The answer must be: these are all made possible by the massive adoption within contemporary culture of the shibboleths of secularism, the denial of which would place one beyond contemporary discourse. But the price of these assumptions is a terrible loss, one that is glaringly obvious if one had the least acquaintance with history, and which loss demands a much more comprehensive explanation of the origins of the secular mind than the ones usually offered.

Let us list some of these principle assumptions, or shibboleths, of the secular mind that we have so far touched upon (in no particular order): religion is monolithic; religion equals 'God'; self arises socially; religion and science are incompatible; religion and critical thought are incompatible; only scientific knowledge is valid; all experience can be understood as brain activity; religion is against progress; religion is obsessed with the afterlife; religion is regressive and compensatory; spirituality means self-denial, and religion is the cause of war. How on earth do these assumptions survive without any serious challenge?

Secularists, if they are bothered to make the least enquiry into these assumptions amongst religionists, find religionists to be (understandably) defensive about the track record of religion. These defensive responses in turn tend to confirm secular assumptions. The most pernicious of all of these is that religion is monolithic, and is defined primarily as a belief in 'God'. The monotheist, aware of nearly three hundred years of atheist attack, is defensive of this central proposition, and entirely colludes in denying the possibility that spirituality is multifaceted and can be open to critical thinking. Monotheism, and the Roman Catholic Church in particular – with its single ultimate source of authority, the Pope – convey to the secular mind that Judaism, Christianity, or Islam is a fixed entity adhering to a single eternal account of religiosity. Hence the secularist is unaware of the history of religion as one of continuous change, experiment and variation. As suggested earlier, the priests of a religion, including the Pope, are usually mere functionaries (mostly conservative at that), and the specialists, those whose life of spiritual practice and enquiry make religion possible, are usually the mystics. It is these individuals who show so clearly how varied the spiritual impulse is, and in the next chapter we explore that variety in depth.

Two:
Articulating Spiritual Difference

Early twentieth-century avant-garde spiritual thinkers like Rudolf Steiner and Carl Gustav Jung, despite their considerable exposure to Eastern thought, were adamant that the systems of Eastern spirituality were unsuitable for Westerners, while the poet Robert Graves was contemptuous of the great Hindu sage Ramakrishna.[46] But this chauvinist position is untenable in the twenty-first century world of global culture – the asymmetry of religious interchange is now more painfully obvious. The missionary thrust of monotheism into the Far East was absorbed into a pluralist spiritual culture there, but the absolutism of the one true 'God' has steadily resisted any reciprocity in the West. When the Inuit took on the Christian 'God' they found no mutuality in it: 'We never told the Christians that they would go to hell if they did not accept our religious beliefs. That is the difference between our spirituality and the white man's.'[47] This quote, from Inuit elder Pien Penashue, epitomises the encounter between monotheism and all other spiritualities, a morally unsupportable position in a pluralist world. The spirituality of the Inuit is a shamanic one, and shamanism is one of the spiritualities we wish to dignify and celebrate in this exposition. But it is a single insight from the Hindu tradition – the most determinedly pluralistic in the world's religious history – that will be the most important component in our system. This insight, as we have mentioned several times already, is that a key difference in spiritual impulse lies between the devotional and the non-devotional, known respectively as *bhakti* and *jnani* in Hinduism.

Spiritual Difference for the religionist, secularist, and New Ager

Celebrating, or even acknowledging, spiritual difference is an alien concept to the three key groups we have invited to the discussion: the rigid religionist (particularly monotheists), the committed secularist, and the more naive New Ager. The committed religionist needs persuading that spiritual pluralism is not the destroyer of faith, or worse, heresy; the committed secularist needs persuading that questions of the spirit are worth examining in the first place; and the naive New Ager needs persuading that questions of spiritual difference are not divisive. But each group also brings special qualities to this debate: the religionist brings historical depth and gravitas; the secularist brings the methods of critical enquiry, and the New Ager brings open enthusiasm. It is only if all three groups cooperate in the forging of a pluralist spirituality that we can avoid religious intolerance, the alienation of secularism, or rampant spiritual superficiality.

The thoughtful and committed secularist in the twenty-first century may examine our initial definition of the spiritual – as a profound connectedness – and respond that it is to be found in quite secular contexts. The first feelings in sexual romantic love, the deepening of these in a long-term relationship, the warmth and intimacy of family life, the sense of community, the sublimity of great art, music, or Nature, and the sense of awe opened up by science on the cosmos; all of these, they might say, are quite sufficient without introducing the problematic sphere of the 'spiritual'. However, as we shall examine in section 3.1.1 *Science and the Century of Alienation*, all of these may be available, yet there may still be a gnawing void in the psyche. And when these secular sites of expanded consciousness fail (through divorce, bereavement, or disability), and the individual is plunged into an alienated and shrunken world, is it the case that secular remedies are sufficient? Is not the secular pride in the sufficiency of its materialist worldview an unfounded narcissism? Deep as all the secular contentments go, high as all the secular exaltations reach, without the spiritual there is no bridge to the sacred.

There are two arguments here: first that however deep the connectedness offered by the secular world, however fulfilled an individual might be in this sense, there is a deeper connectedness and fulfilment available, one that we call 'spiritual', and not at all a contradiction to the best of secular values. It is simply of a different, sacred, order. The second is that when the secular promises fail, then secular remedies may be insufficient. A crisis of alienation may be treated as a psychological crisis but may only really be resolved in the much deeper context of a spiritual crisis – a 'dark night of the soul', for which a more ancient wisdom is required. But a fulfilled secular-minded person exposed to an unsuitable spirituality will only have their anti-spiritual prejudices reinforced, while an unsuitable spirituality will even more deny recovery, in the case of inner crisis, than no spirituality at all.

But all these deeper secular experiences just listed may well draw the secular mind into the world of spirit, without the hardship of personal crisis. The spiritual *may* be a totally separate domain of human experience, or it might start where the psychological leaves off. Or perhaps the *aesthetic* experience segues into the spiritual, or even the *scientific*? We do not need to shut out any of these possibilities. Indeed the analogy with art is useful: personally we may prefer the Impressionists over the Fauvists, for example, but on the whole such a preference does not lead to the view that galleries should display only the one and not the other. Whatever our preferences we accept the idea that the art world is rich and diverse, and that our national collections should reflect that. Spiritual texts often include personal accounts of a struggle to develop along the personal impetus of a particular spiritual orientation, either against the prevailing spiritual climate or with it. These accounts are often the most compelling parts of the spiritual literature and may also call forth an aesthetic response, for example in the medieval mystical text *The Cloud of Unknowing*, or in the poetry of Jellaludin Rumi. The value, just like with art, is that we can contemplate the text as a way of understanding our own spiritual impulse, just as the person who loves only the Impressionists is in some way deepening the understanding of their own aesthetic impulse when contemplating those works. While such a person may never visit an exhibition of, say, the Fauves or installation art, other art lovers will be more eclectic. Those drawn to be artists themselves may likewise wish to pursue a single pre-existing tradition, or pursue a postmodern eclecticism, or in fact initiate an entirely new art movement. Curators of national art collections, and art college teachers, exist to make available and accessible all the great art of the world through its epochs, though there is always a lively dispute as to what qualifies for the category of great art. The untutored response to art may be as dismissive as the untutored response to religion, but nations fund their art collections, and expect them to be representative. The great spiritual texts of the world's religious traditions ought to be regarded as highly, and their differences recognised and articulated. Art is not monolithic, so why should religion be? Pursuing the analogy further: curators may guide public taste to some degree, but it is the great artists that are the specialists – sometimes accorded celebrity status – not the curators. Similarly, in religion it is the mystics who are the specialists, not usually the priests.

But the committed Jew, Christian, or Muslim will find our articulation of spiritual difference a challenge; so may traditional followers of some Eastern religions, though probably to a lesser degree. The committed religionist only too often seeks *certainty* in their religion, glad to surrender their personal choice to a greater principle. The ideas of spiritual difference explored here will show that the major faiths tend to support only a narrow spectrum of spiritual impulse, though often different in each faith. What happens if this analysis demonstrates for the religionist that their personal spiritual impulse is at odds with that fostered by their faith tradition? They may have to exert pressure within their faith for change, or explore another tradition, and these possibilities may be unsettling. Conventional Christians often raise the objection of 'supermarket spirituality' to the possibility of choice, based on an uninformed perception of the New Age scene. This objection is dealt with when the depth and gravitas of the alternatives are properly understood.

For the open-minded New Ager the question arises: isn't the idea of spiritual difference historically divisive? Yes, but we would argue that division has arisen because of *minor* difference, harking back to Freud's dictum, and that real difference has been covered up in the Western tradition. And anyway the modern world is profoundly and positively divided in the sense of its pluralistic culture: a monolithic religiousness, whether of the type 'Christian' or 'The New Age' is no longer possible. The traditional religionist needs spiritual difference in a global multi-faith context, because religious toleration is ever more crucial in global politics. The New Ager needs spiritual difference to go deeper into the spiritual life. Both the traditional religionist and the New Ager have common ground in that the spiritual is already acknowledged as a real and valid part of their lives. But the committed secularist denies this possibility. Somehow a portrayal of the profundity, meaningfulness, and even *ease* of the lived spiritual life is also needed. (A classic Tibetan Buddhist text recognises this in its title: *Profoundly Bent to Ease Us*.[48]) More than ease: one of the great discoveries of the spiritual life is of *religious happiness*, a quiet bubbling source that lies beyond success and failure, beyond any material gain or loss whatsoever.

Spiritual Maps

We shall present a scheme of spiritual difference based on five *historical modalities* of the spirit and four polarities of *personal impulse* of the spirit. In constructing this scheme most of the world's spiritual traditions have been considered, along with existing spiritual maps or schemata. It is worth mentioning some of the most comprehensive spiritual maps that have emerged in contemporary scholarship, along with the reasons they will not been adopted in our analysis. Perhaps the three most highly regarded are that of New Age philosopher Ken Wilber, that of phenomenologist of religion Andrew Rawlinson, and that of Unitarian minister Peter Tufts Richardson. (There are of course thousands of earlier schemes developed over the millennia, but none of them have had the modern context of a global scholarship of spirituality.)

Ken Wilber's spiritual map is represented in his 'two-arc' diagram of spiritual development,[49] postulating stages that begin with the 'pleroma' – the undifferentiated state of infant consciousness – through adult egoic consciousness, through spiritual states we will term as 'esoteric', up to the highest or unitive stage. The last stage in this scheme resembles the first stage in some respects, and hence Wilber can clearly articulate where Freud goes wrong in conflating the transcendent 'oceanic' stage with that of the pre-differentiated stage of the infant. Wilber's enormous contribution to our understanding of world spirituality is undeniable, but his scheme suffers from the simple defect of being *developmental*. While it articulates some of the spiritual difference discussed in this book, Wilber's scheme creates a linear hierarchy privileging 'later' modalities of the spirit as more advanced. This would be sufficient to disqualify it here, but it also neglects what we have already highlighted as the key spiritual difference for understanding the spiritual history of the West: that between devotional and non-devotional spiritualities.

Andrew Rawlinson presents a useful 'quadrant' model of spirituality, but its major shortcoming is that of any quadrant model: it only allows for two variables. In Rawlinson's scheme these two variables are 'hot-cold' and 'structured-unstructured' creating the two axes of his model.[50] 'Hot' in this scheme means external to the aspirant and granted through grace and effort, whereas 'cold' means internal, and intrinsically possessed by the aspirant. 'Structured' means that there is a path to the goal requiring a map or guidance, while 'unstructured' implies that there is no gap between the starting point and the goal. The great advantage of such a model is the notion, implicit in the graph-like diagram that represents it, that different kinds of spiritual impulse may *segue* into each other along a continuum. However, because only two variables are possible in a quadrant scheme, there is an inevitable conflation of many key spiritual differences. In Rawlinson's system this conflation would take place for example between a structured devotional path such as that of Teresa of Avila and a structured esoteric path such as that of Rudolf Steiner, which would both be placed – according to the rules of his scheme – in the top-left quadrant, known as 'hot structured'. (For a similar reason, there cannot be any natural segueing between the two top quadrants.) Our scheme will show that Teresa of Avila and Rudolf Steiner have to be understood as representing radically different spiritual impulses, despite both being Christian.

Peter Tufts Richardson has developed a scheme he calls the 'Four Spiritualities' model, based on Jung's psychological archetypes and the Myers-Briggs typology.[51] The Myers-Briggs Type Indicator (MBTI) is a personality type assessment tool used in business management, based on four polarities of personal impulse: Extraversion / Intraversion, Sensing / Intuition, Thinking / Feeling, and Judging / Perceiving. While there is much to be recommended in Richardson's scheme, it is ultimately too much based in the tradition of psychology, and makes for a sometimes awkward mapping onto the life of the spirit. His Christian background makes him less of a generalist than Wilber or Rawlinson, and leads to some simple but fundamental errors: for example understanding shamanism as a devotional spiritual form.[52]

The Two-Fold Model of Spiritual Difference

We are now ready to introduce the Two-Fold Model of spiritual difference. This model provides a 'baggy schemata' that avoids the rigidity of the models previously discussed, and in particular recognises the tension between a historically located modality of the spirit which acts on the individual from outside, and a multivalent set of personal spiritual impulses that act on the individual from within. The historical component of the scheme could be mis-read as a kind of Hegelian or Wilberian developmental model, if not deployed alongside the personal component. Conversely the personal component alone would create the impression of an acultural free-floating 'self', not influenced by the historical setting.

2.1 *The History of the Spiritual Life*

In this first part of our two-fold model of spiritual difference we look at major modalities of the spirit through the ages. In our scheme the first major religious modality of the human race (wherever people appear on this planet) is animism / shamanism. It is not presented as primitive or lacking in any way, just *different* to the later religious forms. More than that, it happens to offer a modality of the spirit many consider essential to a proper re-engagement with the natural world, without which we are likely to become extinct through ecological catastrophe. Monotheism is also one of the modalities of the spirit presented here, but its position is more complex than any of the others because it is a phenomenon, arising in Middle East, which has dominated the West. There has never been a monotheism east of Iran (Persia), so our historical scheme would need a major variation to deal with the Far East.

The discussion of the earlier modalities of the spirit is informed by anthropology, ethnology, archaeology and the ancient literatures, all of which disciplines are subject to new findings, methodologies or better translations. Simone de Beauvoir, Freud and Jung, to take some examples from the first half of the twentieth century, were profoundly influenced by the anthropology of their day, making many assumptions that would now be discredited by later developments in the discipline. Similarly, some of the approximate dates given here, or even possibly some of the major transitions alluded to, may well have to be revised in the future. However, the modalities of spirit that we are exploring in this historical fashion are not so dependent on the detail of history, but rather their relevance hinges on whether these modalities are archetypally present in our psyches today. We know that contemporary Western city-dwellers actively take up ancient practices (neo-shamanism or neo-paganism for example) or adopt Eastern mystical traditions (neo-Advaita or Zen for example). The flourishing nature of these adoptions illustrates the ability of people to respond at a very deep level to modalities of the spirit that are remote in time or place from their contemporary adoption.

The Five Stages of Religion

We suggest five historical stages in religion as an idealised 'photofit' composite of world spiritual history:

- Animism / Shamanism
- Goddess Polytheism
- Warrior Polytheism
- Monotheism
- Unitive / Transcendent.

These religious stages are shown as a diagram in figure 1. The diagram has been drawn with curved lines to suggest that the boundaries between the five modalities are fluid, even where, in the case of monotheism, it strenuously resists other modalities of the spirit. Arrows have been drawn to indicate the historic progression, with one exception: the arrow from the Unitive / Transcendent to the Animist / Shamanic, which will be discussed later.

Animism / Shamanism

The animistic / shamanistic represents a modality of the spirit that seems to have emerged with humankind itself, everywhere associated with hunter-gatherer cultures. No early hominid traces seem to have been found without evidence of shamanistic practices, which include artefacts – such as fetishes and totems – for rites that revolve around Nature and 'spirits'. The shamanic worldview is predicated on a perception of Nature as imbued with spirit, where the elements of Nature such as rocks, mountains, trees, rivers, animals and skies are inhabited by spirits, and daily life is also filled with the presence of the spirits of the ancestors. This 'spirit world' is both beyond the so-called material world, but at the same time intimately entwined within it: they are not separable in shamanism.

The word 'shaman' itself comes via Russian from the Tungus people of Siberia (though it may have its root in Sanskrit). We are fortunate with shamanism because it has survived at the margins since the earliest of times, generally driven to unfertile or inhospitable territories by later agricultural societies. Hence in mountains, deserts or polar regions there persist to this day shamanic cultures, though increasingly ravaged by their contact with the industrial world, but from which we can assemble a consistent picture. A form of neo-shamanism has recently emerged, particularly in the United States, as people with little previous interest in religion take up shamanic practices under modern teachers or guides. The writings of anthropologist and cult author Carlos Castaneda, and the work of transpersonal psychologist Stansilav Grof have been significant in this revival. Michael Harner, author of the classical work *The Way of the Shaman* comments in the tenth anniversary edition on a 'shamanic renaissance':

> During the last decade, however, shamanism has returned to human life with startling strength, even to urban strongholds of Western 'civilisation', such as New York and Vienna. . . . There is another public, however, rapidly-growing and now numbering in the thousands in the United States and abroad, that has taken up shamanism and made it part of personal daily life.[53]

Entry to the spirit world may be facilitated by the use of dancing, drumming, psychotropic substances, or breathwork, and preferably takes place in natural surroundings. The shaman is usually important as a guide to the novitiate. While a key feature of shamanism is the access or 'flight' to the spirit realm (and hence the imagery central to shamanism of birds and feathers) the physical world of animals,

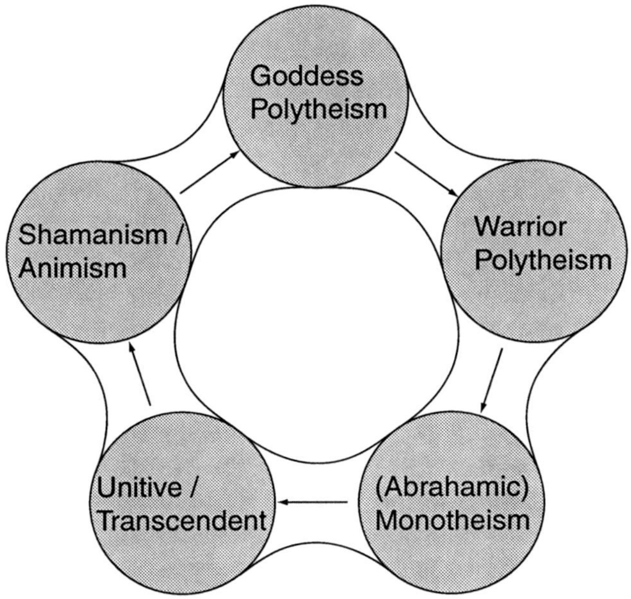

Fig 1. Five religious stages

plants, rocks, mountains and sky is shamanism's other vital pole. The shaman is a 'shape-shifter' not in the sense of science-fiction fantasy, but in the ability to contemplate, say, an eagle, bear or frog, and *become* it. This 'becoming' is not the anthropomorphism that the so-called civilised person imagines when they see the so-called primitive art of half-human, half-animal imagery, rather it is the opposite: a 'zoomorphism', the ability to take on the creature's manner-of-being in the world. The vision of what it is to be human in shamanic terms is a vision of amorphousness, also apparent in a certain gender-fluidity, and gender balance, despite the roles of men as predominantly hunters, and women as predominantly gatherers. Males of shamanic cultures often look a little feminine by modern Western standards, while females may not have our contemporary exaggerated femininity. Neither do these men and women have the individualistic or egotistic natures of Western people; perhaps this has led to the widespread but absurd notion that shamanic peoples have a less developed personal consciousness. A better way to understand a defining characteristic of shamanic peoples is as *self-effacing*. Good portrayals in contemporary culture are to be found in the characters played by Chief Dan George in the films *The Outlaw Josie Wales* and *Little Big Man*, or in the character of Dersu Uzula in the film of the same name by Akira Kurasawa. The well-known Native American actor Gary Farmer plays 'Nobody' in Jim

Jarmusch's film *Dead Man*, vividly conveying the humility at the heart of the Native American way of life, and what is arrogant in that of the white man.

Goddess Polytheism

If shamanism is a contested term, then anything to do with 'goddess' is doubly so. There is in fact a genuine difficulty here: shamanism has survived in the margins, and so is at least in principle recoverable as an ancient practice. But the Goddess religions – to the extent that we now posit their existence – were systematically eradicated by later modalities of the spirit, *and had nowhere to go*. There is growing evidence from archaeology that Goddess cultures replaced shamanic hunter-gatherer cultures in all parts of the world bar the marginal lands, around the period of the late Neolithic and early Bronze Ages. But the interiority of this modality of the spirit is more problematic and less recoverable than the shamanic, because there is no surviving unbroken tradition. Instead, there is a modern revival, led by radical scholars such as Starhawk and women from all walks of life, who seek to imaginatively reconstruct this spirituality. It is the archaeological evidence and the modern revival taken *together* that make the case for this as a major modality of the spirit.

In spiritual terms we can identify two stages in the transition from hunter-gatherer cultures to agrarian ones: firstly to 'Goddess polytheism', and then to 'warrior polytheism'. Both involve an increasing process of abstraction in the conception of the spirits or deities. The hunter-gatherer way of life existed for possibly some three million years, and as a first approximation at least we can associate the shamanic modality of spirit with that way of life. The implication is clear: that shamanism must be deeply rooted in the human psyche if it was present over such huge time scales. Hunter-gatherer societies in general seem to have been non-sexist and relatively non-violent, comprising family groups of about eighty to a hundred individuals, all well known to each other. Some seven thousand years ago two new skills emerged: that of animal husbandry and that of agriculture. Whether as horticulture (small-scale agriculture) or as agriculture proper, the new way of life spread rapidly and pushed the older hunter-gatherer lifestyles to the margins, along with its central spiritual form: shamanism. It is plausible that we can broadly associate Goddess polytheism with small-scale horticultural communities of the late Neolithic and early Bronze Age, and warrior polytheism with large-scale agriculture and societies of the later Bronze and Iron Age.

The growing scholarship on Goddess religions is led by feminist archaeologists and thinkers, whose role as feminists in this is partly to uncover the layers of patriarchal prejudice that have clouded the disciplines of archaeology, anthropology and history up to very recently. For example the huge quantities of goddess figures unearthed in archaeological sites all round the world were routinely dismissed as products of 'fertility cults' by male (Christian) academics.[54] Once the same data is looked at from the recognition that Goddess religions were not marginal cults, but central to thousands of years of human history, a radically different picture emerges. We can say that the word 'cult' is a quick way to dismiss anything non-Christian, and the word 'fertility' a quick way to dismiss anything non-masculine, hence 'fertility cult' conveyed total contempt in the mind of the male (white) Christian. Note that we have emphasised the *polytheism* in Goddess spirituality: this is to counter the Western cultural impetus to merely transpose a male monotheism into a female monotheism, a single 'God' into a single 'Goddess'.

Warrior Polytheism

As large-scale agriculture developed out of small-scale horticulture, methods of creating surplus came into being through the cultivation, drying and storage of grain: this became the first form of wealth and wages. This eventually led to the emergence of the city-state and created a radically new way of life over the small-scale horticultural community. Complex social and economic patterns emerged that allowed a class of society to live removed from the immediate production of food. At the same time there had evolved a new sphere of male human activity: warfare. The Neolithic and early Bronze Age period from approximately 10,000 to 5,000 years ago seems to have been dominated by women, with little militarism or central authority, but the male hunting instinct seems to have been transforming itself during this time into the instinct for war. Perhaps as crops required defending, not just from wild animals, but also from other tribes, defensive and then offensive patterns of aggression developed. With economic surplus and the development of settlements into cities, a military caste came into being, and with it what we are calling 'warrior polytheism'. Society became stratified in a way that was impossible during the epochs of hunter-gatherer societies, leading also to a priestly caste. The shaman might be called the 'priest' or 'priestess' of the shamanic way of life, but his or her powers were in healing and in shamanic flight: the new priesthood became guardians instead of great temples, and therefore also of wealth and power.

Warrior polytheism continues the proliferation of deities, along with the tendency to anthropomorphism, but in the patriarchal pantheon the female deities were demoted or forgotten in favour of new male ones. The new deities were as quarrelsome as the human city states now at war with each other, the ancient Greeks providing us with a good example. The key issue here however is the shift from a female dominated culture and religion to a male dominated culture and religion: from matriarchy to patriarchy. The flowering of Goddess cultures between seven thousand and five thousand years ago was brought to a fairly abrupt end in the Mediterranean world at least, and historians and anthropologists have argued over the causes for more than a century. What brought patriarchy into being is of interest generally, but in terms of religion the shift was nothing short of a revolution. Shamanic spirituality was equally male and female; goddess polytheism was female dominated; but all later religion became male dominated.

City-dwellers needed more symbolic thinking to deal with the increased social, technical and political complexities of their lives, and hence the process of abstraction was necessary. One of the qualities of polytheism is the tendency, possibly inherited from its shamanic roots, to be localised, that is for gods to belong to regions. Hence the Romans, in administrating their conquests, acknowledged the local gods and allowed their worship as long as the gods of Rome, particularly the Emperor, were included. (The Jews were a notable exception in the Empire, refusing to co-operate with this.) In fact the gods in different cultures were rarely so different as to be unrecognisable, Caesar for example had no difficulty in finding the Roman equivalents to the deities he discovered in conquered Gaul (we mentioned this earlier as a process referred to by an early meaning of the word *syncretism*).

Monotheism

In Europe at the time of Christ there was a plurality of religions that encompassed shamanism, Goddess polytheism, warrior polytheism, and the transcendent (yet to be defined), all mixed up, and differentiated mostly across the city / country divide. Also, in one corner of the Mediterranean, there was monotheism. As Jews travelled widely around the Roman Empire their monotheism spread and no doubt influenced a few individuals, but, as the Jews did not proselytise, it would have been a rare religious strand of thought. It was not until the rise of Christianity that monotheism turned from being a regional oddity to a mainstream religion that would silence the earlier religious history of the West. We shall follow the origins and development of Abrahamic monotheism in Chapters Three and Four in great detail, as it is central to the story of Western spirituality and the emergence of secularism. Here we can just sum up some key points in its development: it is geographically unique; originating in the Middle East and nowhere else in the world; it evolved from warrior polytheism, not Goddess polytheism; it is associated with a horrifically violent rejection of what went before; it is patriarchal; and it becomes a religion uniquely associated with the written word (giving rise to 'language mysticism').

Monotheism in one sense may retain the idea of 'God' as a being, a supreme being, analogous with just one of the previous gods, but somehow incorporating the separate characteristics of all of them. Anthropomorphism, that is the natural tendency to project human qualities onto the polytheistic gods, is fiercely resisted in Judaic monotheism, with its prohibition on speaking the name of 'God', and the denial of attributes to him. However, it is inevitable that a single 'God' becomes anthropomorphised in the popular mind, however much this tendency is resisted, and this problem is central to the history of monotheism. Judaic, Christian and Islamic monotheisms are intolerant of other gods, or alternatively they have, for example, read their Judaeo-Christian 'God' into Brahman (supreme cosmic spirit of Hindu philosophy) in a quite inappropriate way. Similarly, Jesuit missionaries in China persuaded themselves that the Chinese 'heaven' was the equivalent to the Christian 'God', though their fellow-missionaries, the Franciscans, thought otherwise and finally convinced the Pope to come down against the Jesuits.[55]

The idea of monotheism seems to have emerged in four possible locations: in fourteenth century BCE Egypt with Akhenaton, in Northern Africa, in Persia (modern day Iran) as Zoroastrianism in the sixth century BCE, and in Israel as an ongoing process of change which may have been influenced by the Egyptian, African or Persian examples. While Egyptian monotheism was rapidly overturned, and Zoroastrianism became a tiny religion on the world stage, it is Judaic monotheism which has had the most impact on the world, through its influence on Christianity and Islam.

The Transcendent / Unitive

In the final development of the religious life, monotheism becomes a transcendent or unitive religion, represented for example by Buddhism and the concept of *nirvana*. However there is no simple example of a monotheistic religion developing into a transcendent one, for example in the case of Christianity and Islam the mystics who pursued this modality of the spiritual life were generally persecuted. Meister Eckhart (1260-1327) is Christian example who was condemned by the Inquisition, though he died before any punishment could be inflicted (see section 4.2.3), while Mansur (Al-Hallaj) is an example in Islam who suffered a horrible martyrdom in 922 CE. In both cases the problem for the mainstream religion in which they were located was that their understanding of 'God' had gone beyond the notion of a separate being or creator-God: their unitive experiences calling for a language of transcendence foreign to monotheism. The position of the mystics in the Judaic tradition is more complex, in that they tended to avoid personal declarations of union, and any popular anthropomorphism of 'God' was balanced by its continual refutation in the writings of Judaic scholars, so the 'transcendent' is generally the most difficult component of the spiritual life to describe, particularly in the West. (Christians please note: the term 'transcendent' is *not* used here as in the usual theological contradistinction to 'immanent'.) The term 'unitive' is equally good, but not as familiar. The East has the well-known concept of 'enlightenment' (*nirvana, moksha* or liberation) describing the goal of the transcendent religionist and a transcendent religion. It is 'unitive' in the sense of 'not-two' (as in Zen and Advaita formulations), which translates in monotheism into the concept of union with 'God'.

Summary

The arrows in our diagram from shamanism to Goddess polytheism and so on can be read as implying a developmental sequence, or even an inevitable sequence. But the final arrow is drawn from from the transcendent back to the shamanic, to counter any sense of inevitable historical development, and to highlight how the shamanic and the transcendent so easily coexist in the East. Tibetan Buddhism is one example, while the coexistence of Zen and Shinto in Japan is another. The arrow linking the unitive or transcendent with shamanism also suggests an *engaged* spirituality: a Buddha who turns again to the world and Nature.

If read in a literal developmental sense these five stages do not map onto the religious history of the world in any simple kind of way. It is clear that by at least 600 BCE all five stages had already emerged onto the religious world scene, though our historical knowledge of this and earlier periods is rather sketchy. In both the Mediterranean and Indian cultures of that period we find evidence that all five strands were present and to one degree or another *available*. This means that individuals, depending on their circumstances and mobility, were able to draw on the support for different types of spiritual life. The fact that an individual has a right to pursue the spiritual life appropriate to them was of course never enshrined in the ancient world either in law (human rights are a recent development), or very often in opportunity (economic and geographical mobility was limited). Nevertheless those who devoted themselves to the spiritual life in the ancient world often did travel large distances to seek out the teachings they could not find locally, and a large part of ancient discourse resulted from such travellers bringing back new teachings (Pythagoras being a good example, or Solon in Plato's *Timaeus*).

2.2 Four Spiritual Polarities

The purpose of listing the five historical modalities of the spirit is not to be accurate or exhaustive, but to be the best possible selection of key modalities of the spirit that can create a *typology*. The list also serves to convey the idea that the monotheism familiar to the Western secular mind – Abrahamic monotheism – is far from the be-all and end-all of religion, and that it can more properly take its place at the table as *one amongst five*. As a typology however the list carries with it the cultural contexts of different epochs and societies; we now describe a second typology that is based on *personal impulse*.

The spiritual life is highly complex, and, if we compare it to the human body, we can profitably slice through it in different ways to reveal different internal dynamics. Just as medical students dissect the body according to the skeleto-muscular system, the cardio-vascular system, the lymphatic system and so on, building up a composite image of the body's workings, so we can dissect the spiritual life. The schemes of Wilber, Rawlinson and Tufts Richardson do this in different ways, as does our historical scheme discussed above. Our second scheme – based on four polarities of personal impulse – is more like a typology of the Jungian archetypes, as in the MBTI scheme. Our four polarities are:

1. esoteric vs. transcendent
2. *bhakti* vs. *jnani* (devotional vs. non-devotional)
3. *via positiva* vs. *via negativa* (orientation to the manifest world)
4. solitary vs. social.

The devotional / non-devotional polarity put forward here

corresponds in some degree to the feeling / thinking polarity in Jungian typologies while our third and fourth categories both relate to the extravert / introvert distinctions made by Jung. However, the spiritual life has a different inner geography to the psychological.

The Esoteric and the Transcendent

The polarity of esoteric vs. transcendent, uses again the term 'transcendent' as deployed in our historical modalities, but here it represents a personal impulse. This distinction has no equivalent in the Jungian tradition because Jung himself did not much understand the transcendent (as we define it), and to the extent that he understood it he believed that the esoteric would lead to it. These impulses are juxtaposed here (despite the fact that some individuals and traditions appear to embody both) because they involve profoundly different paths and practices, and lead to profoundly different spiritual goals. While Wilber recognises the difference between the esoteric ('subtle' in his terminology) and the transcendent or unitive, as a Jungian he has also made the mistake that the one leads to the other. Rawlinson's scheme entirely confuses the distinction, while Richardson fails to recognise the esoteric as a category at all. (As a Christian he inherits a tradition that has written out and vilified much of what we will include under the term esoteric as 'witchcraft'.)

The esoteric (or occult) is defined here as all that pertains to the hidden world of disembodied beings or non-material energies – we called this the 'spirit world' in the context of shamanism. The 'transcendent' or unitive has already been defined, but this can now be a little amplified: it is an inner experience of union in which the sense of a small, separate, volitional self is transcended. We will use the terms 'unitive', 'non-dual' and 'transcendent' interchangeably here, though each arises in different traditions. It is what Rolland was referring to in 'oceanic', and which was so roundly dismissed by Freud.

For those not familiar with figures in the esoteric traditions in the East or West, we can give some examples: William Blake, Jakob Böhme, Emanuel Swedenborg and Rudolf Steiner in the West; and Paramahansa Yogananda (known through his popular *Autobiography of a Yogi*)[56] and Sri Aurobindo in the East. The word 'occult' means hidden, as in 'occluded', but has overtones that may suggest the sinister, superstitious or demonic. The alternative word 'esoteric', also means hidden, and is possibly a less loaded word, but the terms will be used interchangeably here to describe the *disembodied* spiritual life and the operation of non-material energies connected with it, *all* of which lies outside modern science. Clearly the occult or esoteric must have its roots in the shamanic worldview. It was highly active in the Renaissance (as magic and alchemy), and even a hundred years ago was the subject of serious engagement by Western intellectuals and artists. The esoteric movements known as Theosophy and Anthroposophy from 1880 to 1914 had many educated adherents, though such movements were curtailed by WW1 and the subsequent shift in interest amongst the intelligentsia towards socialism. The most scholarly and gifted proponent of occultism in the early twentieth century was Rudolf Steiner, founder of Anthroposophy, and whose books and teachings still have a substantial following today. He was explicit that he taught an 'occult science' the basis of which was knowledge that he gained from access to the spirit world and various kinds of disembodied beings. Steiner tells us that up to the age of thirty he lived more intensely in the spirit world than in the so-called material world. His teachings emerged from the world of the Austrian / Slovenian border in the late nineteenth century, where shamanism probably still lingered, as is suggested in his encounter with a rural herb-collector.[57]

The transcendent or unitive is demonstrated by individuals as far apart in history and geography as Plotinus, Eckhart, Spinoza, the Buddha, Ramana Maharshi and Jiddu Krishnamurti. (We can usefully call these mystics 'buddhas' or 'lost buddhas' in some cases.) Where the occult / esoteric turns its gaze to the manifest and manifold of a hidden (and hotly disputed) world, the transcendent / unitive turns its gaze to the unmanifest, the simple, the infinite and the eternal. Other terms commonly used are: the void, emptiness (*shunyata* in the Buddhist traditions), *nirvana*, the stainless, the imperishable and the unmanifest. The texts and oral traditions of the transcendent / unitive are naturally short on detail compared to those of the esoteric, and much more telegraphic, poetic, and allusional. The unitive impulse is not concerned with the distinction between an *embodied* reality that is manifest and manifold and a *disembodied* reality that is manifest and manifold, and does not set out to map either world. Its orientation is to the whole, not the part. It may pursue knowledge of the manifest as a celebration of its inherent unity, but never as a way of extending the scope of personal or even collective agency. On the other hand the esoteric is usually pursued to gain *power* or *powers* of some kind, whether for the benefit of the individual, of humanity, or the cosmos. (Carlos Castaneda's popular esoteric texts on the teachings of the Yaqui Indian Don Juan focus extensively on the idea of *power*,[58] though how much of his writings are a fiction remains unclear.)

Esotericists are likely to *read the esoteric into* existing spiritual traditions and texts. Hence Jesus is claimed for the occult (by those pursuing esoteric Christianity, for example by Annie Besant), likewise there is an esoteric Buddhism (which would have been anathema to the Buddha himself). Some of the greatest confusion in the history of religion and the spiritual life lies with the failure to distinguish properly between the esoteric and the transcendent. This may be due to the persecution of shamanism, magic and witchcraft which led to secrecy; it may well be that the obscurity of much of the alchemical tradition for example is due to the need to speak in a coded language. It is notable that while occultists rarely warn their followers of the dangers of the transcendent, those of the transcendent path often warn of the dangers of the esoteric. Explicit warnings against the occult come from as far afield as Confucius,[59] Ramana Maharshi,[60] and Jiddu Krishnamurti, whose career after his break with the Theosophists represents a repudiation of the esoteric in favour of the transcendent.

If, instead of attempting to adjudicate between the truth-claims of the two orientations, we explore them as distinctive and equally valid spiritual impulses, then we have a useful tool for understanding many apparent contradictions in the spiritual life. Two great spiritual teachers in the twentieth century that we have mentioned, Rudolf Steiner and Jiddu Krishnamurti, were both part of the Theosophical movement for a time in their lives. Steiner is, however, exclusively esoteric and Krishnamurti exclusively transcendent (more on both teachers in section 3.1.3 *Spiritual Teachers in Modernity*). Both make it very plain that they are not following the path of the other, although events in their lives look similar at times.

Bhakti and *Jnani*

We are now finally ready to make the crucial distinction between the devotional and non-devotional spiritual orientations, the central role of which will unfold in Chapter Four. We can define these two by drawing on ideas from Hinduism and using Sanskrit terminology. Generally one of the biggest obstacles to the wider dissemination of Hindu and Buddhist thought in the West is the large technical vocabulary in Sanskrit and Pali; hence the introduction of the terms *bhakti* for devotional and *jnani* for non-devotional is done with some reluctance. However, no Western terminology exists that can do the same work as these Hindu terms, because the insight behind them never gained momentum in the West, although in Islam there are terms that come close (*'ishq* and *'ilm*). We can speculate that *jnani* is etymologically linked with 'gnosis', but the latter term is too loaded in the Western tradition. (Note that the 'jn' in *jnani* is pronounced as the 'ny' in canyon.)

Bhakti means devotional (a term we are familiar with in the West, though not necessarily in sympathy with), while *jnani* means non-devotional (though not at all the same as atheist or anti-religionist). The non-devotional spiritual inclinations, teachings and traditions of the world have been more or less lost to popular Western consciousness as *philosophy*. The *bhakti-jnani* distinction should be understood more as a spectrum than as a simple dichotomy however, and pertains exclusively to the transcendent, rather than to the esoteric. We can understand the distinction by looking at extremes, for example Teresa of Avila as an extreme of the devotional *bhakti* orientation, and the Buddha as an extreme of the non-devotional *jnani* orientation. Generally, between individuals situated on these extremes, there is a gulf of incomprehension giving rise to more unfortunate misunderstandings in the spiritual life. For the pure *jnani* individual, *bhakti* is mere emotionalism, for the pure *bhakti* individual, *jnani* is mere intellectualism. To those who occupy more of the middle ground falls the important task of trying to reconcile the extremes, though from the centre position it is always possible to miss the importance of the distinction in the first place.

Hinduism, from which tradition the terms *bhakti* and *jnani* derive, has always been aware that the spiritual temperament is roughly divided between these two orientations, though of course many individuals respond to both. One of the best-known articulations of this distinction played out in the lives of Ramakrishna and Vivekananda (late nineteenth century spiritual teachers) and central to their understanding was that *bhakti* and *jnani* are of equal value or weighting in the transcendent spiritual life. The West also has *jnani* or non-devotional exemplars and traditions, Plotinus and Neoplatonism being the best known. However, the West has come to think of the spiritual life exclusively in terms of 'God', a largely *bhakti* way of thinking. A *bhakti* usually creates a language of 'God', a theistic language, or possibly of 'gods', a polytheistic language, as opposed to a *jnani* who tends to develops a non-theistic language. However, we should note that if the prevailing culture is determinedly theistic, then the *jnani* will use the language of 'God' in a particular way, resulting in the 'God of the philosophers'. This term came to prominence when Blaise Pascal spoke so scathingly of it.

The great advantage of a non-devotional spiritual language for the contemporary world is that it allows for a better dialogue with art and science. Neoplatonism, as a Western example of the *jnani* orientation, can be seen as the spiritual engine which kick-started the artistic and scientific flourishing of the Renaissance period, but the Christian mainstream, as we shall see, has been wary of it. This goes right back to St Augustine, who, after rejecting his early faith of many years (Manichaeism), investigated a range of spiritual teachings including Neoplatonism, but settled eventually on the teachings of St Paul. Why? The explanation in the *Confessions* comes down to the simple fact of Augustine's devotionality, in other words he recognised Christianity as a fundamentally *bhakti* religion. Despite Augustine's obvious intellect, he set the tone for the piety of what we now know as the 'Dark Ages', a period that rejected learning in favour of devotion, and which effectively repressed the *jnani* traditions of Neoplatonism. This is explored more fully in Chapter Three.

While the *jnani* element does have a place in the long tradition of Christianity, there can be no doubt that the devotional or *bhakti* has dominated the religion. Put another way, the *jnani* element in Christianity is the vine that grows around the tree of *bhakti*. In Buddhism the reverse is true: the *bhakti* element is the vine that grows around the tree of *jnani*. Although the Buddha himself (as revealed in the Pali Canon) was fiercely non-devotional, the devotional later crept into Buddhism as it became a popular religion. This way of illustrating the difference between the two religions is of course only a first, though useful, approximation.

Many commentators, including F.C.Happold in *Mysticism*[61] and Huston Smith in his classic *The Religions of Man*,[62] have identified the important distinction between *bhakti* and *jnani* in the spiritual life, though not always in these terms. Other important writers on comparative religion, such as Mircia Eliade, ignore the distinction. In this book it will be given prominence, more even than the distinction between the esoteric and the transcendent: the *bhakti-jnani* distinction is nothing less than a Rosetta Stone for the deciphering of complexity in the spiritual life.

Via Positiva and *Via Negativa*

We come now to the third in this set of spiritual polarities, the difference between *via positiva* and *via negativa*. These terms originate in nineteenth century Western theology, and mean approaches to 'God' respectively through affirmation or negation. In the affirmative path the characteristics of 'God' are enumerated as manifest and manifold and to 'His' glory, while in the negative path (originating with Dionysius the pseudo-Areopagite) 'God' is approached through the denial of manifest qualities. More technically these paths are referred to respectively as *cataphatic* and *apophatic*, the latter also referred to as the *negative theology*. Interestingly the *Encyclopedia Britannica* says this about the negative theology: 'It could be called an Eastern idea present and effective in the Occident.' In practice it had a Western Neoplatonist or Gnostic origin.

While leaving cataphatic and apophatic to retain their original meanings, the terms *via positiva* and *via negativa* are borrowed here and slightly modified to extend their application to both theistic and non-theistic spiritualities. *Via positiva* indicates here a spiritual impulse that is outward looking to the manifest world (or world-curious), while *via negativa* is inward looking and contemplative of the unmanifest. The *via positiva* impulse is often found in those with an interest in the arts and science, and an enquiring kind of mind. The *via negativa* impulse is found in the more renunciative temperaments. William Blake is a good example of the *via positiva* impulse, while the Buddha is a good example of the *via negativa* impulse. As with *bhakti* and *jnani*, this tends to be a spectrum rather than an either/or distinction, and individuals may move between these extremes over a lifetime or in a more cyclical way perhaps even during a single day. Again, those adhering to one of the extremes may find it difficult to comprehend the other, or accept it as a valid spiritual expression. We should stress that the cataphatic / apophatic distinction of the Christian tradition is couched in purely *theistic* terms, so it implies a distinction between different types of piety, or different approaches to 'God'. The *via positiva* / *via negativa* distinction as used here in its expanded form can refer to either a devotional (theistic) orientation, or a non-devotional orientation, in other words to either *jnani* or *bhakti* orientations.

One of the best statements ever made in the *via positiva* comes from William Blake: 'eternity is in love with the productions of time.'[63] As we have remarked, the language of eternity is often an indication of a nontheistic *jnani* orientation, in Blake's case *via positiva*, while in the Buddha's *via negativa*. It is inconceivable that the Buddha could have made a statement like Blake's, as there is nothing in all the Buddha's recorded sayings that betray any interest in the 'productions of time'. India produced both some of the most vivid expressions of the *via positiva*, for example in the Upanishads, and also some of the most vivid expressions of the *via negativa*, for example in the Pali Canon of Therevadan Buddhism.

One way to understand the pursuit of what we term the transcendent, as opposed to the esoteric, is in the transcendence of the narrow sense of self, the transcendence of ego. In the *via negativa* one progressively dis-identifies with those elements normally associated with self, such as body, family, opinions, nation, tribe, etc. The Hindus have a term for this: *neti-neti*, meaning 'not this, not that', summing up a spiritual path that has an equivalent in the Western tradition of the negative theology. In the *via positiva* one pursues the opposite method: one progressively identifies with more and more of the manifest universe, thus equally leaving behind the narrow sense of self. Again the Hindus have a term that sums this up: *tat tvam asi*, meaning 'thou art that', and in the West this path is exemplified in Thomas Traherne, William Blake, and Walt Whitman – 'three poets of the *via positiva*'.

The Solitary and the Social

In the last of our four distinctions we look at how the spiritual impulse may tend to the solitary or to the social. The classic account of the spiritual path is of a period of solitude or withdrawal in spiritual searching, followed by a return to the community and the commencement of service or teaching. This means that solitude or social engagement may dominate at different times in a person's life. In the world's spiritual teachings this tension is often found: an exhortation to gather together in prayer or meditation, or other communal ritual, or to carry out good works, but at the same time a recommendation to attend to the inner life. Such tensions are clear in what we know of the teachings of both Jesus and the Buddha, and at a more prosaic level form a dynamic at the heart of our everyday relationships with others: we may be alone and feel the need for company; at a different time we may be in a crowd and wish for solitude. Society generally praises those who are actively engaged with good works or political change for the better, but on the other hand has always (prior to secularism) treasured the accounts of the solitary mystic who reports on states of consciousness clearly not possible in the market place.

In the Abrahamic monotheisms of the West a great deal of emphasis has been placed on the relational, perhaps best summed up in a way suitable for the twentieth century by Martin Buber in his famous book *I and Thou*. For Judaism the ultimate 'I-Thou' relationship is with 'God', but it also represents the highest form of relationship with another human being, one not based on the exploitation inherent in what Buber calls the 'I-It' relationship.[64] In Buddhism there can be no ultimate 'I-Thou' relationship of any kind, only the end of relationship in the state of *nirvana*. However, there is a great emphasis on 'spiritual friendship' meaning a deep sharing in the spiritual life, and a mutual encouragement towards the goal of liberation. In fact, this may be one of the greatest attractions of the spiritual life, a form of friendship deeper than any secular one, yet paradoxically based on the experience of the profoundest solitude. The Buddhist term for a community of spiritual friends is 'sangha', the experience of which can be painfully lacking in the secular life. In all religious traditions a great deal of effort was expended over the centuries regarding the rules of community, both for those in the monasteries and the laity. These rules crucially determine the balance between the social and the solitary. The amount of time spent in withdrawal and solitary contemplation, the amount of time spent in collective prayer and duties around the monastery, the rules over when conversation was permitted; all these were endlessly experimented with for the sole purpose of enhancing the spiritual life.

The Zen Master Hakuin rails in the eighteenth century against the 'unborn' Zennists, a school who 'pursued little discipline, ate much rice, and produced copious quantities of steaming horseflop',[65] while Christians at much the same time were exercised about the 'Quietist heresy'. In both cases the objection was to a spiritual life that seemed overly withdrawn and disengaged. In the case of Eckhart, as we shall see, he combined the exhortation to detachment above all else, with life as an active preacher and administrator with responsibilities to a large region. The Buddha preached a total disengagement with the world, yet drew up highly detailed regulations for the community or *sangha* of seekers that he headed, and served it tirelessly for over fifty years.

Although our typology of spiritual impulses is about the interiority of an individual's spiritual life, and not in the first instance concerned with the form in which they are expressed, we need to say something here about *ritual*. As a general term we can understand ritual to stand for the outer and visible practices arising from the spiritual impulse. When individuals follow the social spiritual impulse, then ritual becomes the key binding element of the spiritual community or *sangha*, more so than belief (this is particularly true in Judaism where observance far outweighs belief as the mark of faith). Ritual may also be the first site of spiritual experience, powerfully affecting an individual long before they can make any detailed articulation of their own spiritual impulses.

2.3 *The Varieties of Spiritual Impulse*

We can now place our entire model in a single diagram (Figure 2).

Note that we have drawn the historical modalities here without arrows, thus suggesting that there is a timeless quality to each (or in Jungian terms that they are *archetypal*) and that there is no developmental ordering. The two parts of our model of spiritual difference are designed to work with each other, to emphasise that a person might be located in a particular epoch and culture – hence subject to social pressures to adopt a particular patterning of the spiritual impulse – but that the person's own spiritual impulses may be otherwise configured. We say 'patterning' or 'configured' because the modalities of the spirit and the personal impulses come in various combinations or clusters, as we shall see in many examples. One more factor plays a part in this discussion, when applying the model to individuals: whether the constellation of spiritual impulses at work in that individual is operating in a nascent or fully developed fashion. We can expect many historical figures to show a nascent spirituality in which our distinctions may not be that apparent, compared to mystics who will have pursued one or more spiritual impulses in depth, and hence make their characteristics more apparent.

The Two-Fold Model Applied to Individuals

We now look at a few examples of how our Two-Fold Model helps us explore and understand spiritual diversity in the lives of well-known individuals.

We will start with two heroes of secularism, Newton (1642-1727) and Einstein (1879-1955). Einstein wrote a delightful book, *The World as I See It*, its popularity a reminder that the secular world takes the private thoughts of its great scientists seriously. While he has been cited as an example of a great scientist with mild Asperger's syndrome[66] – a difficult emotional life – his spirituality has not impinged much on the secular mind. But when we apply our typology to what we learn of his spirituality from his little book we quickly spot that he is transcendent, *jnani, via positiva,* and solitary. Einstein confidently finds in the Jewish tradition that feeling which Freud could not locate within himself as the 'oceanic'. Einstein says of Judaism that it contains 'a sort of intoxicated joy and amazement at the beauty and grandeur of this world . . . [it] is the feeling from which true scientific research draws its spiritual sustenance, but which also seems to find expression in the song of birds.'[67] In other places Einstein vehemently rejects monotheism (seeing it as the

<table>
<tr><th>Historical Modalities of the Spirit</th><th>Personal Spiritual Impulse</th></tr>
<tr><td>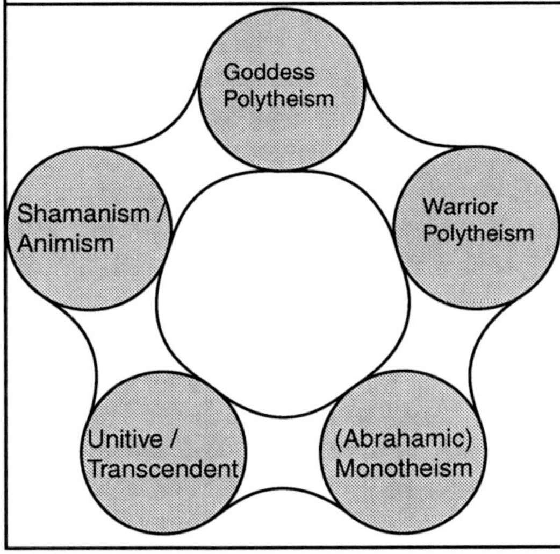
</td><td>esoteric / transcendent
bhakti / jnani
via positiva/via negativa
solitary / social</td></tr>
</table>

religion of fear), being drawn rather to the 'pantheism' of Spinoza. A person who devotes their life to science is a very different person to one who devotes their life to the spiritual, so we are not suggesting that we should see Einstein as an incisive religious or spiritual thinker: his spirituality is nascent. However this brief analysis also shows how little use the word 'atheist' would be to describe Einstein – all it would do is to show that he held a deep-seated objection to some aspects of a specific monotheism: Judaism.

Newton differs from Einstein in a few notable respects: he is likewise *jnani, via positiva* and solitary (possibly suffering from the same lack of emotional development), but his preoccupations are esoteric and Abrahamic rather than transcendent. Newton is more developed religiously than Einstein, but in neither are their spiritual impulses developed to the point where they make a significant historical mark on religion. If we turn to a man in whom the spiritual impulse plays a more public role, William Blake, then we might be surprised to discover that our analysis gives him a similar profile or typology to that of Newton. He is likewise esoteric, *jnani, via positiva,* solitary, and in terms of historical modalities, intensely Abrahamic. The main difference in typology is that he is *also* intensely transcendent, demonstrating a combination of the transcendent and esoteric polarities that arises often enough to have created the confusion between them. Cultural history however has cast him as the arch-enemy of Newton, because, in our terms, his *via positiva,* his engagement with the manifest world (the 'productions of time' as he calls it) is via *art*, rather than science. Culturally speaking Newton stands for science and the Industrial Revolution while Blake, as a father of the Romantic movement, rails against these. In spiritual terms however the two men have far more in common than what separates them, and both fight against Deism, the form of Christian thinking that was ironically triggered by Newton's scientific legacy. If the secular world is only reluctantly facing up to Newton's intense preoccupations with alchemy, astrology and numerology, and also his committed Arianism (a heretical form of Christianity), it has an easier passage with Einstein.

Richard Dawkins has been introduced earlier as a *vituperative* atheist, and we will examine his ideas in more detail in section 3.1.1. But what happens if we read Dawkins on the lookout for signs of a nascent spirituality, as we have for Newton and Einstein? Indeed, what would prompt us to do that in the first place? The answer in fact lies throughout his writings: his joyous and profound awe in contemplating the beauty and complexity of the world, *as revealed by science*. The language he uses and the concepts he conveys in this respect resonate with many great spiritual thinkers of the *jnani via positiva* persuasion. By this characterisation alone we can place him spiritually in the company of Newton and Einstein. His vituperative atheism is targeted against Abrahamic monotheism in its most rigid and authoritarian form, in other words against *clericalism*. But the lyrical prose he devotes to the sheer wonder and beauty of nature, and the efforts he has made to wrest such an activity from the poets (particularly in his book *Unweaving the Rainbow*), attest at least to a nascent *jnani* spirituality. He as much as admits it in the following passage: 'Einstein was profoundly spiritual, but he disowned supernaturalism, and denied all personal gods. Unworthy to lace Einstein's sockless shoes, I gladly share his magnificently godless spirituality. No theist should presume to give Einstein lessons in spirituality.'[68] Indeed. Why should theists give non-theists a lesson in spirituality? But then, equally, why should non-theists give theists a lesson in spirituality? When we can understand a non-theist in this context, not as an atheist, but as a person with the *jnani* spiritual orientation? We will see later that the individual with a *jnani via positiva* outlook fared badly in the medieval Christian context of *bhakti via negativa*; worse still was the fate of those with an esoteric orientation, one that we shall argue has some resonance with the scientific. One then asks: is Dawkins victim of a traumatic historical moment, one long passed, but still active in his and the collective unconscious?

C.G. Jung and Ken Wilber are also both *jnani* and *via positiva* in their impulses, and the balance they show in their emotional lives (as far a we can tell from autobiographical material) suggests that they negotiate the solitary / social spiritual polarity in a creative manner. Where Wilber's instincts are strongly transcendent, Jung's are esoteric however. Wilber's insistence on the non-dual as the goal of the spiritual life, and his own practice as a Buddhist, stand in sharp distinction to Jung's preoccupations with alchemy, the Kabbalah and Gnosticism. Jung's psychological and spiritual instincts are *atavistic,* drawing him much closer to the shamanic, though overlaid with his Abrahamic Western inheritance, while Wilber's instincts are evolutionary, taking him to what he considers are the heights of the transcendent: *nirvana, moksha,* liberation. Hence it is no surprise that while Wilber criticises Freud for falling into the 'pre-trans fallacy' or reductionist fallacy, he criticises Jung for the opposite: the elevationist fallacy.[69] Wilber means by this that Jung takes the deepest parts of the personal and collective unconscious to be the goal of transcendent religion, thus elevating what is 'primitive' to what is 'developed'.

We have not had space in these few examples to adequately justify the assignation of our typologies, and have left details of those with a *bhakti* orientation to a later section. However it is clear that an individual will have a spiritual impulse that can contain any combination of our categories; that they may hold them in seeming contradiction; that the relative emphasis may change over a lifetime; and that each impulse may be present in a nascent or developed form. Without going into any details of how we arrive at the following profiles (most of which are explored later), here are some individuals listed with their spiritual impulses:

The Buddha:	transcendent, *jnani, via negativa,* solitary as regards inwardness, social in the *sangha* and ethics
Walt Whitman:	transcendent, *jnani, via positiva,* solitary in his personal life, social as a democratic visionary
Rudolf Steiner:	esoteric, *jnani, via positiva,* solitary in his access to the spirit world, social in his application of insights so gained
St John of the Cross:	transcendent, *bhakti, via negativa,* mostly solitary
Jellaludin Rumi:	transcendent, *bhakti, via positiva,* solitary in his devotional ecstasy, social in the order he founded

Ramana Maharshi: transcendent, *jnani, via negativa,* solitary to start with, but allowing a community to grow around him

Giordano Bruno: esoteric, *jnani, via positiva,* solitary in his highly personal views, social in the attempt to share them

Richard Rolle: transcendent, *bhakti, via negativa,* mostly solitary

Benedict Spinoza: transcendent, *jnani, via positiva,* solitary in his unique vision, social in his daily life

William Blake: transcendent *and* esoteric, *jnani, via positiva,* solitary visionary, social in the scope of his concerns.

John Muir: transcendent, *jnani, via positiva,* solitary in his Nature Mysticism, social in his vision for preservation of the wilderness

Isaac Newton: esoteric, *jnani, via positiva,* solitary

Albert Einstein: transcendent, *jnani, via positiva,* solitary

Richard Dawkins: transcendent, *jnani, via positiva,* social.

2.4 Pathologies and Correctives in the Spiritual Life

When we try and understand the human being as multi-dimensional, that is as living a number of parallel though interconnected 'lives' or facets, then we may speak of any one of them as being healthy, undeveloped, or pathological, to some degree independent of the others. Hence our working lives may go well but our love lives badly, or vice versa; we may have a vigorous life of the body through sport or dancing but a limited intellectual life through a lack of reading, or vice versa. The spiritual life is presented here in much the same light, as one of these human dimensions or facets. We may have a sex life that is healthy or pathological, or it may be undeveloped or atrophied. By way of a parallel we may have a spiritual life that is healthy, pathological, or, in a largely atheistic secular society, it may be dormant, nascent, undeveloped or atrophied. However in all human activity, pathologies – though dramatic when they arise – occur only in a small proportion of cases. When the field of science and technology goes wrong, for example in the cases of Chernobyl, Bhopal and thalidomide, we don't reject science in its entirety as pathological. We don't necessarily lay the blame for WW1, WW2, Hiroshima, Nagasaki, asthma, obesity, depression, pollution and global warming at the doors of science and technology (though perhaps we should).

Crucial to the origins of the secular mind are the genuine or perceived pathologies within the history of religion, particularly monotheism. However, we need to show briefly that each historical modality of the spirit, not just monotheism, has its risks and dangers. Starting with shamanism, the particular risk it holds is of *stasis* – shamanism is a profound dimension of the spiritual life, but with little dynamic within it for development. Goddess polytheism shares with shamanism this characteristic: great matriarchal goddess-oriented cultures have existed, but their essence was a cyclical rhythm that again may have held little dynamic for development. Warrior polytheism on the other hand, with its heroic/hubristic masculine energies engaged with the outward activities of art, science, politics and warfare, may suffer not from stasis, but from cruelty as its principal spiritual pathology. Monotheism, which we will explore in detail, may extend this cruelty with intolerance and absolutism. Finally, the transcendent stage – epitomised in Spinoza's phrase 'under the aegis of eternity' – may decline into indifference as the mere *facts* of the temporal world lose their power to move the soul.

In terms of our four polarities of personal impulse, some of them will be dealt with in the discussion on Christianity, but the pathologies of the remaining ones can be summarised first. The occult or esoteric impulse, which includes aspects of the shamanic, can be perverted towards 'black' magic: the quest for power can be for harm, while the risk of power in any case is that it can corrupt (as we know in the secular context). The risks of the *via positiva* are that of submergence in the manifest world at the expense of the spiritual (a descent into *materialism*). There is also the possibility of madness, shown clearly in the case of William Blake, and lurking in the background with Walt Whitman (both Blake and Whitman seemed to deal with this through writings of an almost perverse nature, though Whitman's were never published in his lifetime). The risks of the social dimension of the spiritual life are either a degeneration into legalistic morality, or of becoming lost in the social life of the *sangha* (the Buddha once threatened to dismiss his entire spiritual community, complaining that their chatter was as noisy as 'fishermen hawking fish').[70] The dangers of the solitary spiritual orientation are indifference and passivity. But *all* of these dangers live alongside and are far outweighed by the positives in each modality of the spirit.

Control / Indifference

We have suggested that Christianity is predominantly a religion of devotion, hence *bhakti,* while Buddhism is predominantly a religion of *jnani* transcendence. Christianity is largely theistic, Buddhism largely non-theistic. The spiritual impulse of a *bhakti* religion has a societal counterpart in love for one's neighbours (St Paul effectively created this central core of Christianity). The spiritual impulse of a *jnani* religion has a societal counterpart in compassion for one's neighbours (this is at the core of the Buddha's teachings). From this starting point we can make a generalisation about the course of these religions, and of the *bhakti* and *jnani* impulses, when they are corrupted or degenerate. Put simply, when the love central to a devotional religion runs cold it becomes *control.* When the compassion of a non-devotional religion runs cold it becomes *indifference.* This simple statement illuminates both the central spiritual pathologies of West and East respectively, and also the points at which mutual incomprehension comes into being. Central to all the crimes committed in the name of Christianity has been the mania for control, as demonstrated in the Inquisition (as we shall see later). Love in mainstream Christianity (as opposed to within its mystics) is an *active* principle, and the encouragement by Paul of this active principle was central to the early spread of Christianity – it got involved in people's lives (and has been remarked upon as an early example of a welfare system). Its inversion, or its perversion, becomes control. A genuine spiritual love gives rise to a genuine love and care for others, but as the spiritual love hardens into doctrine and dogma, so that active love can degenerate into an active interference and finally an active persecution: a mind police.

In section 1.4 *The Role of Language* we looked at six emotionally charged words from the Christian heritage: idolatry, heretic, pagan, heathen, pantheist and syncretist. We can understand the way that these words acquired their negative charge over time as a manifestation of the absolutist desire for control. In a pluralist and tolerant spiritual context each of these words ought to have a *positive* emotional charge. Christians would do well to repeat the mantra 'idolatry is good, idolatry is good, . . .' as a corrective to their ancient prejudice, for example. As religious scholar Karen Armstrong puts it: 'Despite the bad press it has in the Bible, there is nothing wrong with idolatry *per se.* . . .'[71] But to celebrate spiritual difference in this way has been historically impossible in the Christian tradition because that would have meant a loss of control.

Compassion, as taught by the Buddha, is a less active principle than Christian love, more a profound tolerance and respect (this was the hallmark of the great Buddhist expansions of the Ashokan era in the fifth century BCE). When it derives immediately from the spontaneous unfolding of the *jnani* impulse, then it reaches out just like Christian love, active in its concern for suffering, though expressed more in reference to the mental sources of suffering than the physical sources. The Buddha's teachings on the end of suffering are about an internal *stance*, not about external improvement in society: at no point did he consider ways of reducing the likelihood of physical suffering through medicine, technology, or social justice for example. But as the spontaneity and immediacy of the *jnani* impulse fades, compassion may shrink to a passive tolerance, and finally to indifference. When Christians and Buddhists first encountered each other there was on the one hand a recognition of what was the highest principle in each. On the other hand the pathology of missionary zeal and the accompanying political control were alien to the Buddhists, while for Christians the single most reported impression of their encounter with the East was of the overwhelming indifference to the human condition.

Renunciation

Renunciation becomes a pathology when taken to extremes, and was common in the West, Middle East and the Indian subcontinent, though not much found in China it seems. The perception of renunciation – of family, wealth, power and social responsibility – as at the heart of religion is one of the factors in its contemporary rejection. For example, Jeremy Bentham, the eighteenth-century Utilitarian philosopher, when casting around for a principle which expressed the opposite values to his materialistic political philosophy, chose asceticism. Renunciation favours the next world over this one, and so led to Nietzsche's scathing comment that the 'afterworldsmen' should hurry up and leave this world to those more interested in it.[72] Marx and Bentham saw the doctrines of the afterlife, of a better life to come, only as a means by which religion set out to repress the legitimate demands of the ordinary people for a better conditions in *this* life.

The issue of renunciation is related to the distinction between the social and the solitary explored earlier, and is also a natural expression of the *via negativa*, so we can see that here is a complex issue. Renunciation has a bad name in the Western consumerist outlook, yet at the simple level of logic it is an issue that has to be approached with a new receptivity: obesity at the individual level kills, and obesity at the cultural level – that is absence of voluntary restraint on economic growth – will lead to environmental catastrophe. (These factors could not have been imagined by Bentham in the eighteenth century.)

But the life-denying, renunciative elements within Christianity, Hinduism and Buddhism for example have led many in the modern world to reject religion, not realising that this feature of the religious life is merely an exaggerated *via negativa* or solitary spiritual impulse. There is however a huge difference between the spiritual detachment of the Buddha and Eckhart, on the one hand, and an indifference that is devoid of all human compassion, on the other.

Iconoclasm

We will use the term 'iconoclasm' as a shorthand for all the violence done against idolatry, whether it is the destruction of images (mainly sculpture and painting) – which is the usual meaning of the term – or violence towards the people who hold those 'idolatrous' beliefs. This violence around idolatry is at the root of most religious violence in the world and its most usual justification. To give an example: the destruction by the Taliban of the great Buddha statues at Bamiyan, Afghanistan in 2001, was the expression of an ongoing habit of violence going back to the Old Testament. Another example is the destruction of church art and ornamentation by the soldiers of Oliver Cromwell during the British Civil War of 1642-7. At the heart of Abrahamic monotheism is an inexplicable and arbitrary horror of the worship of idols, present equally in Judaism, Christianity and Islam. In 1757 David Hume expressed his appreciation of Judaism and Islam for their greater determination in the rooting out idolatry than in Christianity.[73] In contrast idolatry is central to Indian religion, and it is a remarkable tribute to the spiritual pluralism of India that it offered to buy the Bamiyan Buddhas from the Taliban to save them from destruction. But India is a largely Hindu nation with only a tiny Buddhist minority, forcing the question: where would we encounter such spiritual generosity in any monotheistic tradition?

Narcissism of minor difference

We can return to Freud's 'narcissism of minor difference' as a useful way to think about group dynamics that generates cohesion within one group at the expense of denigrating another, usually on the basis of a difference that appears quite trivial to outsiders. There are countless examples in the religious history of the world, the 'differences' appearing monumental or trivial depending ones' viewpoint. Hence to an atheist the difference between Sunnis and Shi'ites is minor, but this is not so for those who police Iraq. To a Christian the difference between the Theravada and the Mahayana traditions in Buddhism is minor compared to the more striking fact of the absence of 'God' in both of them. But to Hakuin, and the entire Zen tradition in fact, the difference is not minor at all. Where the monotheistic Western religions differ from the East is that the 'narcissism of minor difference' seems to have led to a *much* higher level of violence than in the Eastern religions of Hinduism, Buddhism and Taoism, and their derivatives. This shows up early in Western religious history with the slaughter of the peoples and priests of religions 'different' to those of the Judaic 'God'. Whether these recorded acts of violence in the Old Testament have any historical basis is unknown, but the Christian and Islamic worlds have certainly used them as a model or justification for later violence in the name of 'difference'. The secular atheist mind can only judge all these differences as minor and the violence as inexcusable.

Teleology / resurrection

Some more peculiarities of monotheism will be outlined here that have their origins in the deserts of the Middle East, and which have affected mostly the West. The first of these relates to the idea of some future salvation or perfection, a discourse known as *teleology*. It starts as a physical idea of the Promised Land, or as a spiritual idea of heaven, but it is always in the future, and may involve a Messiah or saviour. (Such a future-orientation is quite foreign to shamanism, Goddess polytheism, warrior polytheism or the transcendent.) For a beleaguered nation hope is always in the future, and at a practical level there is nothing pathological about such hope: indeed the three Christian virtues include it (faith, hope and charity). However teleology gradually became hard-wired into the Western psyche leading to both positive outcomes and also quite negative outcomes for the spiritual life. The triumvirate of Hegel, Marx and Darwin created a modern teleology linked to a belief in progress that is the secular equivalent of religious teleology. (Evolution in the biological sense has no goal, but the cultural reception of evolutionary theory is profoundly teleological, and led for example to the appalling doctrine of eugenics, the selective breeding of humans.) Teleology can be positive because it avoids the indifference of the Eastern religions described above, but negative because it removes the essential discourse of *now*, the insight that mystics have repeated so often that being in the present moment is the pre-requisite of spiritual practice. (Hence Sartre missed the mystical significance of his protagonist's experience in *Nausea*.) If future-orientedness dominates the psyche, then it becomes difficult to examine the dynamics of the actual world, substituting instead an imaginary (usually fantastical or utopian) teleological future for the present.

At its worst teleology becomes apocalyptical. The concept of a Judgement Day on which 'God' will weigh up the good and bad deeds of all those who had ever lived, sending sinners to Hell and saints to Heaven for eternity, became an elaborate epic in Revelation, planting the idea as a malevolent self-fulfilling fantasy for the West. The tragedies of Ruby Ridge and Waco in late twentieth-century America are the most recent manifestations of this spiritual pathology. In both cases groups created small isolated communities, heavily influenced by Revelation, the last book in the New Testament. In their contempt for the law, in particular US Federal law, they created situations that resulted in armed stand-offs with the authorities, leading to the loss of innocent lives. It was the teleological vision of a Kingdom of Heaven on earth that led both groups to the particular structure and motivations of their communities. These led directly to violence and the self-fulfilling prophecy of apocalypse (the Waco siege in particular led to a hellish conflagration which killed most of the community). The Apocalypse is deeply embedded in American religious culture, as shown in films like *The Rapture* and the *Left Behind* trilogy (based on the best-selling novels by Timothy LeHay). While teleology in the Middle Ages was unrelated to improving the lot of physical life, the Industrial Revolution shifted expectation from the religious sphere to the material sphere. The German Idealists brought the two together in a spiritual vision of the future so powerful that Ken Wilber remarks: 'Henceforth, any spirituality that did not embrace evolution was doomed to extinction.'[74] It is almost as if evolution *is* the new religion.

The second and related fantasy is that of physical resurrection, variants on the theme of a bodily reanimation in an eternal paradise. The idea of rebirth and resurrection in the spiritual life is not confined to Christianity, apparently being part of many traditions in the Mediterranean at that time of Christ. The goal of physical immortality likewise appears as a corruption of Taoist thought in China. The metaphysical essence of these ideas is present more widely in nearly all the world's traditions, as a form of deep inner renewal, though the idea of a *physical resurrection* is peculiarly Western. The East, with its emphasis on reincarnation has not generally taken up the idea of physical immortality, apart from in China. As spiritual ideas in themselves teleology and resurrection may have innocent and valuable origins, as teachings which metaphorically point to essential insights in the spiritual life: those related to the eternal. However, the metaphors become taken literally, and take on a life of their own, with the capacity to trigger the kinds of very visible tragedies outlined above.

Literalism

The Abrahamic monotheisms originated at the same time as the alphabet, and no religion on Earth is more text-oriented than Judaism. A tradition of texts (the Western model), rather than a tradition of teachers (the Eastern model), is bound to lead to the problem of *literalism*: the rigid and literal interpretation of what was originally poetic and allusional. The Gospel of St John opens: 'In the beginning was the Word, and the Word was with God, and the Word was God.' This identification of the highest principle, the most divine, with the 'Word' is unique in the world's religions: Zen in contrast is dismissive of writing and of discursive thought (i.e. thought carried by the vehicle of language). The Chinese language as a pictographic language may be harder to take literally than the alphabetical language originating with the Hebrews (later adopted by the Greeks). Author Leonard Shlain is adamant that the adoption of written language, in particular the alphabetical languages, coincided with, if not directly led to, patriarchy.[75] Whether he is right about this or not, there is no doubt that monotheism is unique as a modality of the spirit that is bound up with the written text.

The Kierkegaard Corrective

Pathologies of the spiritual life are found as extremisms, and are not representative of the flow and adaptation of the religious life in normal circumstances. In most epochs and continents checks and balances to these extremes of the religious life have existed, as they have to extremes in all other aspects of life. The greatest such corrective lies in spiritual pluralism, or the celebration of spiritual difference; hence its articulation is so important. However, such pluralism is what has been mostly absent in the monotheistic religions, and so we need to identify where the core corrective for these religions truly lies. Clearly it is in religious love.

In casting around for a relatively modern exposition of religious love, the writings of Kierkegaard recommend themselves, in particular an essay called 'Works of Love'. Soren Kierkegaard (1813-1855) was a Danish writer and philosopher often described as the father of existentialism. His thought and personality are complex and contradictory, and he shares with Blaise Pascal three characteristics:

he is oddly modern yet deeply Christian; he is an intellectual with a strong feel for the devotional; and he often writes in an aphoristic style. In 'Works of Love', written in 1847 he puts forward as the central Christian idea the commandment given by Jesus in Matthew 22:39: 'But the second commandment is like it: Thou shalt love thy neighbour as thyself.'

Kierkegaard points out that this contains within it the presupposition that every person loves themselves, but that 'it is Christianity's intention to wrest self-love away from us human beings.'[76] He goes on to say: 'If the commandment about loving the neighbour were expressed in any other way than with this little phrase, *as yourself*, which simultaneously is so easy to handle and yet has the elasticity of eternity, the commandment would be unable to handle self-love in this way.' He keeps returning to the idea that eternity lives between self-love and the love of the neighbour *as yourself*, in other words that selfish love is narrow and claustrophobic, whereas the love of others is the straight road to eternity. The very use of the term eternity is typical of the *jnani*, yet Kierkegaard's repeated return to the theme of love in his work shows a remarkable and unusual empathy for *bhakti*, as is found in Blaise Pascal.

We can call this precept to love one's neighbour 'The Kierkegaard Corrective' when understood as a central idea in all three monotheisms, and which, in the daily practice of all three religions, mitigates against the very specific pathologies of the spiritual life found within them. This is no illusion: even a sceptic attending a Jewish, Christian or Islamic service in full swing cannot fail to notice a sense of religious love that goes far beyond personal friendship or the ties of family. And the inextricable link between charitable work and religion cannot be brushed aside: even left-wing secularist and *Guardian* writer Neal Lawson says: 'I don't care if they are Muslim, Catholic or Church of England – if they preach the cause of the poor and needy in our bloated materialistic world, then they are my people.'[77]

In 'Works of Love' Kierkegaard shows that the literary for him lays far below the religious in its significance; that the romantic love of the poets is precisely the secret self-love which is merely the erotic, not the eternal. This is quite consistent with his attacks on the German Idealist tradition in philosophy – particularly Hegel – which draw on the Romantic tradition. For example, Werther's all consuming love, in Goethe's highly influential *The Sorrows of Young Werther*, is painted as sublime for the elevation of its object – Charlotte – to the role of transcendent inspiration and muse, and leads to Werther's dramatic suicide. Kierkegaard's point here is quite precise however: the erotic love, which appears to place its object so central as to lead the poet to a state of self-abnegation, is in fact purely selfish. This is simply because this love is based on preference, as it is in friendship for that matter. To love one's neighbour *as oneself* is to love all neighbours, with no preference: after all it is easy to love the beautiful, the generous, the attractive, but what about the poor, the halt and the lame? Kierkegaard is well aware of the enormity of this challenge, but his instinct for eternity as a deeply religious man makes him sober enough to cast off the false coin of the Romantics.

If we return to Matthew we can remind ourselves that the first commandment, from the same passage, is: 'Thou shalt love the Lord thy God with all thy heart, and with all thy soul, and with all thy mind.'

This helps us understand that in the transition from romantic love to the expanded love that includes within it all humanity, it is the love of 'God' that is the catalyst, the method, and the goal. In our terminology this love defines the *bhakti* impulse, but which at the same time requires no particular 'God' or even any 'God': it is a profound spiritual impulse of the heart. This impulse may be entirely inwards, in those we characterise as 'solitary', or it may find outward expression for the neighbour in those we characterise as 'social': in practice the poles of this tension create a continuous movement between them.

We also suggest that in 'love thy neighbour as thyself' we have the spiritual corrective not just to the pitfalls of temporal love, but to certain pathologies of the spiritual life that Jung termed 'inflation'. To come close to 'God', to attain union with 'God', or to attain to the *jnani* equivalents: eternity, infinity, *nirvana* and so on, is to pose all the time the risk of ego-inflation. 'The Kierkegaard Corrective' – not of course originating with him, but so exquisitely thought through by him for the modern world – is the corrective to all forms of narcissism: we stop looking at ourselves and look at others instead. To love the neighbour as oneself is an expression of that profound connectedness that we define as at the heart of spirituality. By attaching Kierkegaard's name to it, as a serious, even gloomy philosopher, we also indicate that the realisation of it is arduous, and is part of a personal struggle on the other side of which is alienation and depression in a very modern sense.

The Kierkegaard Corrective also translates well into the core motivation of secular ethics, and in particular of Marxist ethics. We look at Marx in more detail in section 3.2.4 *The Death of 'God'*, but here we point out that Kierkegaard and Marx were born within five years of each other and were both inheritors of the German philosophical tradition. Where Kierkegaard was deeply religious Marx was totally atheist however, yet in Marx too we find the 'neighbour' at the heart of his ethics. In *Capital* Marx shows that a commodity has a value based on an equal regard for every person's labour, an insight that Aristotle failed to uncover because of the slavery central to the ancient Greek economy.[78] Where Marx believes that he has shown this 'scientifically' we would argue that it is an *a priori* belief which no empirical investigation could prove: it is an ethical standpoint of egalitarianism which places the neighbour's economic equality as paramount. As the old Jewish saying goes: 'my neighbour's material problem is my spiritual problem.'

2.5 The Problem of 'God'

The word 'God' has been placed within quote marks in this book so far without explanation. We suggested in section 1.4 *The Role of Language* that Wittgenstein's view on language should be made more precise: that the meaning of a word *for any major user-community* is the use of that word *in that community*. We further suggested that if the user-community for an English word is the entire English-speaking world, then that word has an undisputed currency. Everyday objects like 'table' would fall into that category; however, 'God' would not. We are now in a position to explain this properly. The English-speaking world may be growing exponentially, as English becomes a global language, but when a Chinese person learns the English word 'God' there is a problem: it is untranslatable into Chinese (so in fact is the word 'religion')[79] 'God' as a concept has no proper currency outside of monotheistic cultures. But even in the earlier Latin-speaking world of the Enlightenment its use had become highly contested, as we shall see in later sections. In the secular world 'God' was emptied of meaning, though the husk of its presumed earlier usage remained. Hence we see that 'God' had multiple meanings in pre-atheistic Europe, a shrunken meaning in secular culture, and no meaning at all in the original vocabularies of the Orient, at least anywhere east of Iran. In terms of our historical modalities of the spirit it has meaning in only one out of five of them – monotheism. Such widely conflicting usage – and non-usage – destroys any common meaning: one ought endlessly to qualify the word with 'for such-and-such a user-community.' The quote marks round 'God' have simply been used to flag up this point.

Defining 'God' for Religionists, Secularists and New Agers

While 'God' as a word can have no universal currency for the reasons given above, what if we were to find within its spectrum of denotations a meaning that worked for the three Western user communities this book addresses: the religionist, the secularist and the New Ager? The common ground that we have been establishing so far for this three-way dialogue is the idea of personal spiritual impulse and its variety. In an important sense this approach is like that of Feuerbach's (further discussed in section 5.1.4) whose position is summed up in the dictum: 'God is made in the image of man, not man in the image of God.' Although this sounds like the overturning of religion, Feuerbach merely opposed the literal, realist notion of 'God', making him instead a necessary conception of man as a religious being. It is only Marx and his tradition that took Feuerbach's radical religiosity and made it into radical atheism, by insisting that 'God' was an *empty* construct. For the purposes of the later discussion we need a definition of 'God' that recognises the term's origins in human impulses, but which does not permit such an emptying out.

To say that 'God' arises out of the monotheist impulse is simply a tautology, but within our polarities of the spiritual impulse we can identify one that could lead to 'God'-type conceptualisations: the *transcendent*. In terms of the other polarities, the issue of *via positiva* or *via negativa* is secondary (though greatly colours the conceiving of 'God'), as is the issue of social or solitary, but what radically changes the character of 'God' in a discourse of transcendence is the *bhakti-jnani* distinction. Hence we are led to the following definition: *'God' is the objectless object of the transcendent* bhakti *or* jnani *impulse.'*

If secular thought is prepared to consider that there is a transcendent spiritual impulse in the first place, a drive to a sacred connectedness, a genuine sense of the oceanic, then that impulse will have a direction, effectively to some object. But that object is transcendent and so beyond any material description; hence the term 'objectless' is used to indicate that. This definition should hence pose no problem to any of our three communities. The religionist will find in it a tension (between *bhakti* and *jnani*) which goes to the heart of the Judaic origins of monotheism, and still plays out in all contemporary theology. For the secularist it emphasises the non-realist position (that 'God' is not a thing in any ordinary sense); the only assumption required is that there is such a thing as a spiritual impulse in the first place. For the New Ager, the idea of 'God' as arising out of either love or wisdom is readily acceptable, though perhaps prompting the question: why not both? But the rigid religionist will reject the anti-realist stance of the statement; the rigid secularist will find, with Francis Wheen perhaps, that 'objectless object' is just mumbo-jumbo; and the less serious New Ager might prefer to skip off amongst the daisies. For those still left we will put this definition to work.

The Two 'Gods': Pascal and Hume

The implications of our definition of 'God' will be worked out in full in following chapters, but here we begin to show how the definition recognises what has remained obscured in the spiritual history of the West: that there are effectively two 'Gods': one the result of the devotional (*bhakti*) impulse, and the other the result of the non-devotional (*jnani*) impulse. We can illustrate this with the examples of Blaise Pascal (1623-1662) and David Hume (1711-1776). Both men were intellectuals leaving behind influential writings that have been studied over the centuries. Pascal was practical, mathematical and scientific, where Hume was sceptical, theoretical and philosophical. Pascal has had a highly visible impact on cities in every corner of the globe: he invented the omnibus. He is also brief in his style, which is aphoristic, and also sceptical in a curiously modern way. Pascal, a Jansenist, is the natural inheritor of St Augustine's religious thought; Hume was a Calvinist and natural inheritor of the Protestant Reformation. But, when examined closely, their writings reveal the key difference for us: Pascal has a typically *bhakti* spiritual impulse while Hume has a typically *jnani* spiritual impulse. The spiritual impulse is less developed in Hume, and it is this in combination with his *jnani* orientation – for which there is no framework of understanding in the West – that led to accusations of atheism levelled at him (though any attentive reading of his book *The Natural History of Religion* will show that the charge of atheism is absurd). Pascal however provides us with the key Western terminology for understanding Hume: the notion of the 'God of the philosophers'. We have come to inherit this from Pascal in the most dramatic of ways: on 23 November, 1654, he underwent a profound spiritual experience, a revelation lasting from half past ten in the evening to half past midnight. His own brief account of it was so important to him that he transcribed it onto parchment and sewed it into his clothing, to be found only after his death.[80] Pascal, after his intense religious experience, now conceived of 'God' only in terms of *love*, and considered the emerging Deism of the time no better than atheism. (Deism is dealt with in more detail in section 5.2; for now it should be understood as a non-devotional *jnani* religiosity.) Hume on the other hand saw in the notion of the love of 'God' something fit only for primitive peoples and women:

'Ignorance is the mother of Devotion', he said,[81] a typical *jnani* response to the *bhakti*.

For Pascal 'God' is the objectless object of the devotional (*bhakti*) impulse, while for Hume 'God' is the objectless object of the non-devotional (*jnani*) impulse. Regardless of how developed those impulses were in the two men, they are strong enough to represent the total abyss of understanding between their versions of 'God'. For a Hindu these two versions of 'God' would be no more difficult to reconcile than the existence of male and female genders. *For the West we will show that this difference was irreconcilable, and shaped the entire trajectory of the Western tradition.*

Anthropomorphism and the Trap of Theistic Thinking

The central problem of monotheism is the trap of absolutist theistic thinking. Because of the monolithic nature of a supposed single 'God', theism struggles to serve both the *bhakti* and the *jnani* impulses: to be simultaneously the 'God' of devotional practice and the 'God' of the philosophers. Judaism, Christianity and Islam all attempted different solutions to the problem, but in India and the Far East the problem didn't arise.

It is nearly two hundred years since the German philosopher Arthur Schopenhauer was introduced to the Indian scriptures that he claimed were central to his own development as a philosopher (he read the *Upanishads* in the winter of 1813-1814), yet no Western philosopher has engaged with Eastern thought since that time in any serious way. As a result the striking fact that the religions of the East are largely non-theistic has had no serious recognition in the West. We find then that at the most trivial level of popular culture religion is discussed exclusively in terms of 'God' – perhaps not too surprising – but also that the intellectual elite including Continental, British and American philosophers do the same – actually rather shocking.

When *jnanis* find themselves in the historical context of Abrahamic monotheism, they vigorously resist the image of 'God' as in any way human: in other words they resist the anthropomorphism of 'God'. David Hume is adamant on this, finding the Jews and Muslims better at it than the Christians. However, when we turn to a religious thinker with a much more developed *jnani* impulse, like Meister Eckhart, we find that the resistance to anthropomorphism goes so far as to come close to dispensing with 'God' altogether. Those with a *bhakti* (devotional) impulse within a monotheistic setting very naturally tend to anthropomorphise 'God'. Even so, they are adamant that 'God' is unlike a human being in the profoundest of ways. But the love they feel is directed towards its goal, and even though we have used the phrase 'objectless object' to indicate that 'God' is not a *thing*, the goal of their love has to be described in human language, often borrowing from romantic or even erotic imagery. Because Christianity has been dominated by *bhakti* thinkers rather *than* jnani thinkers, anthropomorphism was never resisted strenuously enough or publicly enough, as it was in Judaism and Islam. Hence even in the popular secular mind, the term 'God' is irrevocably understood in terms of a 'being' – the 'Supreme Being'. This is the realist position on 'God' which creates the realist atheist: one who denies the anthropomorphic 'God'. The atheist has never in fact denied the other 'God' of our definition.

2.6 *The* Bhakti / Jnani *Distinction*

Bhakti is the 'love that dare not speak its name' – at least in the modern world. This phrase was used to describe lesbianism, though Queen Victoria refused to enact legislation to make it illegal (like homosexuality) on the grounds that it didn't exist. While the religionist or New Ager might accept at least the principle of a spiritual love, the secularist is likely to ridicule it or decide that it doesn't exist. Even in the eighteenth century Hume had declared that 'devotion is the daughter of ignorance'; the secularist now is even more likely to give it a wide berth. A defining axiom of secularism states that the unquestioning submission to any external authority, particularly religious, is the hallmark of feudalism, and is opposed to every democratic principle. Yet 'Love, Devotion, Surrender' (to quote the title of a Santana / Mahavishnu John McLaughlin album) in the spiritual domain has a radically different meaning than in the political domain. To press the point: 'Volunteered Slavery' sounds absurd, and even more so when we know that it is an album title by the American jazz saxophonist Rahsaan Roland Kirk – a black Afro-Caribbean, probably of slave descent. Santana and McLaughlin drew on the *bhakti* tradition in Hinduism; Kirk on the *bhakti* tradition in Islam (which anyway means 'submission'). Secularists buy their music in droves, and see in these musicians not just great artistry, but an integrity implicit in their non-commercial styles – usually without thinking through the implications.

However, to further engage the secular mind with the concept of *bhakti*, we will need to adopt what we have identified as secularism's greatest strength: critical enquiry. So far we have simply made unsubstantiated claims that key historical figures have either a *bhakti* or *jnani* impulse, and that this affected their thinking, and so the course of history. We therefore need to provide a rigorous method for identifying the *bhakti* or *jnani* hallmarks in an individual or tradition. First we need to examine some Western sources for possible guidance on this. The Greeks used an interesting distinction for three sorts of love: 'philia' for the love towards things generally, 'eros' for sexual love, and 'agape' for brotherly (spiritual) love. (Sanskrit has ninety-six words for love, apparently, while Greek has three, and English only one.) Agape has since been adopted by Christianity to mean 'Christian love' and hence is the closest to *bhakti* that we have in the West. However its use for the love of 'God' is muted by its origins as the love between fellow religionists, mainly men. It was the Romantic period that generated a new discourse around eros, elevating it at times to a near-spiritual quality. The Sufi tradition in Islam uses the terms *'ishq* for *bhakti* and *'ilm* for *jnani* in distinguishing between the two spiritual orientations. Shaykh Zulfiqar Ahmad (a leading contemporary teacher of Islam in the Tasawwuf tradition) says for example that the (spiritual) sustenance of the heart is love (*'ishq*) and the (spiritual) sustenance of the mind is sacred knowledge (*'ilm*).[82]

We might expect to find some systematic method for characterising the devotional and non-devotional impulse in the work of C. G. Jung. In his key typology of the ego, he has two pairs of opposites: thinking / feeling and intuitive / sensing. Would not thinking and sensing belong to the *jnani* orientation, and feeling and intuition belong to the *bhakti* orientation? As a distant analogy it is useful, but Jung describes his opposites in largely secular terms. Also, when we apply Jung's own typology to himself, we discover a thinking type at work, or in our terminology a *jnani*; furthermore his interests lie in the esoteric, as we saw. As such his instinct for the transcendent is undeveloped, and his insight into *bhakti* even more limited.

Worse still, while the West is saturated with *jnani* thinkers, there is even less insight into the *jnani* orientation than the *bhakti* orientation. That which was *jnani* in the ancient Greeks is now understood as merely speculative philosophy. We turn instead to the Hindu tradition, and in particular the great spiritual teachers Ramakrishna (1836-1886) and Vivekananda (1863-1902), mentioned earlier. While Ramakrishna wrote nothing, there is an extensive literature on him, including the *Ramakrishna Gospel*,[83] a diary kept by a close disciple, providing a fascinating insight into Indian life during that period, but also many valuable references to the Indian concepts of *bhakti* and *jnani*. Vivekananda, the more intellectual of the two (he was educated in a Western-style Indian University), was the *jnani*, while

Ramakrishna's intense devotionality made him an outstanding example of the *bhakti* type. Romain Rolland, the friend and correspondent of Freud, chose to start his account of Ramakrishna's life with this salutation:

> Greeting to the feet of the Jnani [seeker on the path of awareness (knowledge)]! Greeting to the feet of the Bhakta [seeker on the path of devotion]! Greeting to the devout who believe in the formless God! Greeting to those who believe in God with form! Greeting to the men of old who knew Brahman! Greeting to the modern knowers of Truth. . . .[84]

The terms used in Rolland's text are Jnani and Bhakta (a slight variation on *bhakti*), and his own definitions for a Western audience are placed in the square brackets. The reference to 'God' with and without form is natural in the pluralistic tradition of India: both an anthropomorphic *bhakti* 'God' and a *jnani* one with no form at all are equally valid. (The Western 'God' as a concept had arrived centuries earlier with the Islamic invasions.) Rolland had instinctively picked out Ramakrishna's key quality: Ramakrishna moved endlessly from one modality of the spirit to another, including experimentations with Christianity and Islam at one point in his life. But the key polarity that had been with Ramakrishna all his spiritual life was the dynamic between *jnani* and *bhakti*: he was instinctively a *bhakti*, but from early in his life as a priest of Kali on the banks of the Ganges he had been introduced to the profundity of the *jnani* orientation, and grappled to master it. Later he became the guru of the much younger Vivekananda who said of them both: 'Outwardly he was all Bhakta, but inwardly all Jnani . . . I am the exact opposite.'[85] The exposition of *jnani* and *bhakti* in the lives, relationship and teachings of Ramakrishna and Vivekananda is prominent, but it was also part of the Hindu tradition stretching back to the *Bhagavad Gita*, in which Krishna instructs the recalcitrant Arjuna on both paths.

There is no central authority in Hinduism to turn to, for validation of the insights put forward by Ramakrishna and Vivekananda (Westerners forget that even the concept of a central authority in religion is a cultural accident: its most conspicuous occurrence is in Roman Catholicism). Hence the choice of Ramakrishna and Vivekananda may seem arbitrary as originators of the key analytical concepts deployed in this book. Yet they are universally revered in India, probably only second to Gandhi in the recent spiritual pantheon of the subcontinent. They were also household names in Europe and America in about 1900, due to the impact that the young Vivekananda had in his brilliant exposition of Hinduism at the 1893 World Parliament of Religions in Chicago and in subsequent lecture tours.

Our system is more complex of course than a simple devotional / non-devotional dichotomy, though we place that at the heart of the analysis. The presence of these two orientations is greatly refracted by the overall modality of the spirit (which is largely monotheism in the period of Western history under consideration) and the issues of *via positiva / via negativa,* social / solitary, and whether the esoteric plays a part. In other words, while we are taking a Hindu conception of spiritual interiority, we are consciously operationalising it for the West. We end this chapter with a checklist of *bhakti* and *jnani* characteristics that we will be able to use later on, when we need a systematic method for distinguishing these impulses in key religious figures and traditions.

The *Bhakti / Jnani* Checklist

Bhakti characteristics:

1. love;
2. devotional impulse towards a range of possible recipients: idols, images, gurus, saints, deities, or the transcendent objectless object we have defined as 'God';
3. devotional behaviour including the construction and tending of idols or images, acts of devotion to a living guru, devotional writings, visible signs of love including swooning, transports of joy and ecstasy;
4. tendency to anthropomorphise the recipient of the devotional impulse;
5. emphasis on submission and surrender;
6. tendency to sexual metaphors and language to describe the impulse and experience;
7. activity lies in preparation for the 'beloved', to make worthy of sacred love, to receive grace;
8. unlikely to be drawn to the esoteric.

Jnani characteristics:

1. compassion;
2. an impulse towards knowing, 'seeing', knowledge, wisdom, learning, awareness, insight;
3. enquiry, possibly doubt, as core behaviour;
4. instinctively anti-anthropomorphic;
5. emphasis on the will;
6. 'infinite' and 'eternal' are key words;
7. may be drawn to materialist or esoteric science.

Three:
Returning to the Roots

In this chapter we explore two factors in the emergence of the secular mind: the 'double whammy' of the rise of science, and the legacy of religious cruelty. The analysis works backwards through Western history in a zigzag fashion: starting with the secular world view of today, events are identified in the recent past that led to it, then a jump made further back to show how the events of that period led up to the assumptions of the recent past, and so on, reaching to the very roots of Western thought. This means a slow acclimatisation to the more and more distant past, a little like diving without the 'bends'. Crucially, we use the spiritual typologies developed in the last chapter to expose the historical forces acting on religion. The Enlightenment, its legacy, and events leading up to it are considered, and we ask if its great thinkers were persuaded by the new rationalism and the desire to end superstition and religious cruelty, to actually want the end of religion itself.

3.1 The Twentieth Century

3.1.1 Science and the Century of Alienation

The twentieth century as a whole brought to fruition all the forces of the previous three centuries that were ranged against religion, but to call the twentieth century atheistic would be to ignore the fact of religion's survival, albeit in emasculated form. A better epitaph for the twentieth century would be to call it *the century of alienation*. We identified *control* as the principal pathology of Western religion; we identify *alienation* as the principal pathology of Western secularism. Those for whom control is not the ultimate terror retreat to old religion; those for whom alienation is not the ultimate terror cling to secularism. Between the extremes lies the New Age, with its own set of problems.

A Vocabulary of alienation

The twentieth century was a century of democratic progress that gave each family its 'nuclear' isolation, based on material goods undreamed of by the Victorians, but progressively decoupling citizens from each other through the independence of the motor car, the supermarket and the television set. In the philosophy, psychology and literature of the period we find a groping towards a *vocabulary* of alienation, new words that would serve to name vague feelings of dread, boredom, indifference and uprootedness. The *Encyclopedia Britannica* suggests five meanings for alienation: (1) powerlessness, (2) meaninglessness, (3) normlessness, (4) cultural estrangement, (5) social isolation, and (6) self-estrangement, of which it finds the last to be the hardest to define. Marx used the term alienation in his early work to mean the estrangement of the worker from the means of production, though Marx was also interested in the concept of 'self-estrangement'. It is this last sense of the term alienation that we are pursuing here, though with the much expanded notion of self that our definition of spirituality implies – the self capable of a sacred connectedness with existence, but also facing the possible suffering born of a profound disconnectedness. The 'means of production' – the nature of the workplace – might be boring or exhausting, but its meaninglessness is not the question here, because those who do own the means of production are no less prone to an alienation of the spirit than their workers.

Increased material wealth amplified the whole cluster of 'alienations' and spread them to a far wider population than the effete and wealthy aristocrats of the Romantic age – whole new strata of society had education and time on their hands. Christianity had an ancient term, 'acedia', meaning spiritual torpor, apathy or not caring, but new phrases were emerging. Alienation occupied a prominent role in Hegel's *Phenomenology of Spirit* in 1808; Kierkegaard invented the phrase 'sickness unto death', in an 1849 essay; and Durkheim coined the term *anomie* for normlessness in 1893. Heidegger took from Kierkegaard the term *angst*, a German word for fear, redefining it in his 1927 book *Being and Time*; and Sartre used *nausea* and 'iron in the soul' in the 1930s to describe a *new* form of inner suffering that was unknown in previous eras. The developing discipline of psychology had a field day with 'schizophrenic', 'shadow', 'subconscious' and so on. *Ennui*, which is etymologically derived from 'to hate', gained a new usage which suggested listlessness and dissatisfaction rather than the active passion of hatred. The theologian Paul Tillich, writing in 1952 said: 'Today it has become almost a truism to call our time an 'age of anxiety.' '[86] Kierkegaard had anticipated him a hundred years earlier when he talked of an 'age of disintegration, an esthetic, enervating disintegration.'[87] Max Weber, writing around the time of WW1 , came up with the term 'disenchantment' (a translation of his German term 'entzauberung') to mean a society abandoned by its gods. Late nineteenth century psychiatrists were called 'alienists', reflecting an insight that the root affliction of mental patients was alienation: we can only speculate that this nomenclature was dropped as alienation became so endemic as to be invisible.

If this alienation has its roots in the three pillars of secularism: the social, the scientific and the psychological, then it is also mirrored and reinforced in Western cultural production: the arts. We have suggested that the arts, where they reflect or reinforce this alienation, instead of providing as they do at times a re-enchanted vision of humanity, can be thought of as producing a kind of cultural autism.

Scientism

We suggested in section 1.1 that scientism underpins the assumptions of today's secular outlook. Scientism may be populist in the sense of it being widely held but poorly articulated, as is often shown in optimistic television documentaries on science, or it may have a scholarly underpinning, particularly from Logical Positivism. This philosophy was put forward by the group known as the Vienna Circle between the wars, and stated that not only was scientific knowledge the *only* legitimate form of knowledge, but that metaphysical questions could not be answered from personal experience: *they were meaningless questions*. In fact philosophy had increasingly focussed its attack on metaphysics, a closer inspection of which suggests that this is really an attack on religion. The contemporary philosopher Brian Magee puzzles, as many have done, on how Wittgenstein – whose writings are often rather mystical – came to be so closely allied to the Vienna Circle. What is clear is that a mutual and instinctive dislike for metaphysics forged this temporary alliance.[88] To attack religion directly during the first part of the twentieth century was 'old hat' for the philosopher, but metaphysics – questions of being, divinity and eternity – seems to have been a legitimate target.

Logical Positivism appears to be the natural end product of the Enlightenment project, a massive shift from the religious worldview to an emasculated rationalism. It is also a reaction to the overblown rhetoric of German Idealism. Because the emphasis on the scientific method gives results that are quantifiable, and hence carry the argument against all expressions of 'emotionalism', other voices went unheeded. Questions of spirit are meaningless, says Logical Positivism, creating the atmosphere in which the great minds of the twentieth century ignored those questions, allowing our interiority to be claimed by psychology, or shallow New Age romanticism, or a turn to the dark philosophies of Nationalism and Fascism (possibly via German Idealism), or the interminable circumlocution of postmodernism. The grip of Logical Positivism on the universities of the West in the second half of the twentieth century set the agenda for education and research which shaped its culture.

Scientism on the one hand and the 'linguistic turn' on the other kept the popular pundits and the scholars busy explaining to the general public that subjective world of personal experience and feelings was no conceivable resort for making sense of the world. If one broke down under the meaninglessness of this twin onslaught, then there was psychiatry – perhaps your relationship with your father or mother was at fault – or, as the last resort, there were anti-depressants. Tragically the last resort has become increasingly the first resort, and, thanks to the Internet, dangerous anti-depressants are now often self-prescribed, in isolation even from a doctor, and avoiding regulatory frameworks.

Antidepressants as Indicator of Alienation

Whereas 'melancholy' (a disposition supposedly caused by an excess of 'black bile') was more often considered a personality trait than a mood, 'depression' was the new mental condition for the century of alienation. Melancholy had good cause: to take just one example, no parent in earlier times could automatically expect all their children to survive into adulthood. For example, both Charles Darwin and Thomas Huxley were devastated by the early deaths of their daughter and son respectively,[89] as was Marx devastated by the loss of his daughter. Disease and accident could take any person in their prime without warning, life was far more unpredictable, and a natural melancholy might be the response to any such loss. Only the tiny aristocratic and leisured classes could afford to dwell on their melancholy however (and Romanticism gave voice to that), but in the twentieth century with dramatically increasing wealth and the securities offered by science and technology, an uncaused 'iron in the soul' became a large-scale phenomenon and required a new name: 'depression' or even 'clinical depression'. Psychotherapy is of course the preferred treatment, but the cost of the 'talking cure' and its thousand derivatives is prohibitive, not in a material sense, but in the time of highly-paid professionals. Drugs have stepped into the breach for the family doctor, now prescribing anti-depressants on a massive scale, and, even more alarmingly, increasingly to children.

Anti-depressants are entirely consistent as a materialist solution to a problem: depression, say the materialists, is caused by chemical imbalances in the brain, and these can be chemically fixed. The idea of 'downward' causation, that the soul or will can affect a material system like brain and body, is anathema to science. This means that science cannot tolerate the idea that the 'self' as a non-material entity is experiencing an unhappiness that *subsequently* appears as a measurable imbalance in brain chemistry. For science, the imbalance is the cause, not the effect (as neuroscientist Thomas Damasio would probably say). At best the scientific world might permit the psychoanalyst or cognitive-behavioural therapist their reductionist language of interiority, and give them a shot at 'curing' the patient, but, for a busy family doctor, with scarce State resources, the easy route is to write out a prescription for drugs.

According to *IMS Health*, a company providing pharmaceutical market intelligence to the US administration and industry, new prescriptions written for six of the most widely prescribed antidepressants during 2000 were:

Sertraline (Zoloft) 10.7 million
Paroxetine (Paxil) 10.49 million
Fluoxetine (Prozac) 10 million
Citalopram (Celexa) 5.29 million
Venlafaxine (Effexor) 4.2 million
Nefazodone (Serzone) 2.34 million.[90]

This is a staggering total of 43 million *new* prescriptions in the year 2000. The drugs prescribed to children in the US include all of the above, often despite lack of licensing for paediatric use. Other studies show antidepressants to be firstly no better than placebos, and secondly to cause suicide. The UK has banned prescription of a number of antidepressants to children, but still there were 50,000 children as young as six on antidepressants in the UK in 2003.[91] A study by the UK health authorities suggested that Paxil (known as Seroxat in the UK) increases suicidal tendencies in the under-18s by 1.5 to 3.2 times that over receiving a placebo.[92] Paxil has six month sales figures of a billion dollars, indicating the enormous financial temptations to steamroller clinical trials, and the resources available for marketing. Paxil produces intense and emotionally painful withdrawal symptoms, but the company is allowed to claim that it is not addictive because it does not produce all the medically-defined aspects of addiction, including the necessity for continually increased dosage for the same effect. In a TV interview a family doctor, when confronted with the ordeal of a young woman attempting to get off the Paxil he had prescribed, explained pleasantly that the withdrawal symptoms were useful because it reminded the patient to go on taking the drug.

A jury in the US found GlaxoSmithKline guilty of negligence and awarded $6.4 million compensation after a man who took Paxil for just two days shot and killed his family,[93] but the drug has not been withdrawn, and field trials went ahead on children. Parents were induced to put their children forward for free trials by an advert that asked: 'Do your children suffer from Social Anxiety Disorder?' This so-called condition has well-meaning web sites devoted to it, in which the reader finds that Paxil has received approval from the US Food and Drug Administration as a treatment: it is an easy leap to then order the drug over the Internet. The enormous sales, according to the defenders of antidepressants, are due to their effectiveness, though some studies contradict this.

Prozac: for the 'Worried Well'

The best-known antidepressant of the SSRI class is Prozac (its name now stands in effect for the whole class of selective serotonin reuptake inhibitors), and a considerable literature has grown out of its status as a social phenomenon. This includes psychiatrist Peter Kramer's ground-breaking survey *Listening to Prozac* (1993), Elizabeth Wurtzel's autobiographical account *Prozac Nation* (1995), and a collection of essays called *Prozac as a Way of Life*, edited by Carl Elliott and Tod Chambers (2004). The latter volume expresses the near-consensus amongst healthcare professionals and commentators that the use of antidepressants in cases of 'clinical depression' is more than justified because it can be a life-threatening condition. (At the same time they also acknowledge trials which suggest that SSRIs have no impact over placebos.) But as Elliott argues in his introduction to *Prozac as a Way of Life*, the real debate is over the term that Kramer coined: 'cosmetic psychopharmacology' – the prescription of Prozac as an enhancement technology, a drug for the 'worried well', a drug to make the client 'better than well'. It is Kramer's essay in *Prozac as a Way of Life* that perhaps best approaches the *spiritual* implications of Prozac, argued in terms of alienation. He says:

> Imagine we have defined possible elements of existential alienation, such as spleen, anomie, angst, acedia, vertigo,

> malaise, emptiness and the like. . . . Elliott's worry is precisely that if a medication replaces pessimism with optimism, anxiety with assertiveness, and diffidence with gregariousness, it will have robbed us of a tendency to remain at a critical distance from our own existence. . . .
>
> Our sensibility has been largely formed by melancholics. Much of philosophy is written, and much art has been created, by melancholics or the outright depressed as a response to their substantial vulnerabilities. . . . If loss were less painful, the good life might be characterised not by ataraxia [serenity] but by gusto.[94]

Kramer's argument is that society has valorised the melancholic over the sanguine, and that this is the only source of the moral anxiety over Prozac. This is a quintessentially American viewpoint, a context within which the masculine heroic model of individuality would naturally find the intellectual and melancholic personality a little suspect. If Prozac helps the alienated individual live a life of 'gusto', what could be wrong in that? Kramer says: 'If Prozac induces conformity, it is to an ideal of assertiveness, but assertiveness can be in the service of social reform of the sort ordinarily understood as nonconformity or rebellion.'[95] This is the secular American ideal to which Prozac helps the 'worried well' conform. It is true that for the Romantics, their alienation was almost elevated to the erotic, that suicide was noble, and that melancholy was the only natural response to existence. But for the great spiritual thinkers we have been drawing on, the indulgence in either gusto or melancholy misses the point. It is precisely the issues which confront one with one's own alienation that can also lead to its overcoming in the spiritual life. Susan Squier, in her essay in *Prozac as a Way of Life*, confirms this. She had taken Prozac after a depression:

> But even with the depression under control, I told my psychiatrist, I still felt a hollowness at the core of life: a loss. A Quaker, she understood what I was describing. She knew that this was an experience I needed not to control or to defend against, but to encounter. Had I ever tried meditation, she asked?[96]

Squier was lucky to be treated by a psychiatrist who was also grounded in the spiritual life, who recommended meditation as the spiritual path of *encounter* with the roots of her alienation. (Squiers reports that she has now stopped Prozac but continues with Zen meditation.) Antidepressants may or may not help in cases of clinical depression, but on their own, and particularly when used as 'cosmetic psychopharmacology' merely shift the client from melancholy to gusto, escaping any meaningful confrontation with the roots of alienation. If alienation is understood in the first place as a spiritual issue then there is no need either to insist that 'gusto' is unspiritual. Kramer's American model of 'self-reliance' is ultimately a false model however, because the successful individual operating out of this worldview merely ignores the hollow premise of this individualism, that one is separate. Alienation exists because the ultimate connectedness of self to existence has not been discovered, whereas in that connectedness lies *real* gusto, along, of course, with the acceptance of *real* loss. Prozac as a *philosophy*, just as much as a treatment, removes responsibility from the individual for spiritual enquiry and examination of the real nature of self.

We can use the phenomenon of depression as an indicator of a spiritual loss, a loss of meaning or connectedness, and the attempt to use chemistry to alleviate it as indicating the response of scientism. Writers on science who extol the prevailing scientism include Daniel Dennett (who sees Darwinism as a universal acid that eats away all competing explanations), Francis Crick (who sees us as 'nothing but a pack of neurons'), Richard Dawkins (introduced earlier), and Richard Atkins (who believes in religious intolerance – that is an intolerance of religion on scientific grounds). We will look at three of these figures, whose writings appear on the popular science shelves around the world.

Daniel Dennett and the Argument from Design

Daniel Dennett is a philosopher and author of many books and articles including *Consciousness Explained* (1991) and *Darwin's Dangerous Idea* (1995). He co-edited *The Mind's I* with Douglas Hofstadter in 1981 (a book that is surprisingly open to the 'metaphysical'). His thinking as a materialist philosopher in respect of the spiritual could be characterised as 'benign dismissive', rather than 'vituperative atheist'. Nevertheless his approach is instructive regarding a broad strand of scientific secularism which makes contact with religion at the level of 'explanation'. Once he has shown, as it is easy to do, that religion's power of explanation with regard to natural phenomena is flawed, there is no further interest. He doesn't seem to be incensed, as Dawkins is, that religious fundamentalists ignore all demonstrations of religious fallibility.

In *Darwin's Dangerous Idea: Evolution and the Meanings of Life*, Dennett devotes brief sections to the philosophers John Locke and

David Hume, in order to introduce the distinction between 'revealed religion', i.e. a religion based on divine revelation, and 'natural religion', which he defines as a religion based on rational deductions from evidence (this is a narrow way of describing what is known as 'Deism'). This is useful for our discussion because it flags up a stage in the development of Christianity that made it so vulnerable to attacks from scientism: reliance on the proofs of the existence of 'God'. The key proof in the emerging age of science in the eighteenth century was the Argument from Design, an appeal to behold the exquisite order in the natural world as evidence of the hand of 'God'. Dennett quotes a passage from Hume's *Dialogues Concerning Natural Religion* as an eloquent expression of the Argument from Design, finishing with this line: 'By this argument a posteriori, and by this argument alone, do we prove at once the existence of a Deity, and his similarity to human mind and intelligence.'[97] Dennett is anxious to establish that the protagonists in Hume's *Dialogues* demolish the Argument from Design, through the imagination and reasoning power of both sides as they put forward argument and counter-argument. Indeed the *potential* flaws in the argument are exposed from every conceivable angle. The fact that in the end the Argument from Design is *agreed on* in the *Dialogues* is seen by Dennett as a caving in by Hume,[98] but this is a perverse deduction, as a reading of Hume's *The Natural History of Religion* would confirm. For example Hume writes in the introduction to that book: 'The whole frame of nature bespeaks an intelligent author; and no rational enquirer can, after serious reflection, suspend his belief a moment with regard to the primary principles of genuine Theism and Religion.'[99] But Dennett, following popular prejudice, concludes that the *Dialogues* cave in to the Argument from Design because Hume lived prior to the discoveries of Darwin. Dennett chooses this passage to mark Hume's capitulation:

> From this enquiry, the legitimate conclusion is, that the causes have also an analogy: and if we are not contented with calling the first and supreme cause a GOD or DEITY, but desire to vary the expression; what can we call him but MIND or THOUGHT, to which he is justly supposed to bear a considerable resemblance?[100]

Dennett has read the *Dialogues* as an attack on religion, but which failed to carry through because Hume could not imagine another *explanation* for the design of the natural world. Dennett does not comment on the fact that Hume offers alternative terms for the 'deity' such as MIND or THOUGHT, but is convinced instead that Hume has attempted and failed in an exercise in cosmogony (an account of the origins of the universe). It is true that cosmogony features largely in sections of the *Dialogues*, but what Dennett's philosophical scientism fails to spot is that Hume is engaged, not with a dispute against religion, but a dispute *within* religion: Hume is arguing for natural religion (Deism) against revealed religion. In other words Dennett applies a secular bracketing out of the spiritual content of the *Dialogues*, as is often exercised against all the philosophers of the early Enlightenment.

Dennett concedes that the *Dialogues*, which Hume prevented publication of in his own lifetime, was written without devices to hide its meaning. Hence we can take seriously Hume's interesting suggestion that the 'first and supreme cause' might be given a name other than 'God'. We will see later on that this is the natural instinct of the pre-modern philosopher, who had for centuries offered the 'God of the Philosophers' as an alternative to the anthropomorphic 'God' of the devotional impulse. MIND and THOUGHT are typical terms arising from the *jnani* impulse (which philosophers generally exhibit), avoiding the anthropomorphism that Hume rails against in the *Dialogues*. Dennett, inheritor of the mainstream anthropomorphic Christian tradition that he is anxious to discredit, believes that once the Argument from Design of natural religion is disproved, then religion is done for. However, as Hume puts it: 'The question is not concerning the BEING of God, but the NATURE of God.'[101] How could this possibly be the thinking of an atheist? But both religionists of the time and secularists of the modern age promote Hume's atheism as an indisputable fact, effectively denying the possibility that Hume's clear engagement with the Argument from Design is a valid religiosity.

We will return to Hume and the issue of natural religion later, but for now it is interesting to note that by the end of the eighteenth century it was increasingly common to understand the natural world as a mechanism. This raised serious questions about the meaning of life, and was the origin of the alienation that eventually became endemic in the twentieth century. Dennett attempts to bridge the gap caused by the loss of the spiritual with an *evolutionary* account of meaning, but in both *Darwin's Dangerous Idea*, and the subsequent book, *Freedom Evolves*, his secular scientism seems to offer cold comfort. As a philosopher he may try to add purpose or meaning to evolution, but the biology as a science is clear: mutations are random and purposeless.

Francis Crick

Francis Crick (1916-2004) was the British biophysicist, who, with James Watson and Maurice Wilkins, received the 1962 Nobel Prize for Physiology or Medicine for the discovery of the molecular structure of DNA, a great twentieth-century scientific breakthrough. After the triumph of the DNA discovery Crick sought a comparable scientific challengein the field of consciousness research. He published his early findings in 1994, as 'work in progress', in a book called *The Astonishing Hypothesis*, which opens with the following statement:

> The Astonishing Hypothesis is that 'You', your joys and your sorrows, your memories and your ambitions, your sense of personal identity and free will, are in fact no more than the behaviour of a vast assembly of nerve cells and their associated molecules. As Lewis Carrol's Alice might have phrased it: 'You're nothing but a pack of neurons.'[102]

Crick has unconsciously used the very phrase that many thinkers have adopted to criticise the type of analysis presented in his book: 'nothing buttery'. Yet Crick's work is representative of an essential component of twentieth-century alienation, the idea that one is 'nothing but' something, where that 'something' is variously a pack of neurons, a collection of atoms, a cellular collective for the propagation of the selfish gene, peptides, and so on. Hence both his work and his stance regarding religion are instructive.

Crick is outspokenly anti-religious. He says: 'I went into science because of these religious reasons, there's no doubt about that. I asked myself what were the two things that appear inexplicable and are used to support religious beliefs: the difference between living and nonliving things, and the phenomenon of consciousness.'[103] In 1961, Crick resigned his fellowship of Churchill College, Cambridge, over its proposal to build a chapel. Sir Winston Churchill, in whose honour the college was founded, wrote to him, pointing out that 'none need enter [the chapel] unless they wish', Crick replied that, on those grounds, the college should build a brothel; he enclosed a cheque for ten guineas towards its construction.[104]

Like Dennett, Crick tends to approach religion at the level of *explanation*. Crick is explicit about this:

> The record of religious beliefs in explaining scientific phenomena has been so poor in the past that there is little reason to believe that conventional religions will do much better in the future.[105]

For a scientist this point is meaningful, and supported by

overwhelming historical evidence. However from the spiritual point of view, religion is, in the first place, nothing to do with explanation, and certainly not about explanation of 'scientific phenomena' which, by definition are observed from the *outside*. Crick's neurologising of self is both the expression of alienation and contributes to it, making of man a (biological) machine.

Richard Dawkins

If Dennett is mildly dismissive of religion and Crick merely outspoken, then we must turn to Richard Dawkins as the most truly vituperative of the scientific atheists today. He is the Salk Professor for the Public Understanding of Science at Oxford. His commitment to this role is exemplary, and his energy and imagination in making science come to life is a great service to it. However he has also taken upon himself to publicly attack religion at every opportunity. Whether it is his scientific writings alone, or whether his anti-religionist stance has greatly contributed to it, *Prospect* magazine voted him its top intellectual in 2004.[106]

Dawkins's first book *The Selfish Gene* went to the best-seller lists after its publication in 1976 and its thesis remains controversial and hotly debated to this day. It is a starkly reductionist vision where Darwinian survival is no longer understood at the level of individual or species, but at the level of gene. The life of a human being is determined by genes that have no investment in either the individual human or the human race, but only in ensuring their replication in new carriers. Dawkins also coined the term 'meme' – meaning an idea or cluster of ideas that persevere in human culture – by way of analogy with gene, and suggested that memes replicate and survive in a similar way. The 1976 edition of the *Selfish Gene* made little direct reference to religion, other than to give 'God' as an example of a cluster of memes which survive tenaciously. Dawkins writes: 'Another member of the religious meme complex is called faith. It means blind trust, in the absence of evidence, even in the teeth of evidence.'[107] In the second edition, which is expanded with Endnotes to each chapter, Dawkins responds to readers who objected to his characterisation of faith, by repeating his claim that faith is a belief without evidence.[108]

Dawkins's second book, also a best seller, is called *The Blind Watchmaker*, and the title itself is a reference to the science-religion debate, though the book makes even less direct reference to religion than the *Selfish Gene*. Dawkins explains that the 'blind watchmaker' refers to the illustration given by the eighteenth-century theologian William Paley for the Argument from Design.[109] Paley sets forth in his *Natural Theology* a series of arguments for the existence of 'God', including the idea that both the natural world and a watch imply the existence of a maker. Dawkins disagrees in the case of the natural world (just as Dennett does): no 'God' is required, but merely the workings of Darwinian evolution.

The argument between Dawkins and established Christian religion has developed over a number of years, to the point where Dawkins is now seen as the principle opponent of the 'creationists', those who insist on a literal Biblical account of the origins of the natural world. Yet Dawkins himself tells us again and again of his wonder and awe for the natural world, feelings such that he finds himself with more in common with Paley than with some secular atheists. Dawkins counts himself as an atheist of course, but says that: 'although it might have been *logically* tenable to be an atheist before Darwin, Darwin made it possible to be an intellectually fulfilled atheist.'[110] Dawkins explains that Hume's attack on natural religion was logically correct, but that Hume was unable to provide any suitable account for the miraculous complexity and coherence of the natural world. It was Charles Darwin who eventually provided that intellectual fulfilment. Because Hume is assumed in contemporary culture to be an atheist, Dawkins is able to suggest that Hume would have agreed with him, and would have been 'intellectually fulfilled' by Darwinism. That is debatable, because of Hume's profound scepticism towards science, but not the point here: the real point is that to be intellectually fulfilled is not what makes a complete human being.

Dawkins has become the standard bearer of scientific objection to religion, being willing to offer an opinion on any perceived defect of mainstream religion, and finding publishers and news media ever willing to convey his views. After the September 11 attacks on New York and Washington in 2001, the British newspaper *The Guardian* published a response from Dawkins, which was effectively an attack on Islam.[111] To the secular mind Dawkins' stance is rational and reasonable, and the vitriol of his attacks on religion is simply consistent with his passion for science and his gifts as orator and writer. Crucial to the thesis of this book is the opposite view however, that Dawkins's stance is contradictory and only made possible by the accidents of the development of the *Christian* religion, not religion per se.

We assessed Dawkins in section 2.3 as a person with a nascent or undeveloped spirituality – according to our Two-Fold Model – with these characteristics: transcendent, *jnani, via positiva,* and social. We can now apply the *jnani/bhakti* checklist of section 2.6 in support of that assessment. Of the eight *bhakti* characteristics we enumerated, he demonstrates a leaning to only one: **unlikely to be drawn to the esoteric**. Of the seven *jnani* characteristics we enumerated he falls in with four of them, which include: (2) **an impulse towards knowing, 'seeing', knowledge, wisdom, learning**; (3) **enquiry, even doubt, as core behaviour**; (4) **an emphasis on the will**; (5) **instinctively anti-anthropomorphic**. All of these indications must be considered in the recognition that a nascent spirituality obscures clear manifestations of personal spiritual impulses. Yet a clear *jnani* transcendent *via positiva* portrait emerges. Secularists could argue that, far from a spiritual sense of the transcendent, Dawkins is merely a competent scientist who happens to have a poetic, even rhapsodical streak in him. But the historical development of the Western mind out of medieval Christianity and through the Enlightenment (elaborated fully in following sections) places Dawkins in a specific context. If we read his anti-religionist stance in the direct lineage of Hume, and understand it as expressing the same anti-anthropomorphism, then everything changes. Dawkins' argument – like Hume's – becomes an argument *within* religion, not against it. In fact he is most usefully attacking a product of religion's core pathology, control, though culturally lacking the intellectual tools to understand that his own version of the Argument from Design – which his writings represent – would support a non-theistic, *jnani, via positiva,* socially-oriented religion. He is in fact brought to consider the possibility that 'scientific awe' is the same as religion in a series of published exchanges, but dismisses it on the grounds that it would make for a very 'flabby' definition of the word 'religion'.[112] What Dawkins calls a 'flabby' definition of religion is what we are calling spiritual pluralism, and in its articulation we can find a reconciliation that his scientism forces him to reject. When Dawkins uses the word 'religion' he means only monotheism, and only its worst aspects at that. If he were to attend a synagogue, church, or mosque – one which had any religious vitality – for a period of time, he would discover what is not easily obvious to the non-participant: the everyday operation of what we are calling the Kierkegaard Corrective. An attack on religion made in complete unawareness of this corrective, which acts both as a moderating principle on the majority and as a very genuine source of happiness, is like the non-musical attacking music, or the non-sporting attacking sport.

Religious happiness is like no other, having a source beyond all external factors. But it is not known to all, particularly in the modern age, or perhaps some people have always been 'tone-deaf' to religion and spirituality. Jean-Jacques Rousseau, writing in 1762, thought that the difficulty of trying to explain the deep source of religion within personal experience was like trying to convince the deaf of the existence of sound.[113] In his analogy the deaf were the materialists who denied the evidence of their own unbounded inner life, and equated themselves only with the bound material of their bodies.

Explanation, proof, control, power

Atheistic scientists attack religion, as we have seen, at the level of explanatory power, unlike say, Gore Vidal or Polly Toynbee, whose atheism starts at a social or cultural level and is a protest at the oppressive character of old religion. The attack on religion on the basis of explanatory power resonates widely with a Western secular audience, and very little defence seems to be mounted using the argument that religion *isn't* primarily about explanation. (After all people enjoy art with no expectation that it will explain anything: so why should religion?) Could it be because the Western mind has learned a profound lesson since the seventeenth century: that scientific explanations, however impenetrable or counter-intuitive they may be to the lay person, have the power of accurate *prediction*? And that accurate prediction means control of the environment, and that control means protection from the forces of Nature, and the acquisition of power and wealth? This surely is the unconscious equation that drives such widespread acceptance of the views of scientists like Dawkins, Crick and Dennett. Francis Bacon is the originator of the saying 'Knowledge is power', appropriate for the man credited with kick-starting the empirical revolution in science. It is perfectly natural to want control or power over one's circumstances, but the secular mind has internalised this stance in a way never seen before.

It has become impossible for the atheistic mind to conceive that surrender to the will of 'God' is a metaphor with a genuine reality behind it, or that seeking harmony could be more important than seeking power (as in Taoism). Westerners are just as baffled when encountering the non-devotional (*jnani*) equivalent of surrender to 'God' in the Eastern religions: recognition of the emptiness of 'self', *shunyata* in Buddhism, for example. To exist in a state which the materialists have characterised as fatalist appears as a self-abnegation with only a pathological reading. Free will, the will to power, choice, freedom, self-expression, self-determination, are all attributes of self that are central to the secular mind (and if not, prescribe Prozac). Science can underpin their fulfilment; religion can only underpin their denial.

The argument here isn't that the manipulation of Nature through science may lead to an ecological catastrophe (though it might). Instead it is suggested that an attack against religion on the basis that its 'explanations' are inferior to that of science rings true to the uninformed because 'explanations' have become highly valued. The link between scientific explanations and mastery of the natural world is proven a million times. The benefits of technology are everywhere apparent to the Western secular mind, but the negative and insidious corollary is rarely perceived: that the world has therefore become a machine. If all natural phenomena are to yield to scientific explanation and technological manipulation, then all natural phenomena are mechanisms, and so, eventually, is the self. We even have a specific science that tells us so, psychology. And if self is a mechanism, *then self is alienated from itself*. William Blake understood this well when he said: 'the bounded is loathed by its possessor.'[114] A machine cannot do more than repeat the 'same dull round', and even the entire universe would soon become 'a mill with complicated wheels'. Blake was truly prophetic of the sickness of the modern era.

The very word 'explanation' has an etymology meaning 'to flatten out'. The scientific worldview removes the mystery from the world, disenchants it, and provides a safe, technologically mastered environment; gated communities emerge within which individuals are safe from hunger, the elements, disease, ageing – and other people. An insidious cocoon grows up around the technologically rich to protect them from the technologically poor, and, once this alienation from Nature and others produces the inevitable psychopathologies, the ultimate technological intervention is to hand: the anti-depressant. If the self is also a mechanism, then it can be fixed like any other machine, or at least patched up to some cultural criteria of normalcy. Love, which many would argue is the only true restorative, is scientifically understood as neuronal, hormonal, or a side effect of the selfish gene. It is an emergent property of the psychologised, neurologised, peptidised, genetic self. The love of 'God' is a superstition, surrender to his 'will' a fatalist delusion, mastery of the elements is everything. But how much Prozac, Ritaxil, Valium, Mogadon, Miltown – the brand names come and go, making depression as datable as hairstyles – does it take to hold that sense of alienation at bay?

3.1.2 Presecular Religion in the Twentieth Century

In the twentieth century old religion – in particular Christianity – had to make a profound accommodation with the secular culture in which it was located. Effectively a 'toleration pact' had grown up between old religion and secular society, articulated quite differently in America, Britain, France and Turkey, for example, as we showed earlier. The New Age as the third element had no significant access to the official state roles of old religion (such as providing the rituals of birth, marriage and death, or ceremonies for remembrance and times of national mourning), or to the levers of power and cultural production held by the secular institutions. The relationship between old religion with its feudally-constructed mindset, and secularism with its core of alienation was, since the seventeenth century, increasingly a stand-off, with New Age thinking unable to bridge the gulf. At arms length, the worlds of old religion and secularism, which have potentially so much to give each other, led parallel lives. The painfully slow and incomplete modernisation of church institutions in blank incomprehension of the rapid advances within secularism is mirrored by the painfully slow growth of interest in religion from the secular world and the more likely blank incomprehension from vituperative secularists.

Effectively then we have a polarisation of society, between those who hold views that are secular, prone to scientism, and atheistic (ranging from a casual Monty Python atheism to a vituperative atheism), and those who retain the religion of their forebears, and amongst whom we can count a small proportion of fundamentalists and extremists. The American Religious Identity Survey (ARIS) conducted in 2001, with a sample size of 50,000 Americans, suggested that 76.5% of the population were Christian, and 13.2% secular or nonreligious, the remainder being divided amongst other religious groupings.[115] If only 13% of the US population is secular, then why is it that the secular component of society runs the engines of cultural production? Professor David Martin at the London School of Economics has proposed a 'centre-periphery' theory to explain this, which says that large cities and their universities concentrate a liberal elite which tend to atheism, while the majority living outside of these cultural centres are still religious.[116] (Martin also points out that *explanation* is not the purpose of religion: he says, 'Christianity is exclamatory, not explanatory.'[117]) The liberal elite write books, control TV stations, radio and journalism, according to this theory, ignoring religion or even actively denouncing it. There must be some truth in this, but it ignores a vital feature of contemporary media: that it operates – at least in principle – within a free market where demand intimately shapes production. The question remains then: why isn't there a demand for cultural productions that deal creatively, critically, and intelligently with religion? We have constructed a phrase to help explain this: that religion flies below the cultural radar of the West. This means, without wishing to denigrate the presecular religions in any way, that the creative, literary and artistic intellectuals of the West rarely engage with religion, *to the considerable detriment of both religion and secular culture*. Religion has become an intellectual and creative ghetto, even if statistically it claims vastly more adherents than secularism. A collective failure of the imagination means that the religious majority has been unable to demand of the cultural elite the films, novels, poetry and plays that properly reflect the status of religion. This may be because the religious majority in the West are only weakly religious, whereas the liberal arts minority are vociferously atheist.

3.1.3 Spiritual Teachers in Modernity

If old religion and secular modernity had reached a détente in the twentieth century, the balance of which is now on the move, what then of the New Age? Why has it been unable to bridge the gulf between them? In two key scholarly works on the New Age – *New Age Movements*[118] and *The Spiritual Revolution – why Religion is Giving Way to Spirituality*[119] – the outer, more visible aspects of the New Age are well explored. We focus on a key aspect of the New Age, the spiritual teacher, because of the specific problems this raises, and because it gives us access to the interiorities of spirit within the New Age.

The Problem of the Teacher

A central issue in any East / West dialogue remains the different attitudes to the spiritual teacher or guru. The very word 'guru' raises an ambivalent set of responses in the Western mind, and yet just means 'spiritual teacher'. The West has such a monolithic concept of religion that the idea of the autonomous spiritual teacher is almost unknown and so we use the Indian word 'guru' for this unfamiliar person. The nearest Christian equivalent is the 'spiritual director', used within the monastic tradition as a term for a senior monk or nun who acts as guides to the novitiates, but not widely known outside.

Karen Armstrong in her book *A History of God* suggests that East and West diverged as long ago as the eighth century BCE in its attitudes to the teacher:

> The gods were no longer very important in India. Henceforth they would be superseded by the religious teacher, who would be considered higher than the gods. It was a remarkable assertion of the value of humanity and the desire to take control of destiny: it would be the great religious insight of the subcontinent. The new religions of Hinduism and Buddhism did not deny the existence of gods nor did they forbid the people to worship them. In their view such a repression and denial would be damaging. Instead, Hindus and Buddhists sought new ways to transcend the gods, to go beyond them.[120]

Armstrong makes the point well, but we might challenge the implication that the West did not share this insight at all, did not place value on the spiritual teacher. As she focuses on the history of 'God' it is natural for her to follow the monotheistic strand from its localised genesis in Israel to its domination of the West. But parallel to the Judaic development were a plethora of other spiritual traditions in the Mediterranean lands and Middle East, stretching back for millennia (just as was the case in India), including the Hermetic, the Gnostic and the Neoplatonist. The Western world at the time of Jesus was a melting pot of spiritual traditions and it is not all clear that the spiritual teacher at this time was valued less than in India. Indeed, Jesus becomes more accessible to the modern mind if understood as an ordinary man who underwent a spiritual awakening, and became a spiritual teacher. We can then understand him by working backwards from great contemporary spiritual teachers (whose lives are well documented), rather than by attempting to understand contemporary teachers by comparison with a mythologised, supernaturalised figure from the ancient past. What is undeniable in Armstrong's analysis is that the *subsequent* religious development of the West into a monolithic and exclusive monotheism did marginalise the spiritual teacher, and placed him or her as subservient to 'God', tradition, and the machinery of the church.

Although the differences between East and West are much greater than this, it does serve to illustrate one way in which new spiritualities developed in modernity. Christianity has had many new thinkers and movements in the twentieth century, but outside of its deeply conservative framework we find a plethora of non-Christian spiritualities emerging. Some fall under the umbrella of New Age religions and spiritual movements, while some are direct imports from the East. The common factor is usually a gifted spiritual teacher.

But the relationship between seeker and teacher is problematic in the West if that teacher instructs under their *own* authority, partly because the West is inclined to extremes in this relationship of either determined hostility or uncritical devotion. The uncritical following of charismatic figures can lead to abuse, which leads to a hostile reaction or over-reaction. Hence we find amongst the new spiritual movements accusations of abuse, of the kind rarely found in the East, even though the origin of the teacher may have been Eastern. Buddhism is no exception to this, with the Tibetan teachers amongst the most colourful and controversial of those teaching in the West, subject to such accusations and even lawsuits.

Andrew Rawlinson's *The Book of Enlightened Masters – Western Teachers in Eastern Traditions* provides one of the most comprehensive surveys of spiritual teachers in the West. His spiritual taxonomy (mentioned in Chapter Two) was developed so he could describe the huge array of teachers and traditions his book covered. The sheer wealth of material in the book conveys both a richness and excitement that alternative spiritual movements have provided since the nineteenth century, and demonstrates how much they have drawn from the East.

Christianity and the New Age

Mainstream Christianity has been forced to pay close attention to the New Age as its potentially greatest rival for the loyalties of the younger generation. In 2003, for example, the Catholic Pontifical Council for Interreligious Dialogue published a paper titled 'Jesus Christ The Bearer Of The Water Of Life – A Christian reflection on the "New Age" ' which thoughtfully reviewed the phenomena and its potential for conflict with Christianity. It concludes with the passage from Luke 16:13: 'No servant can be the slave of two masters: he will either hate the first and love the second, or treat the first with respect and the second with scorn.'[121] The sentiment expressed is familiar: it is an 'either-or' situation, never a 'both-and' situation, and is true to the 2,000-year tradition of Christian absolutism. Rawlinson's book shows however that there is a great appetite in the West for pluralism (though his taxonomy is not detailed enough for our purposes). The well-known passage from Luke cited above is a quotation from Jesus which concludes: 'You cannot serve God and Money.' Yet the contemporary Western experience is that one 'serves' many masters, 'Money' being just one of them. Any attempt to lead an entirely one-sided life is not just a human impossibility, but increasingly unrealistic in the modern world. The spiritual life of the West will inevitably reflect this multifaceted experience of daily life, but if we only had Western models of religion we might find ourselves with no religious precedent.

The spiritual teacher or guru in the West often run into accusations of malpractice, as mentioned above. A key factor in this situation was the timing of many of the emerging Western teachers in Eastern traditions: during the sexual revolution of the 1960s and 1970s. India had developed a highly repressive attitude towards sex, and when Indian teachers went to the West, or encountered Westerners in India, the unfamiliar sexual openness sometimes resulted in inappropriate behaviour (the story of the Beatles and their guru is a classic example). Sexual abuse is the most common accusation against spiritual teachers, though it seems to be, in legal terms at least, mostly between consenting adults. The story of spiritual teachers and their fallibilities is explored in two valuable books: *Holy Madness* by Georg Feuerstein[122] and *Feet of Clay – A Study of Gurus* by Anthony Storr.[123] Feuerstein is generally sympathetic to gurus, while Storr is generally hostile, though both provide plenty of material that demonstrates gurus' various weaknesses. Another recent survey of the guru scene, by Mariana Caplan, is tellingly titled: *Halfway Up the Mountain – the Error of Premature Claims of Enlightenment*.[124] Caplan's book represents a midpoint between enthusiasm and caution, and draws on many writers in the field. However it makes no attempt at a taxonomy of the spiritual, other than narrowing the field to 'enlightenment' as a category of spiritual experience.

Andrew Cohen

We can look at the example of spiritual teacher Andrew Cohen to illustrate some of the issues we have raised so far about the autonomous spiritual teacher in a modern Western setting. He was born in 1955 in New York. The details of his life are set out in his *Autobiography of an Awakening*,[125] in which Cohen tells us of a relatively unhappy, though relatively privileged childhood, marked at the age of sixteen by a profound spiritual experience. He travelled to India where he met both his future wife Alka, and his future guru, H.W.L.Poonja, also known as Poonjaji. This spiritual teacher was a direct disciple of the great Hindu sage Ramana Maharshi, and is considered by many to be fully self-realised. Cohen had found a teacher of sufficient stature to match his enquiry, and in turn Poonjaji found a student of unusual capacity, claiming that he had only seen the emptiness of self-realisation in the eyes of three people in his lifetime: Maharshi, himself in the mirror, and Cohen. Poonjaji was soon talking of Cohen as his spiritual 'son', a statement that is very significant in the spiritual life in India.

After a period Cohen broke with Poonjaji, and now teaches under his own authority, and has an international seminar schedule and following. He is founder of *What is Enlightenment* magazine, which has a considerable international readership, and which reports on many new spiritual teachers, as well as his own teachings. Since the fall/winter issue of 1997 which featured interviews with leading writers Ken Wilber and Georg Feuerstein, Cohen has increasingly been associated with Wilber's ideas of evolutionary consciousness. Cohen's initial teachings of enlightenment underwent an early shift of emphasis with the tag 'impersonal' and a second shift to 'evolutionary' since the collaboration with Wilber. The idea of evolutionary enlightenment draws on a wider field, however, including Teilhard de Chardin and Sri Aurobindo. This shift in Cohen's teaching aligns him with many New Age concerns, principally the core belief that a profound change in the whole of humanity is immanent.

Three of the most controversial gurus of the twentieth century might be G.I.Gurdjieff, Bhagwan Shree Rajneesh (Osho), and Chögyam Trungpa, at least if we take note of Rawlinson, Feuerstein and Storr. However, the saintly Jiddu Krishnamurti (who we look at shortly) doesn't completely escape criticism, being the subject of accusations that he slept with the wife of his closest assistant.[126] Cohen has publicly insisted on the highest possible standards for a spiritual teacher, but, in a book by his estranged mother (Luna Tarlo) called *The Mother of God*, she provides a damning indictment of his behaviour.[127] The book is alarming because of the very standards Cohen insists on in his own writings:

> The delicacy of this is beyond measure and it is indeed terrifying to perceive the karmic implications of what it means to claim Enlightenment, for if one would dare to do so, one must indeed be able to manifest and BE that perfection without hesitation and without fear. . . . That which is sacred is indeed sacred. No one is exempt from the absolute reckoning of one's entire being. Most of the world-renowned masters, gurus and prophets of our era have left a legacy of, at its worst corruption, and at its least confusion, only for this reason.[128]

History will judge however whether Cohen is perceived as he sets out to be, or as his mother tells it. At the time of writing Cohen remains a largely respected teacher with a substantial following.

However there is a small but growing group of disaffected pupils, whose collective voice might yet discredit him totally, or at least show that he lacks the self-awareness which would caution him against proclaiming his superiority over other gurus.

Cohen's spirituality, according to our Two-Fold Model, is clearly of the *jnani* orientation, demonstrated throughout his teachings and also in his lineage from Ramana Maharshi through Poonjaji. A parallel lineage exists in the teachers Nisargadatta Maharaj, Ramesh Balsekar and Wayne Liquorman, striking in the fact that both lineages start with Indian gurus and end in American ones. Both lineages are also in the Advaita tradition of India. It is useful to describe Liquorman and Cohen as representing a Neo-Advaita, i.e. a version of the Advaita that is adapting itself to the West through Western teachers. The Advaita itself (meaning 'not two') is one of the clearest examples of a pure transcendent, *jnani, via negativa* tradition, in which we find no trace of the esoteric impulse. The Neo-Advaita includes not only Liquorman and Cohen, but such teachers as British-born Tony Parsons, and the German teacher Eckhart Tolle. Despite their popularity, the Advaita does not travel that easily to the West, falling prey to accusations of indifference in the same way that the Buddha's teachings can. Where the Buddha built in substantial correctives in the shape of his strictures on compassion, precepts of monkhood, and on the life of the *sangha* (spiritual community), both the Advaita and the neo-Advaita have only the inherent corrective of the *via negativa* to draw on. In India the *via negativa* is intimately bound up with renunciation and poverty, which safeguard against many kinds of mystical inflation. But when the Advaita travels to the materialist West, particularly when located within what religious scholar Paul Heelas calls the 'Prosperity Wing' of the New Age,[129] risks multiply.

The Neo-Advaita is an essential component of the New Age scene, but, as we flagged up early on, the New Age has a reluctance to distinguish between different spiritual impulses within its umbrella. While Rawlinsons' research on spiritual teachers in modernity is invaluable, the principal shortcoming of his quadrant scheme is that it assumes that the esoteric /occult is a path to enlightenment, an assumption that is widespread in the New Age. As made clear in section 2.2, enlightenment is the goal of the transcendent / unitive path, and has nothing to do with the shamanic / esoteric / occult. We now explore this proposition in more detail by looking at two twentieth-century teachers: Rudolf Steiner as exponent of the occult, and Jiddu Krishnamurti as exponent of the transcendent.

Rudolf Steiner

Rudolf Steiner (1861-1925) was arguably the greatest and best-known occultist of the late nineteenth and early twentieth century. He is explicit that the movement he founded, Anthroposophy, was based on an *occult* science, one that was put forward in opposition to material science. Steiner represents the most remarkable updating of the shamanic impulse into the modern world that has ever been attempted: a meeting of the world of spirits with the world of science and philosophy. Steiner was educationally advanced, and would sit in his high-school class apparently reading his dull set text, but between its covers he had inserted works by Fichte, Schelling or Hegel and was correcting them. There are good biographies of Steiner by Hemleben[130] and Lissau[131] and also an autobiography, *The Course of my Life*,[132] which cover between them his contribution to medicine, philosophy, religion, education (including Waldorf schools), architecture, art and eurhythmy. His occultism is often described as an esoteric Christianity, which usefully indicates that he did not draw on Eastern traditions. Yet no account of the spiritual life of the twentieth century in the West and of its great spiritual teachers could be complete without Steiner.

Spiritual teachers who lie outside of tradition, whether their impulse is towards the esoteric or the transcendent, derive their insights from within. The Western mania for a lineage of thought in all spheres requires Steiner to be a follower of Goethe or Theosophy, but any careful examination of his life shows that his insights were firstly his (deriving from esoteric experience), and only later cross-referenced against the religious and philosophical heritage of the West. His reading was extensive and so was his acquaintance with the leading thinkers of his time in mainland Europe: we find in his autobiography an irenicist approach to philosophies even if he completely disagreed with them. All the more interesting then to find in chapter eleven of *The Course of my Life* a short but clear account of why he did not hold himself to be a mystic. (Note that Steiner adheres to the meaning of 'mystic' as one on the unitive path, like, for example Meister Eckhart or St John of the Cross.) In doing so he articulates the distinction between the occult and the transcendent, as an occultist who frankly finds the mystic (on the unitive path) to be a poor relation. He says:

> The mystic seemed to me to be a person who failed to establish a right relation with the world of Ideas, in which for me the spiritual was manifest. I felt that it was a deficiency in real

> spirituality when, in order to arrive at inner satisfaction, one wished to plunge with one's ideas into an inner world void of ideas. In this I could see no road to light, but rather a way to spiritual darkness. It seemed to me an impotence in cognition when the mind seeks through an escape from ideas to reach spiritual reality, which although it does not itself weave in ideas, can be experienced through ideas.[133]

This is an important passage, and its significance in terms of the esoteric / transcendent distinction needs teasing out in detail. Steiner finds the world of the mystic to be a dark one here, and in the following passages he adds cold, too. Yet the mystic feels the reverse, that esotericism / occultism is a dark cold world devoid of the ecstatic light and warmth of the unitive state of being. This gulf of incomprehension makes it almost impossible to explain the one impulse to someone of the other impulse, but much is at stake to make it worth the attempt. Earlier we said that the transcendent / unitive impulse leads to the goal expressed in Eastern terms as 'enlightenment' or variously, *moksha, nirvana,* liberation and so on. Amongst the theistic mystics of the West the goal was articulated as union with 'God', most properly an expression of the *bhakti* orientation. But what is the goal of the occultist / esotericist if it is not this (and for Steiner it most emphatically is not)? The definition given in Chapter Two is that it is about an expansion of personal agency through esoteric knowledge and esoteric powers. That expanded personal agency may be put to social purposes, as in the case of Steiner, or be a simple commitment to learning, or, as in the case of so-called 'black' magicians may be used for selfish personal ends. Whatever the exact ends of the occultist, they are unrelated to the mystic's goal of transcendence of self. Also, as Steiner makes so clear in the above passage, the esotericist engages with the world of ideas and imagination. In the New Age this is referred to as the 'imaginal' or the 'imaginal cosmos', and its exploration represents a complementary but quite alien spirituality to the Neo-Advaita. Where Steiner uses the term 'impotence' to reject the transcendent state of the mystic, this is also a rejection of the *via negativa,* and resonates with Jung's characterisation of *nirvana* as an 'amputation'.[134] Both were, at heart, esotericists.

Jiddu Krishnamurti

Andrew Rawlinson only devotes an appendix entry to Krishnamurti in his book on Western Masters, but tells us that he represents '*Cool*

Unstructured teachings in perhaps their most pronounced form.'[135] Krishnamurti was born in 1895, and in the terminology used here is a teacher of transcendence of the *jnani* temperament with an interesting balance between *via positiva* and *via negativa*, and between the social and the solitary. His mother may have had some presentiment that her future child would hold spiritual significance: she chose, against the explicit religious and caste traditions regarding birth, to deliver Krishnamurti in the *puja* (shrine) room of her small house.

As a child Krishnamurti was not considered unusual in any way, but was discovered in 1909 by Charles Leadbeater, a leading member of the Theosophical Society. It had as its stated goal the preparation for a new World Leader (the second coming of both Christ and the Buddha), and before long it declared that it had found it in the person of Jiddu Krishnamurti. He was prepared for his role through occult initiations at the hands of Leadbeater and Annie Besant, a process that involved communications with so-called disembodied 'Masters', and ultimately the excruciatingly painful preparation of his body to become the vessel for the (Buddha) Maitreya. But he gradually shook off the ministrations of the Theosophical Society, and in a dramatic gesture dissolved the Order of the Star, which was the organisation founded to support his work. He could not shake of his destiny however, and entered a life of teaching that lasted fifty years. The teachings were his, however, and could be summed up in one phrase: *choiceless awareness*.

Krishnamurti simply jettisoned the whole of Indian religious history (as well as all other religious apparatus) and talked for fifty years on the pristine state of a silent mind that lives with this choiceless awareness. His emphasis on no-mind borrows nothing from Zen Buddhism, and he seems to have taken no interest in any mystical figure or teaching, however similar to his own: he was reputed to read detective novels or watch Clint Eastwood movies by way of relaxation. Although he showed moderation in material things, he liked to dress smartly, and enjoyed sports-cars, his dogs, gardening, reading and films; his Indian instinct for renunciation only showing itself in a vague dismissal of human affairs such as war and politics, and a general distaste for the coarser sides of life. The latter is in indication of the *via negativa* in his orientation, but he held a profound love of Nature which is more *via positiva*, and in complete contrast to what we know of the Buddha (who is otherwise a good comparison).

Christ-comparisons were also made throughout his life – here are some comments from contemporary figures:

> George Bernard Shaw called Krishnamurti 'a religious figure of the greatest distinction', and added, 'He is the most beautiful human being I have ever seen.'
>
> Henry Miller wrote, 'There is no man I would consider it a greater privilege to meet . . .'
>
> Aldous Huxley, after attending one of Krishnamurti's lectures, confided in a letter, 'the most impressive thing I have listened to. It was like listening to the discourse of the Buddha – such power, such intrinsic authority. . . .'
>
> Kahlil Gibran wrote, 'When he entered my room I said to myself, 'Surely the Lord of Love has come.' '[136]

It is a commonly held view that Krishnamurti was enlightened, but a poor teacher, yet over the years of his teaching many thousands from all over the world attended his talks, attracted by his reputation and his words. The list of cultural celebrities who held him in high regard includes not only figures named above (George Bernard Shaw, Henry Miller, Aldous Huxley, and Kahlil Gibran) but also many Western spiritual teachers of the next generation, including Andrew Cohen. Krishnamurti did not recommend outward forms of self-discipline and he was outspoken in his condemnation of all gurus, teachers, and Masters. His antipathy to the whole context of master and disciple may have been a result of his early training, but it placed him in a paradoxical position as a teacher. More than any other teacher of his time he attempted to give the aspirant independence and self-reliance from the outset, but his own nature and background and the nature of seekers in general made this maddeningly difficult. It is apparent in the conversations between Krishnamurti and the physicist David Bohm, where Krishnamurti *is* the master and Bohm *is* the disciple, but the dialogue has the outward form of equality, or even an inversion of roles. David Bohm was in fact an important figure in the twentieth century revolution in physics (his 'implicate order' is a key New Age science concept), but did his knowledge of physics warrant such an apparent inversion of the guru-disciple relationship? Krishnamurti's cautiousness was established long before the publication of Feuerstein's and Storr's works and the wider concern about gurus, so Krishnamurti would not have known about the mounting catalogue of accusations against such teachers.

But what is it that makes Krishnamurti's spirituality so vehemently transcendent, as against Steiner's vehement esotericism? It is epitomised in a single statement found in Krishnamurti's *The Only*

Revolution: 'Imagination in any form destroys truth.'[137] Although Krishnamurti's writings contain within them poetic and rhapsodical accounts of Nature, they are the antithesis of any Romantic imaginings. Where Steiner finds an 'impotence of cognition' in the escape from ideas, Krishnamurti asserts the opposite: freedom lies in rejecting the world of the imagination, of word-constructions, of the 'imaginal cosmos'. They are rejected because they so easily become ego-constructions in which truth can be avoided, in which the fiction of a separate self can be maintained. As with the Buddha and the Advaita, Krishnamurti's path to truth is an austere one, but more than that, all these accounts of transcendent truth (in the singular) are radically incommensurable with Steiner's account of esoteric truths (in the plural).

In Paul Heelas's analysis of the New Age he makes the useful distinction between the 'prosperity wing' and the 'counter-culture' wing.[138] The former uses its spirituality as an aid to material success, and sees success as an endorsement of its spirituality; while the latter pursues a spirituality that withdraws from or resists the mainstream. This is a useful division of the external and visible phenomena of the New Age, but our division between transcendent and esoteric addresses the spiritual interiority of its practitioners. And it is the esoteric which more naturally carries forward the key New Age idea: of an imminent transformation of humanity spoken of and influenced by the imaginal cosmos. In terms of the *jnani* / *bhakti* distinction, the New Age is predominantly *jnani*. On this basis we can see why the institutional Church, as instinctively *bhakti* and with a historical antipathy to the esoteric, would be hostile to the New Age and its teachers.

3.2 The Enlightenment

3.2.1 The Special Role of the Enlightenment

How is that spiritual teachers in the twentieth century like Steiner, Krishnamurti, and Cohen had neither any impact on the mainstream, nor on the ghettoised Christian world? How is it that religion in the secular West shrank to occupy the cultural margins of society? And how exactly did twentieth-century thought, with its adherence to scientism, come into being in the first place? If we go back three or four hundred years, then Europe and America had no conception of science as an argument against religion; could not conceive of religion as anything less than at the centre of all cultural life and exchange; and, rather than ignore new spiritual teachers, rushed to react for or against them in the most public of ways (the reaction to the extraordinary work of George Fox, founder of the Quakers, is a good example). What was the revolution in thought that made these two worlds, that of the twentieth century and that of the seventeenth century, so unrecognisable to each other? To put it more baldly, what caused the complete rupture of tradition – a tradition going back to mists of time in which religion was the centrepiece of human living – and the emergence of the secular mind? The conventional answer is: *the Enlightenment*.

What had been previously joined as one integrated human endeavour: religion, philosophy and science, now were set on the road to become three mutually uncomprehending realms of thought. The Renaissance had left its mark with the reintroduction of Platonic thought; the Reformation and counter-Reformation had left its mark with the great schism between Catholics and Protestants; new religious sects sprang up everywhere; and science was born, leading to the Enlightenment and the Industrial Revolution. As we seek the origins of the secular mind within the Enlightenment period we need to explore how a key shibboleth of secularism emerged: that science is opposed to religion. Which of the Enlightenment scientists were atheists? Which of the Enlightenment philosophers advocated the end of religion, or the birth of atheism in anything like a modern form?

We will see that a key concept that will help understand this period is the distinction between atheism on the one hand, and anticlericalism on the other. A real atheist is one who denies the existence of 'God' and who demands the end of religion as mere superstition. Anticlericalism means instead the objection to Church authority and its power both over daily life and of doctrine. Anticlericalism means an end to superstitious beliefs and practices, the beginning of toleration between sects, and above all the separation of Church and state. Of these two, atheism or anticlericalism, which properly represents the thinking of the key Enlightenment figures?

3.2.2 The Rise of Science

Science clearly plays a role in the origins of the secular mind, and it is Sir Isaac Newton (1642-1727) who is universally recognised as the single most important scientist of the Enlightenment. Richard Dawkins – who believes that science is irrevocably pitted against religion – considers Newton to be 'one of the greatest thinkers of all time'.[139] The word 'science' is however rather loosely used, often covering ancient disciplines from as far afield as China and the Arab world. What arose in the seventeenth century to be more precise was 'physics' in the modern sense of the word, a discipline understood by few then or now. However, it was a genuine breakthrough in human thought and discovery, and it is important to understand that its birth took place nowhere else in the world or at any other time. 'Physics explains chemistry explains biology' is a simple mantra that college students may hear, but is not always acknowledged. Nobel laureate in physics, Murray Gell-Mann, explains this hierarchy well in chapter nine of his popular science book, *The Quark and the Jaguar*.[140] Lord Kelvin famously remarked: 'all science is either physics or stamp collecting.' Physics, chemistry and biology are known as the hard sciences, led by physics, and it is these which underpinned the Enlightenment and Industrial Revolution, and provided the economic basis for all the political developments which were to follow.

The Birth of Physics

If the death of 'God' took place in 1885 (at the hands of Nietzsche) then the birth of science took place in 1676 (at the hands of Newton). The new science of mathematical physics was born in the Royal Society premises in 1676, on 27 April to be precise. This is proposed in a biography of Newton by science writer Michael White where he details the *experimentum crucis* carried out by Newton's arch-rival Hooke to demonstrate a key theory of the *Opticks*: the splitting of light by a prism.[141] This date is not universally accepted as the birth of science because Newton's work was built 'on the shoulders of giants' (i.e. earlier and contemporary experimenters), but does mark a very important landmark in science, the establishment of *verification* of a theory by duplicating experiment, the process which establishes its consensual nature. Newton had reached the propositions regarding colour and light that he wrote up in his *Opticks* by a series of brilliant hunches and experiments, but this is insufficient for theories to pass into 'science' in the modern sense. It is only when the scientific community considers the propositions worthy of testing, independently tests them, and finds confirmation, that the theories becomes scientifically proven. Even then they are in perpetuity open to falsification (as Sir Karl Popper asserted in the twentieth century that they must be); in practice this rarely happens, other than through a process of refinement. Given Newton's present-day status as cultural hero, we might expect that the Royal Society just rubber-stamped the findings in the *Opticks*, but of course at the time Newton's reputation was not yet made, and he had in Robert Hooke a rival and the bitterest of enemies. This is what gives this moment such significance in the context of the seventeenth century – that a new human endeavour was born where personal animosity counted for nothing in establishing truth. (Note that some consider Newton's famous quote about the 'shoulders of giants' to be part genuine tribute and part a vicious jibe at Hooke, who was so stooped and physically deformed as to resemble a dwarf.)

To bring home how revolutionary this moment was we can consider how agreement in other spheres of human activity is much harder to come by. For example Mullah Omar, the leader of the Afghani Taliban most closely associated with Osama Bin Laden, might put forward a set of theological arguments on the subject of *jihad* or holy war. Imagine then that the Pope is brought to an International Society of Religion to demonstrate the *validity* of Omar's findings. He could only do so if his vested interests in opposing Mullah Omar were overcome by incontrovertible evidence that Omar was right: clearly impossible. In our analogy Hooke is to Newton as the Pope is to Omar. Hooke had everything to gain by proving Newton wrong, but the methods of the radical new science – physics – persuaded him that Newton was right against his emotional investment. The history of the hard sciences since that time is littered with great names that resisted certain theories – including at times even their own – until the empirical evidence emerging in favour of them forced a total u-turn. Conversely, even the Pope-like status of Einstein today doesn't prevent the scientific community from disagreeing with him over his rejection of indeterminacy in quantum theory: the evidence is massively against him.

In 1676 the Inquisition still cast its long shadow over the Continent and religious dissent was everywhere an action that could have terrible consequences. The Cavalier Parliament passed the Act of Uniformity in 1662, resulting in the 'Great Persecution', which, though it did not lead to deaths, resulted in the trauma to nearly 2,000 Puritan ministers of losing their jobs. Newton himself had to keep his interests in alchemy secret, and even more of a liability for a public figure, and even more vital to hide, was his adherence to Arianism, a heretical Christian belief with its origins in the fourth century. The *experimentum crucis* in 1676 was truly momentous in its implications for the secular revolution to come because it gradually became apparent that there were certainties to be obtained in the physical world that did not arise from authority or revelation, and which the bitterest of rivals had to yield to via an arbitration beyond question. Of course it did not impinge on the seventeenth-century consciousness overnight, and we find for example that the eminently reasonable and capable scientist, Christiaan Huygens, chose to disagree with Newton on his *Opticks* as a matter of principle rather than to test the propositions in the laboratory.[142] The Royal Society was the exception in that day, Huygens the norm, but this position changed in the following decades as scientific momentum grew in pace. The experiment at the Royal Society had a *consensual* basis, that is the scientific facts under investigation provided a repeatable, distinct, and precise impression on the *senses* of those present. The deductive leap from those sense impressions (nearly always visual in physics) to the scientific facts was compelling, particularly as those results were expressed mathematically. The *consensus* reached by the distinguished gentlemen was of a type never reached before in any human field of enquiry, and certainly not in the field of religion. As Voltaire put it: 'There are no sects in mathematics, in experimental physics.'[143]

Physics and Philosophy

We are attempting to make a link between physics and its impact on a culture of repressed dissent that may not be immediately obvious. Physics itself deals with knowledge that ultimately has no relation to the important issues of meaning that religion and politics for example deal with. How relevant is the relationship between the curvature of a lens and its focal length to the relationship between citizen and state, man and wife, or layperson and clergyman? How does the fact that the period of oscillation of a pendulum remains constant over a range of amplitudes of swing help us understand our subjective relationship to time? (As so poignantly investigated by, for example, Bergson and Proust?) Of course the first 'fact' helps us create microscopes, telescopes and eyeglasses, and the second in the construction of clocks, but where is the passion of human life in that? People today who take close interest in such 'facts' are dismissed as 'nerds', so why did the proceedings of science have such a huge impact on the eighteenth-century mind? Long before the technological windfall became ubiquitous in people's daily lives?

The Industrial Revolution took off on the back of the new sciences, but it was the philosophical and intellectual world that was galvanised by them into the rationalism of the Enlightenment (as we explore in the next section). Scientists didn't write much that directly shaped this revolution in thought; it was the broader intelligentsia of the time that did the work. The philosophers didn't often cite the 'facts' of the new science (indeed they often didn't understand them at all, as Locke admitted after attempting to read Newton's *Principia*) but interpreted its methods as a dawn of rational thought. But let us go back a little deeper into the origins of modern science. The ancient Greeks created part of the mindset necessary for physics to develop, yet it did not emerge in its modern form in that culture. Rational thought itself and the beginnings of mathematics and geometry were present, yet did not add up to science because of an ambivalent attitude to the physical world, represented by the opposing views of Plato and Aristotle. Plato was distinctly anti-empirical, for example chiding the Pythagoreans for their efforts to understand sound by measuring vibrations in strings,[144] and telling us that Nature was of no interest, or at least nothing outside of the city gates was worth looking at, only the realm of men and politics within the city's walls.[145] Aristotle placed a greater emphasis on the observation of Nature, and created the first taxonomies of plants and animals, in fact the basic structure of his system survives to this day. But his *mechanics* was so mistaken as to hold back its development for nearly 2,000 years. Hence a systematic empirical instinct was not part of the Hellenistic outlook, and did not surface in Europe until Leonardo da Vinci, who opposed the Greek revival, favouring instead the principle of *observation*. (Michael White usefully explores the significance of Leonardo for science in *Leonardo: The First Scientist*.[146])

The Problem of the Planets

What there was of the scientific mind in the Greeks was lost to Europe in the so-called Dark Ages (though not to the Middle East) until it re-entered in the thirteenth and fifteenth centuries via the Arab world, and ushered in the Renaissance. What the Greeks bequeathed to Europe, and which engaged its intellectuals from Plato's time (about 500 BCE) until Newton was 'the problem of the planets'. The very word 'planet' comes from a Greek word meaning 'to wander', and it was this erratic movement of the planets, as opposed to the fixed stars, that presented a problem in understanding. Richard Tarnas in the *Passion of the Western Mind* cites the 'problem of the planets' as being the central question that hung over the mind of intellectuals in Europe and the Islamic world for over two thousand years after the time of Plato.[147] Saint Augustine demonstrates this point. After a profligate youth he turned to religion and became a devoted follower of Manichaeism for about ten years. He eventually left the religion partly because a Manichean bishop called Faustus failed to satisfy him on matters of astronomy, matters which could only have been related to the erratic motion of the planets.[148] (More on Augustine in section 4.1.3.)

The solution to the problem is now well charted – the key figures in this drama being Copernicus, Galileo, Kepler, Tycho Brahe, and Newton. Prior to the solution put forward by Newton (with the help of many leading scientists of the time) an ad hoc model for understanding planetary motion had developed over the centuries, called the Ptolemaic model. This model had to be clumsily modified to account for each new observation of the movements of the planets, and its predictive power was erratic. However the revolution that Newton brought about was a completely new basis for understanding natural phenomena – which we now call physics – underpinned by mathematics. The solution to the problem of the planets is expressed in a formula known as Newton's inverse square law of gravitation, and it is not an exaggeration to say that its discovery changed the world forever. It is also true to say that unless one has some training in maths and physics it is almost impossible to grasp its significance. However, some points are not difficult to make, the first of which is that, by studying the planets, scientists were able to properly understand the motion of objects free from friction and air resistance, such as they are constrained by on Earth. More important still, the behaviour of the planets approximate well to what is known as a 'two body problem', again, not easily obtainable on Earth. By conquering a seemingly impossible, ancient and well-known problem, the science that arose out of it – mechanics – was put on a firm footing and could be used on Earth by factoring back in the effects of air resistance and friction, or, alternatively, by excluding such influences in the laboratory. The invention of the vacuum pump assisted in this and also allowed further confirmations of the new mechanics. In fact the laboratory then became the place where the scientist strove above all else to eliminate external influences on the subject of experimentation: a narrowing down of focus.

Jnani, Occultism and Science

Let us return to the person of Newton, for whom we provided a provisional spiritual profile according to our Two-Fold Model of Spiritual Difference: esoteric, *jnani, via positiva,* solitary. We also suggested that, while he made no significant contribution to religious thinking, his spiritual impulse was more than nascent or undeveloped. He was in fact, even by the standards of his day, a deeply religious man. So how could that fit with his momentous and founding contribution to science? We need to investigate a little more how science and religion co-existed in general at this time in history.

The ideas of Aristotle entered European culture in the thirteenth century via the Arabs, and found expression in the treatises of the Schoolmen (also known as Scholastics), culminating in the works of Thomas Aquinas. Scholasticism can be seen as the West's second attempt at a science, the first being that of the ancient Greeks, both dying away before the events that led up to birth of science proper in 1676 at the Royal Society. We have suggested why the Greeks failed to establish a real science, but the reasons for the Scholastic failure are slightly different. They shared with the Greeks some of the ambivalence which combined scepticism towards Nature with a renewed interest towards it, but the main problem here was of motivation. It was the Scholastics who took the idea of 'proving' the existence of 'God' to new heights, and this gives the clue: their efforts are best understood as acts of piety. A rationalist and genuine interest in Nature was in servitude to a spiritual impulse, to the glory of 'God'. Even so, the work of Aquinas was not at first accepted by the Church, and he was charged with heresy. We suggest that this is because his work had an element in it that was genuinely foreign to Christianity as it had evolved to this point. It had an element of *inquiry* that was alien to the devotional tradition largely fostered by St Paul and St Augustine, and it had an element of the non-devotional, non-theistic *jnani* spirituality that we can define more sharply as we progress. For now we can say that neither a science could develop out of Scholasticism, because it had religious goals, nor could a new non-theistic spirituality grow out of it because the Church was intolerant, in its deepest instincts, to such a development.

The *jnani* spiritual impulse includes an element of enquiry. In the Buddha this became an entirely inward exploration, because, in addition to the *jnani* impulse he also demonstrated a powerful impulse to the *via negativa*. However where the *jnani* impulse is coupled with a more *via positiva* approach, as, we suggest, in Aquinas, a scientific outlook can arise. As we shall see in section 4.2.4 *The Renaissance and Reformation*, Plato provided an impulse to science in the West through the loose collection of traditions we call Neoplatonism, and which perpetuate in Western history the powerful influence or mindset that we call occultism. Its origins, we have suggested, lie in shamanism, but result from the retention of shamanic or mediumistic contact with the spirit world within and alongside the culture of the agrarian city state. In Plato, generally upheld as one of the great Greek rationalists, we find many of the powerful occult ideas that are maintained in the Neoplatonist tradition: an invisible world of 'forms', astrological influence, reincarnation, and the myth of Atlantis. The origins of astrology and alchemy are lost in the mists of time, but are ideas fully in keeping with the shamanic worldview, and outlive the destruction of the shamanic way of life at the hands of monotheism.

When Dawkins characterises Newton as one of the greatest thinkers of all time, we are left with a puzzle. Does Dawkins not know about the religious and occultist sides of Newton? Or if he does, is he assuming that Newton was a divided personality, part man of his superstitious times, part exemplary scientist? Were the religious and occult sides of Newton just a foible, perhaps the result of the enormous strain of his forced labours, not found in the other pioneers of science? We need to look therefore at the other scientists of this period to answer these questions.

Galileo and Other Early Scientists

Disregarding any possible sarcasm, Newton's astonishing achievements were truly built on the shoulders of giants, including Copernicus (1473-1543), Johannes Kepler (1571-1630) and Galileo Galilei (1564-1642). Copernicus was a canon in a Polish monastery serving under his uncle, pursuing his mathematical and astronomical interests alongside his quite conventional religious life. He waited to publish his key work *De Revolutionibus* until 1543, shortly before his death, an indication that he was anxious about its reception. It was finally placed on the Forbidden Index (the Roman Catholic Church's list of banned books) in 1616 subject to revision, along with several other books that defended the Copernican system. The Roman Church was slow to act because Copernicus was a canon and he dedicated his book to Pope Paul III, though the counter-Reformation was increasingly making unorthodoxy more problematic. However, there is nothing we know about Copernicus to suggest that he rejected any part of his religion, other than its adherence to the geocentric model of the universe, and the Aristotelian / Ptolemaic orthodoxy in astronomy.

Johannes Kepler developed the ideas of Copernicus, playing a major role in the eventual solution to the 'problem of the planets'. His life epitomises the intellectual upheavals of the seventeenth century as he contributed to the birth of modern physics, published works in astrology and philosophy, was religiously syncretic and irenic by temperament and fell victim to the Inquisition twice over, but in ways quite normal for the times. Firstly his mother was accused of witchcraft (his aunt had already been burned at the stake for the same offence), so that he had to devote considerable time and resources to her defence, and secondly a number of his books were banned. Kepler's *New Astronomy*, his *Epitome of Copernican Astronomy* (1618-21), and his *World Harmony* (1619), a Pythagorean-influenced treatise, were all placed on the Index of Forbidden Books. Kepler originally studied theology, and his astronomy was intimately mixed up with his astrology: in other words he was similar to Newton in pursuing religion, occultism, and science.

During the proceedings against Galileo in 1633, his *Dialogue Concerning the Two Chief Systems of the World* was placed on the Index, where it remained until 1824. His prosecution by the papacy is still famous today as the epitome of religious intolerance in the face of scientific advance. Copernicus, Kepler and Galileo all had their books banned by the Roman Catholic Church, and Galileo more visibly was victimised for his science through the threat of torture during his trial, and his subsequent house arrest. It is therefore reasonable to say that the Church attacked these crucial early scientists, but is it reasonable to say that this was a mutual hostility? Did the scientists attack religion per se? Clearly not in the case of Copernicus, and only marginally if we consider the syncretism of Kepler as an attack on religion (as we saw earlier, syncretism tends to raise Christian suspicions). In Galileo's case all we find – apart from his insistence on scientific truth – is a poor handling of the politics of the day in his attack on Aristotelian 'science'. (His *Dialogue* could be construed as presenting the Pope as an Aristotelian simpleton.) Although the early Christian church was opposed to all of Greek learning, including its 'proto-science', there had been a slow shift since the time of Aquinas in the thirteenth century which had not merely rehabilitated Aristotle, but almost canonised him. His pronouncements on the natural world had slowly become Christian orthodoxy, despite their being no correlation at all with any Biblical passages, but sadly, the same dogmatism was applied to his ideas as to all Christian thinking. Hence to contradict Aristotle was effectively heresy.

For the new science of physics to emerge, Aristotle *had* to be overthrown, which was a much easier job in countries like England. This was because 'Thomist' philosophy (Thomas Aquinas' Aristoteleism) was anyhow abandoned. Newton was also very shrewd politically, and in any case lived in the slightly more tolerant overall Anglican context. Taking the four men together, it seems that while their science offended religion to varying degrees, religion did not offend them at all. *We can find no possible source of the secular world view in the life and works of these key early scientists*. Clearly by the time of Crick, Dennett, Dawkins and Atkins we find scientists considerably offended by religion, so we need to search for scientists after Newton who might have created the precedent for this, or some other cause altogether.

Boyle, Laplace, Lavoisier, Franklin, Priestley and Faraday

Newton, though pre-eminent amongst scientists of his time was not in other ways unique: Robert Boyle (1627-1691) was another pioneering scientist of his day and founder member of the Royal Society. He also assiduously pursued alchemy, assisting Newton with the construction of furnaces, and made a bequest in his will to fund an annual sermon to prove the Christian religion 'against notorious infidels, viz. Atheists, Theists, Pagans, Jews and Mahometans.'[149] (Note a difficulty with terms here: Boyle presumably meant 'Deist' when he used the term 'Theist'.) Boyle is also reputed to have said that one could only be an atheist if one had *not* studied nature,[150] a sentiment perhaps standing for all the scientists of the time, and indicating the complete lack of any perceived antagonism between science and religion from the viewpoint of the scientists.

The case of a French scientist, Antoine-Laurent Lavoisier (1743-1794), is similarly devoid of atheism. He showed no hostility or scepticism towards religion despite being to chemistry what Newton was to physics. His monumental contribution to chemistry is, however, another example of the struggle against Empedocles and Aristotle: this time against the 'four elements'. The classical conception of matter as constituted from only four elements was as big an obstacle to the proper development of chemistry – which was to discover over a hundred elements – as Aristotelian mechanics was to physics.

Pierre Simon Laplace (1749-1827) took Newton's astronomy and substantially built on it, and is genuinely deserving of the epithet 'The Newton of France'. He is also regularly seized upon by atheists for his famous retort to Napoleon on the apparent absence of 'God' in his astronomy: 'Sire, I have no need of that hypothesis.' Perhaps we have found the first of the founding physicists to have seen in their science a reason for atheism? Apparently not. An examination of his life shows him to have been as religious as the previous four examples, summoning not just one priest to his deathbed but two. There does not even seem to have been any alchemy, astrology, numerology, Arianism or syncretism in his religiosity that would place him at odds with his religious tradition. In saying that he did not need the hypothesis of 'God' he most likely meant that he did not need it for his *astronomy*. For a religious person 'God' is not an entity of hypotheses: we have suggested rather that it is the objectless object of either the devotional or non-devotional spiritual impulse. But the atheists who would like to claim Laplace as one of their own do have a point: Newton, standing at the dawn of modern physics could not shake off the earlier unity of science and religion. Despite his brilliant successes, he did not yet have that modern sense of faith in science, that everything would ultimately yield to a mechanical explanation. For him, the questions still left about the solar system, for example why it was that the planets rotated in the same direction, were simply proof of 'God's' beautiful design. Laplace, in offering the nebular theory of planet formation, not only removed 'God' from that question, but began to make it possible that *all* scientific questions would be resolved through mechanistic explanations. But that did not in the least make him an opponent of religion.

Looking at three other great scientists born in the eighteenth century, Benjamin Franklin (1706-1790), Joseph Priestley (1733-1804), and Michael Faraday (1791-1867), we find them all joyously religious in their life and writings. And no scientist born in that century – perhaps ever – has shaped modern life as intimately as Faraday, who invented both the electric motor and the generator which would supply it with electrical power. Can one conceive of life today without electrical appliances? Even the petrol-driven motor car would fail to operate if it did not have an electric motor and generator, and the diesel engine, which *can* run without electricity, would have to be started with a blow-torch. Imagine life without the washing machine, the vacuum cleaner, and without all the electronic means of media and communication. Yet Faraday was a deeply religious man, who saw no contradiction between religion on the one hand, and the intimately bound pursuits of pure science, mathematics, and applied technology on the other. Crucially, all these scientists held some variant of a *materialist* view of the world, but a materialism which did not in the least imply atheism.

How could Physics lead to Atheism?

We have drawn a blank in searching amongst the key founders of modern science for any hint that they found their science to be a contradiction to their religion. Some possibilities now remain. Firstly, was it just the blindness of an era, prior to the overwhelming triumph of the rationalist Enlightenment, that enabled them to hold science and religion within a single mindset? The examples of deeply spiritual twentieth-century physicists like John Polkinghorne and Erwin Schroedinger suggest otherwise; the greatest nineteenth century physicist, Lord Kelvin, was also quite conventionally religious. In fact it is hard to find an atheist physicist in any period, or at least one who has in any way accelerated the course of atheism. All that took place was the gradual divorce of science and religion, with no intention that one of the two partners should then die.

But what about Dawkins's suggestion that 'although it might have been *logically* tenable to be an atheist before Darwin, Darwin made it possible to be an intellectually fulfilled atheist'? Although the examples of Polkinghorne, Schroedinger, and many others, rather suggest that one can be a religiously fulfilled scientist, Dawkins is giving us a clue: perhaps we should be looking to *biology* rather than *physics*. We will do so, in particular Darwinism, but first we need to look at a possible early source of secularism outside of any of the hard sciences: philosophy.

3.2.3 Enlightenment Philosophy

There is no doubt that science, philosophy and religion diverge in the Enlightenment period. Could it be that the scientists were not philosophically astute enough to understand the contradiction between science and religion? Was it the philosophers, responding to the new science, who became the atheists of that time? The scientists did not generally pick a fight with the Church, but it is easy to find philosophers of the period who did, and none more visibly than Voltaire. It is the French *philosophes* and Encyclopaedists who made direct attacks on the Roman Catholic Church, and who suffered most in this period by way of retaliation. But can we be sure that their attack was on religion itself, or was it on the powerful institution that it had become and the oppressive nature of its role in the affairs of State? In other words, were Enlightenment philosophers atheists, or merely anticlerical? And, if such atheism existed, had it derived from the new science? For one thing is clear: that philosophy at that time was utterly contingent on the scientific revolution, taking its cue effectively from the ethos surrounding the *experimentum crucis* of 1676.

He Who Hid Well, Lived Well

In Antonio Damasio's book on Spinoza and neuroscience he points out that despite the haven of tolerance that Holland provided in the seventeenth and eighteenth centuries (compared to France for example), it was still not possible to speak one's mind. He cites the epitaph that Descartes had intended for his tomb: 'He who hid well, lived well', to suggest that Descartes for one was keenly aware of the dangers of free thought in that period.[151] Hume scholar Paul Russell shows that Hume, living in *relatively* tolerant Britain, is nevertheless careful to conceal his true thoughts. Russell suggests that the epigram from Tacitus: 'seldom are men blessed with times in which they may think what they like, and say what they think', which appears in Latin on the title page of Hume's *Treatise of Human Nature*, acknowledges the dangers facing any radical thinker of that time.[152] One could face a ban on printing (as in the case of Hobbes) or repeated imprisonment (as in the case of Voltaire), though no philosopher of this period was tortured or killed by Church or State (Lavoisier was executed in the Terror for entirely political reasons). What this means is that the writings of the philosophers cannot be read like the thinkers of today: we have to expect that to one degree or another they hid their true meaning. At the same time this explains the often strange manner of writing in that period, using a variety of devices such as 'Dialogues', circumlocution and ellipsis, the use of verbosity, the use of aphorism, or, in the case of Spinoza, a pseudo-Euclidean structure. Dialogues were a useful device because, though the opinion of the author was usually voiced by one of the protagonists in the debate, there could be no certainty as to which. (Galileo's attempt to use this device was not totally successful however.) It is quite possible that the habit of philosophers to be unnecessarily prolix originated in this period.

The necessity of Enlightenment thinkers to hide their ideas has led secularists to assume that where their thought is obviously religious, this was merely as a sop to the authorities. But this approach is too often an easy way out, allowing secularists to bracket out what is massively and genuinely present in the writings of Enlightenment thinkers, their profound *engagement* with religion. That this engagement was often critical is what caused their contemporaries to accuse them of atheism, a prejudice carried through into modernity.

Bacon, Hobbes and the Big Six

Before we look at the French *philosophes* and Encyclopaedists who openly attacked the Roman Catholic Church, we examine eight key Enlightenment philosophers for evidence of atheism: Bacon, Hobbes, Berkeley, Locke, Hume, Descartes, Spinoza and Leibniz, bearing in mind that to some degree all of them would have exercised caution in expressing their views. Of these Bacon, Hobbes, Hume and Spinoza were accused of atheism in that period, and some secularists today still hold to this opinion. The word 'atheism' in the seventeenth and eighteenth centuries was highly charged or cathected as Freud would say, and none of the eight philosophers listed here wanted to be labelled as such. But we need to be precise about what the term 'atheism' means in this context, recalling that it would mean nothing at all in the contexts of Buddhism, Taoism, Yoga or the Advaita (as we shall see in Chapter Four). In Europe of the early Enlightenment 'atheist' meant any attack or criticism on Christian thought, in particular the principle of revealed truth handed down through scripture. In this sense all eight could be charged with atheism to one small degree or another, merely on the basis of a commitment to rational thought.

We are interested in other implications of 'atheism' however. Firstly, to what extent are these thinkers merely anti-anthropomorphic or non-realist in their conceptions of 'God'? Secondly, to what extent are they casual or disengaged atheists like Marx, who simply dismiss the entire religious worldview? Thirdly, to what extent are any of them vituperative atheists like Crick, Dawkins, Vidal or Toynbee? And, most crucially, to what extent did any of them want the end of religion?

Taking Francis Bacon (1561-1626) first, we encounter a philosophy that rejects Scholastic subservience to Aristotle. Bacon exhorts reason on the one hand, and empirical investigation on the other (his failure to attain the status of Newton is because he did not grasp the necessity for a mathematical modelling of empirically gathered data). His desire to separate reason from the revelation of Scripture and allow each to go their own way was typical of all the Enlightenment philosophers, and would open him to the charge of atheism of the first type, that at least he did not support tradition. As to his own opinion of atheism, he wrote a well-known essay attacking it, of which this extract sets the tone: 'It is true, that a little philosophy inclineth man's mind to atheism, but depth in philosophy bringeth men's minds about to religion.'[153] He is anti-anthropomorphic or anti-realist in a very typical manner of these philosophers: while 'God' may not be active in each moment to keep the world turning, the world is most definitely his creation. There is nothing in him of modern atheism: neither disengaged like Marx nor vituperative like modern angry atheists. Whether the spiritual impulse in him was particularly developed is hard to say, but, if it was, then it probably followed a Hellenic modality: potentially *jnani,* and probably *via positiva.*

Thomas Hobbes (1588-1679) was often thought of as an atheist in his time, though he wrote nothing that we could take as a direct attack on Christianity. He shared with Bacon the emphasis on reason and on the empirical, though went much further in his materialism, the real source of attacks on him. But in his most famous work, *The Leviathan*, there are, however, substantial sections presenting his understanding of the Christian faith, the Christian Commonwealth and the Kingdom of God. He denied his atheism in a most instructive manner in a response to the accusations of Bramhall, saying that 'God' was neither incorporeal nor immaterial, as these words were not found in Scripture.[154] Similarly, it has been argued that his views on mortality betray a fundamentally theistic belief.[155] But when taken in the context of the religious writing of the *Leviathan*, we can say again that this is an argument within religion, not against it (though possibly not a very interesting argument). Hobbes was in trouble because he supported monarchy, not on the old principle of the divine right of kings, but through rational argument. He shared with Newton the conviction that, as one pursued the causes of things down their chain, one inevitably comes to: 'one First Mover; that is, a first and an eternal cause of all things; which is that which men mean by the name of God.'[156] As to a nascent or developed spirituality, it is hard to judge, beyond a vague *jnani* and *via positiva* leaning, subsumed within the Abrahamic religiosity typical of Protestantism.

In contrast Bishop George Berkeley (1685-1753) managed to be both conventionally religious and a deeply interesting spiritual thinker. Along with Locke and Hume he makes up the triumvirate of British philosophers known as 'empirical' because they held that there was nothing in the mind that was not first in the senses. At first glance this would rule out religion entirely (unless, like the Buddha, we counted 'mind' as a sense organ), but, in Berkeley's case there is an extraordinary caveat: 'save the spiritual.' At least this is the summary that the *Encyclopedia Britannica* gives of Berkeley's thought. Berkeley takes his 'empiricism' from John Locke (1632-1704), who was deeply influenced in this by Bacon. Yet the step that Berkeley takes from what is originally a materialist empiricism – one that a physicist would understand – to an idealism which denies the physical world any independent reality, is remarkable. Only 'God' is independently real, and we are ideas in his mind. Whether the thinking of Berkeley and Locke was radical or not in a religious sense is open to debate, but in no sense at all can either of them be charged with atheism.

But we saw earlier that David Hume (1711-1776), the third of the British 'empiricists' was often charged with atheism, and this reputation lives to this day. While Berkeley may be the more interesting religious thinker, it is Hume that is more relevant to us here: does he epitomise Enlightenment thinking? He is perhaps the most virulently anti-anthropomorphic in his approach to 'God', but shows absolutely nothing of the modern vituperative atheism. Although it would have been provocative to the authorities of his day, and hence remained unpublished in his lifetime, Hume's *Dialogues* places him not just as a thinker of the *jnani via positiva* temperament, but close to the Deist position. This is confirmed in his unpublished *Natural History of Religion*. This was scandalous because it suggested that a crude monotheism was no better than a crude polytheism, but there can be no doubt that what Hume was defending was a *true* theism, which required a 'first principle of mind or thought'. Those, like Dennett, who believe that Hume had demolished the Argument from Design in his *Dialogues*, would be forced to think again when encountering *The Natural History of Religion*. Hume is supremely confident of 'God', of a true theism, expressed as a first principle, and proved endlessly by the beauty of creation. He was attacking what he thought was the corruption of religion, the vulgar superstitions of those whose belief was not founded on reason:

> Even at this day, and in Europe, ask any of the vulgar, why he believes in an omnipotent creator of the world; he will never mention the beauty of final causes, of which he is wholly ignorant. . . .[157]

Hume continues with a description of the human hand and 'the suppleness and variety of joints in his fingers' – a habit of demonstrating the existence of 'God' through the wonders of Nature that was common to nearly all thinkers of the time. This habit was termed 'natural religion', and is so central to Deism, that the two terms are best considered synonymous.

Turning now to the Continental 'rationalists', René Descartes (1596-1650), Benedict Spinoza (1632-1677), and Gottfried Wilhelm Leibniz (1646–1716), we find an even more remarkable *engagement* with religion. Descartes was cautious about publishing and moved to Holland to avoid censure; Spinoza never published his major work the *Ethics* at all; while Leibniz seemed to avoid attacks through a highly astute political sensibility. It is Spinoza who was systematically charged with atheism however; even Hume described him as an infamous atheist, and his ideas as a 'hideous hypothesis'.[158] Paul Russell considers that Hume was merely covering his tracks as a follower of Hobbseian and Spinozist atheism, but from our perspective Hume and Spinoza are equally interesting in their anti-anthropomorphist conceptions of 'God', making neither men properly atheist at all. In fact we can say of all four men, Hume, Descartes, Spinoza and Leibniz, that they stand outside of their Christian tradition, are anti-anthropomorphic, highly engaged with religious questions, and demonstrate not a trace of either the Monty Python or vituperative atheism of the twentieth century. And none of them wanted the end of religion.

We are not directly concerned with the formal contribution of these eight thinkers to philosophy per se, and so have placed the terms 'empiricist' and 'rationalist' – usually used to distinguish the British from the Continental philosophers – in quote marks. From a spiritual point of view all eight are rationalist, because of the weight they give to the power of reason, and from a scientific point of view of none of them is 'empiricist' in a way that the emerging discipline of physics required. We have seen that all could be considered 'atheist' only in the sense of original religious thinking, and not one of them denied the existence of 'God' in any modern manner. Yet clearly their emphasis on reason collectively denied the older principle of religion as revealed through the text of scripture, and it is this that underpins the later genuinely atheistic developments. However, before we move to the later period we need to answer the question posed at the beginning of this section: did any of the Enlightenment philosophers take the new findings in physics, chemistry and biology to demonstrate a contradiction with religion where the scientists plainly did not?

Physics and Philosophy

The philosophers clearly took their cue from the new sciences, but we need to ask: in what way? Modern atheists like Dawkins are adamant that science and religion are incompatible, so did the Enlightenment philosophers, in their study of the science of their day, make such a conclusion inevitable? It is Hume that gives us the answers to these questions. Firstly he is famous for the open ambition to be a 'Newton of the mind', subtitling his key work *A Treatise of Human Nature* with this phrase: 'An Attempt to Introduce the Experimental Method of Reasoning into Moral Subjects.' We have suggested before that Newton's *experimentum crucis* reverberated through the Enlightenment: here is its most direct citation. The astonishing success of Newton's work, based on the 'experimental method of reasoning', created the imperative for all the thinkers of this age to attempt this approach. Yet here is the contradiction: Hume and

Newton interpreted Bacon in radically different ways. Hume not only had no understanding of the scientific method of Newton, but Hume's philosophical scepticism undermines the very foundation of it. Hume was sceptical about both causality itself and of 'induction', through which term he effectively dismissed Newton's entire and revolutionary foundation of the discipline of mechanics.[159] Famously, Hume considers the case of billiard balls striking each other, and proposes that what we understand as cause and effect is merely a habit of thought derived from 'constant conjunction'.[160] Newtonian mechanics was not then taught in schools as it is today, and Hume did not realise that we can know with extraordinary precision the momentum (expressed in $kgms^{-1}$ as a vector quantity) of both balls before and after impact, the energy lost in friction (measured in Joules and calculated through coefficients of sliding and rolling friction, and through the coefficients of elasticity of the balls) and so on. His opinion is, regrettably, that of the 'vulgar' who have no training in physics (to turn his own phrase upon him).

More than any thinker of that period Hume's scepticism allowed the philosophical tradition, so galvanised at this time by Newton, to strike out in a completely unrelated direction. It is no exaggeration to say that Hume severed for ever the disciplines of science and philosophy: his natural successor, Kant, merely put further 'clear blue water' between them. Enlightenment philosophy may have been inspired by the *idea* of the emerging science, but in practice its response was to *recoil* from it. The fact is that philosophers are temperamentally disinclined to the empiricism and pragmatism required of a scientist, though they readily invent speculative systems of thought with those names. Empiricism in science means measuring the 'primary' qualities (reducible to mass, length and time) of things and events, while pragmatism means accepting that any measurements, and the theories supported by those measurements, can only be known to so many decimal places.

What of the other 'empiricists'? Berkeley was in fact a known opponent of Newton, and Locke was honest enough to admit that he did not understand the *Principia* of Newton – his friend – and had to ask Huygens his opinion of it.[161] While all three to some degree believed in the 'experimental method of reasoning', none of them were in any position to pit either the method or the results of Newton against the domain of religion. Newton himself ends the *Principia* with a eulogy about 'God' which conveys the piety of a deeply religious man, and no conceivable reason to suggest a contradiction with science. The three Continental 'rationalists' were in a better position to understand the physics of Newton, but none of them made any significant contribution to the discipline or found within it the least opposition to their conceptions of 'God'.

While not wishing to anticipate our discussion of Descartes too much, we can suggest that he gives us the best handle on the issue of physics and religion: his terminology of *res cogitans* (thinking stuff) and *res extensans* (extended stuff). What the scientists of the period had discovered was a reliable method of investigating 'extended stuff' that leads to sure knowledge of inanimate matter, and upon which our modern scientific and technological world is built. Galileo had indicated the basis of this when he identified what he called the 'primary qualities' for science, those attributes of extended stuff that could be measured. What Hume and others who attempted to apply the 'experimental method of reasoning' of Newton to the area of morals and other abstract spheres did not understand is that it only works for 'extended stuff' – the material world. For mind or spirit – all that has no mass or extension – an application of the experimental method of the scientists is a category mistake. This is because science, drawing on the measurements provided by physical instruments, is *consensual*. The gentlemen of the Royal Society could agree because the optical instruments in front of them occupied a consensual physical reality: each one repeatedly and reliably received the same visual sense impressions from the apparatus. 'God' and all the other descriptions we give to the experiences that arise from the spiritual impulse involve no sense impressions that are repeatable and consensual. But the philosophers of this period, transfixed by the glamour of Newton's success, wished to make the same application of human reason to intangibles. In the process they found no evidence against 'God' at all, but ended up elevating reason to his status. They also parted company with science.

The French Experience

We now turn to the French *philosophes* or Encyclopaedists. These were vituperative enough, but were they atheists? *Britannica* defines the *philosophes* as follows: 'the movement was dominated by Voltaire and Montesquieu, but that restrained phase became more volatile in the second half of the [eighteenth] century. Denis Diderot, Jean-Jacques Rousseau, Georges-Louis Leclerc de Buffon, Étienne Bonnot de Condillac, Anne-Robert-Jacques Turgot, and the Marquis de Condorcet were among the *philosophes* who devoted their energies to compiling the *Encyclopédie*, one of the great intellectual achievements of the century.' It was a compendium of scientific and liberal thought of that time, and an intellectual precursor to the French Revolution. As such it represented a profound challenge to the Roman Catholic Church and its authors were targets of suppression, arrest, exile and book-burning.

France in the eighteenth century contained an explosive dynamic: on the one hand a repressive hierarchy which supported a reactionary monarchy, and on the other hand some of the leading and most outspoken thinkers of Europe. This group is well represented by the case of Voltaire (1694-1778), the pen name of François-Marie Arouet, of a middle class family, and educated by Jesuits. He acquired from them a taste for the classics, but only a distaste for conventional religion as it was taught by these most religiously militant of Roman Catholics. Voltaire was not a philosopher as such, dismissing the speculative systems of thought of the philosophers as 'metaphysico-theologico-cosmolonigology' (as it is ridiculed by Pangloss in Voltaire's best loved work, *Candide*). It was as a man of letters and public life that he became the famous symbol of freedom and revolution for France, and the model for Thomas Paine's call to freedom in America. Like other French intellectuals of his day, he was impressed by the relative freedoms and religious tolerance of countries adjoining France: England, Holland and Germany. It was in The Hague that Voltaire first met refugees from religious persecution, and their plight was the first of many causes that fired in him his revolutionary literary talent. His published works soon brought him into conflict with the authorities, and the year 1717 saw his first imprisonment in the Bastille, the most notorious symbol of French repression. He was returned to the Bastille in 1726, but Voltaire agreed to exile rather than imprisonment, leading to a stay in England.

It was in the relative tolerance of England that Voltaire sharpened his criticism of French tyranny, and on his return published works exhorting similar religious tolerance. We have seen that England in the eighteenth century was, compared to the twenty-first century, a place where extreme caution and circumlocution was required in published works: the examples from Hobbes to Hume attest to that. However, France was orders of magnitude worse, due principally to the dominance of the Roman Catholic Church in public life. Those who did not conform were denied access to the professions, and were regularly tortured and judicially murdered for their beliefs. In 1762 Voltaire championed the cause of those who were unjustly persecuted, such as Jean Calas, a Huguenot shopkeeper in Toulouse who was falsely accused and convicted of murdering one of his sons (allegedly because his son wanted to become a Roman Catholic). Calas was tortured, broken on the wheel, taken to the scaffold, his limbs broken, and after two hours of agony was finally strangled to death by the executioner, having at no point broken down and confessed to any alleged crime. Voltaire took up the case and eventually secured an overturn of the verdict.

Most of Voltaire's writings were banned, resulting in their illegal circulation. The Church and State attempted the utmost control over the thought of French citizens: no criticism of either religion or authority was permitted. In the twenty-first century we regard the works of reason in politics, philosophy and science to be bloodless and probably tedious, at least compared to the fictional outpourings of the contemporary imagination: who would even know where to find Newton's *Principia* today? Yet Newton's works were banned in eighteenth-century France, and in their illegal circulation created a fascination we now find hard to imagine. The repressive regime of France produced in its highest form the exact opposite of what was intended: revolutionary rationalist thought. Voltaire's life and work epitomises this process, and his battle-cry was *Écrasez l'Infâme* – eradicate infamy. His targets were twofold: 'superstition' and repression, and though he did not live to see the Revolution of 1791, his remains were returned to the Panthéon as its cultural hero.

So what kind of atheist was Voltaire, if at all? His attack on Christianity was as vituperative as that of Dawkins et al. His exposure of 'God' as forcing Mary into adultery, and countless other scandalising interpretations of biblical dogmas, was for his time unthinkable, though the same ideas in the hands of vituperative atheists today have lost all power to shock. In this sense Voltaire is the progenitor of Vidal, Toynbee, Dawkins, Crick and Atkins, but a major difference remains. He *believed in religion itself*. In his search for religious tolerance – or spiritual difference as we call it – he drew on many sources, including the East. J.J. Clarke has documented how European intellectuals of this period were influenced by the religious and political ideas from the East, which returned with Jesuit missionaries.[162] Voltaire was highly impressed by both the Indian religion and Confucianism, detecting in the public life of India and China a religious pluralism unknown in the France of his time. Voltaire most tellingly discovers early Hindu texts which appear to be the equivalent of the Old Testament for the Indian peoples . . . but which palpably omit slaughter as the necessary accompaniment to state- and religion-building.

The Enduring Puzzle of Enlightenment Philosophy

In our survey of eight of the leading Enlightenment philosophers and a consideration of Voltaire as the key thinker of pre-revolutionary France we find a common move towards *reason*. To varying degrees this developing facility of rational thought, triggered by the Newtonian revolution in science, was deployed to attack the backwardness of Christian institutions, a stance understood as anticlericalism. However, not one of these thinkers, including the most outspoken of them, denied the existence of 'God'. At worst, Hobbes wanted to describe him as 'First Cause' while Hume wanted to rename him 'MIND' or 'THOUGHT', a move that we can understand picturesquely as the deification of reason. More crucially we discover in all of these thinkers a spirituality that we characterise as non-devotional or *jnani*, which naturally tends in a theistic milieu towards an abstract, non-realist, anti-anthropomorphic 'God'. Their 'God' was the objectless object of the non-devotional or *jnani* impulse, and as such it was simply the 'God' of Christianity viewed through a certain temperament. The disappearance of cant and superstition, along with repressive systems of government was their shared goal. The disappearance of 'God' was emphatically not their intention. We have shown that the arguments of the scientists were not against religion but against Aristotle, and that the arguments of most of the philosophers were not against religion, but *within* it.

If neither the work of the scientists, nor the thinking of most writers and philosophers of this period sought the death of 'God', then what exactly led to the dramatic break that laid the ground for atheism and secularism? For in the vision of this period there was the potential for an end to superstition, an end to religious intolerance, and the creation of a religious alternative centred on the 'God of the philosophers'. But in the next section we document, not such a possible expansion of humanity's vision of 'God', but his complete dying.

3.2.4 The Death of 'God'

Thinkers began to emerge in eighteenth-century France who proposed the next step: the abolition of religion in its entirety. Where Voltaire saw infamy in the exercise of religious power, the new thinkers saw infamy in religion itself. Where Voltaire saw superstition as a part of religion to be excised, the more radical thinkers saw religion as entirely a construction of superstition, and to be excised *in toto*. Two such thinkers were Julien Offray de La Mettrie (1709-1751), and Paul Henri Thiry, Baron d'Holbach (1723-1789). Both were responding to the Descartes' ideas: the division between *res extensans* and *res cogitans*. We will show that Descartes' *res cogitans* – thinking stuff – has been misunderstood, but for now what is significant is that his method of doubt led him to be sure of thinking stuff, and sceptical of extended stuff (matter). The new thinkers were materialists however, drawing on the tradition of Bacon and Hobbes, and hugely impressed by the advances in science, particularly in biology. Hence they turned Descartes upside down by denying thinking stuff altogether and accepting only extended stuff. To start with La Mettrie and Holbach were in a scandalising minority, but in time their thought became the standard for scientism. (Holbach was in fact supported in his work by the key *philosophe* Denis Diderot, who seems to have made an interesting journey from theism, to Deism, to atheism.) To put it another way, these were truly the thinkers who initiated the dying of 'God', a process that was later completed by Marx, Darwin, and Nietzsche, and capped by Freud.

It was Nietzsche who most dramatically announced the death of 'God', but he merely gave artistic voice to a process already completed. Here are six dates in the dying of 'God':

1748 La Mettrie: *Man a Machine*
1749 Diderot: *Lettre sur les aveugles* (*Letter on the Blind*)
1770 Holbach: *System of Nature*
1844 Marx: '*Contribution to the Critique of Hegel's* Philosophy of Right'
1859 Darwin: *On the Origin of Species*
1885 Nietzsche: *Thus Spoke Zarathustra*

La Mettrie

La Mettrie's *Man a Machine* posits a mechanistic view of the universe, requiring no 'God' at all. Here was a humanist ethics and a rhapsodical, holistic vision of nature in which the human being took its place along with the other animals. La Mettrie lived in Paris, but after an earlier controversial publication in 1745, he was forced to flee to Leiden where he completed *Man a Machine,* the publication of which aroused even greater hostility and forced him to seek refuge in Berlin. His thesis is not so much a refutation of 'God' as an early appeal to a materialist doctrine that much later became the hallmark of twenty-first century biologists. His finds no need for Descartes' dualism, that is no need for two 'substances', one material and the other immaterial:

> Let us conclude that man is a machine, and that in the whole universe there is but a single substance differently modified.[163]

La Mettrie anticipates by 250 years the arguments against dualism in Daniel Dennett's *Consciousness Explained*. Dennett says 'The prevailing wisdom, variously expressed and argued for, is *materialism*: there is only one sort of stuff, namely matter – the physical stuff of physics, chemistry, and physiology – and the mind is somehow nothing but physical phenomenon.'[164] He goes on to add: 'It is not that I think I can give a knock-down proof that dualism, in all its forms is false or incoherent, but that, given the way dualism wallows in mystery, *accepting dualism is giving up*.'[165] Dennett believes that it was the philosopher Gilbert Ryle's 1949 attack that finally and convincingly discredited Descartes' dualism. La Mettrie has not completely given up on dualism however, being rather an agnostic who simply doesn't find much cause to retain 'God':

> I do not mean to call in question the existence of a supreme being; on the contrary it seems to me that the greatest degree of probability is in favour of this belief. But since the existence of this being goes no further than that of any other toward proving the need of worship, it is a theoretic truth with very little practical value. Therefore, since we may say, after such long experience, that religion does not imply exact honesty, we are authorized by the same reasons to think that atheism does not exclude it.
>
> Furthermore, who can be sure that the reason for man's existence is not simply the fact that he exists? Perhaps he was thrown by chance on some spot on the earth's surface, nobody knows how nor why, but simply that he must live and die, like the mushrooms which appear from day to day, or like those flowers which border the ditches and cover the walls.
>
> Let us not lose ourselves in the infinite, for we are not made to have the least idea thereof, and are absolutely unable to get back to the origin of things. Besides it does not matter for our peace of mind, whether matter be eternal or have been created, whether there be or be not a God. How foolish to torment ourselves so much about things which we can not know, and which would not make us any happier even were we to gain knowledge about them![166]

La Mettrie is very modern in his indifference to the 'infinite' and the acceptance that we 'must live and die like mushrooms.' (We will see on the contrary that Descartes was sure that we have a *better* idea of the infinite than the finite: a true sign of the *jnani* spiritual orientation.) His genuine love of nature is reminiscent of Dawkins, and his ethical optimism is reminiscent of later secular humanists such as Bradlaugh and Ingersoll. But, like all the agnostics and atheists who derive their thinking from science, it is not at all clear how 'a single substance differently modified' – Descartes's extended stuff – should generate either beauty or morality, and the recent attempts by Dawkins et al to read these qualities into matter, conceived of as a machine, are unconvincing.

Diderot and Holbach

Denis Diderot, renowned as the principal editor of the *Encyclopédie*, is considered by Michael Buckley to be the first true atheist. Buckley argues that Diderot's *Letter on the Blind* 'introduces a critical transition in Western thought with his dismissal of transcendence and assertion of the virtualities of dynamic matter.'[167] This may well be true, but a better-known work, both then and now, is the *System of Nature*, written by Holbach with support and assistance from Diderot.

Baron d'Holbach anonymously published the *System of Nature* in 1770, and it circulated clandestinely in France. It rapidly became known as the 'atheists' bible'. Holbach was interested to refute Descartes on two counts, his 'proofs' of the existence of 'God' and the 'Cartesian split', the notion of two distinct substances. In this sense Holbach is more actively atheistic than La Mettrie, attacking Descartes on the grounds of a belief in 'God'. Holbach did not allow his name to appear on his works, preferring instead to remain a central part of the Parisian intellectual circle formed around the Encyclopaedists. Holbach's argument against religion is very modern, that body and mind cannot act upon each other, as all action takes place through energetic exchanges of *matter*. This was a direct consequence of the triumph of Newtonian mechanics (though would have been anathema to Newton himself). Holbach's motivations are summed up in this 'advertisement' at the start of his work, and are largely similar to the motivations of Crick, Dennett and Dawkins, and has pre-echoes of Marxist thought:

> Mankind are unhappy, in proportion as they are deluded by imaginary systems of theology. Taught to attach much importance to belief in religious doctrines, and to mere forms and ceremonies of religious worship, the slightest disagreement among theological dogmatists is oftentimes sufficient to inflame their minds, already excited by bigotry, and to lead them to anathematize and destroy each other without pity, mercy, or remorse.[168]

Dennett believes that a retreat to dualism is *giving up*, by which he means giving up in the attempt to explain natural phenomena. The founding belief of scientism is that *in time* all things that are now unexplained will yield to scientific investigation. In the meantime there should be no recourse to the supernatural, as Holbach explains:

> If in the chain of these causes we sometimes meet with obstacles that oppose themselves to our research, we ought to endeavour by patience and diligence to overcome them; when it so happens we cannot surmount the difficulties that occur, we still are never justified in concluding the chain to be broken, or that the cause which acts is supernatural.[169]

Both men have fallen victim to the idea that spirituality or religion is founded on *explanation* of the mysterious or miraculous (Dawkins also insists on this as the basis for his attacks on religion). A cool look at science since Newton shows the obstinate fact however: Holbach's 'chain of causes' never terminate. Newton's gravitation moved the mystery of the planets one step along a chain of scientific explanation to the mystery of attraction at a distance. This is now 'explained' in terms of *gravitons*, fundamental particles of the class boson which are massless in themselves but carry the force of attraction between objects. Unfortunately no gravitons have yet been detected, and their behaviour would still require the reconciliation of relativity with quantum mechanics. In the quest to 'explain' gravity enormous scientific progress has been made, and the predictive power of what we do know is massively accurate, but we cannot be sure with Holbach and Dennett that the 'chain is not broken'. The 'particle zoo' of quantum mechanics gives no hope yet of a termination in any ultimate explanation (no Unified Field Theory yet exists); indeed Murray Gell-Mann suggests there may be an infinite number of sub-atomic particles.[170] The ungrounded optimism in a science that will inevitably terminate is as naïve a faith as that of theistic realism.

Marx

Atheism in the mid-nineteenth century was still contentious and problematic for its adherents. We mentioned in section 1.1 that Charles Bradlaugh helped to establish the National Secular Society in 1866. Bradlaugh had been in trouble for his atheist views, and was three times elected to the British parliament, and three times refused entry because he would not swear allegiance on the Bible. Bradlaugh's opposite number in America, Robert Green Ingersoll, was known by the tolerant as 'The Great Agnostic' and by his enemies as 'Injure Soul'. The journey from *Man a Machine* in 1748 and *The System of Nature* in 1770 to the thought of Marx, who was working and writing at the same time as Bradlaugh and Ingersoll, is remarkable. In the nineteenth century atheism lived within an intellectual climate that was profoundly interested in religion. But Marx is remarkable because his atheism was complete, final, and casual.

Karl Marx (1818-1883) was born to Jewish parents, but after the conversion of his father to the Evangelical Established Church, Karl was baptised. As a teenager he showed Christian devotion and a longing for self-sacrifice on behalf of the welfare of others. Yet by the age of twenty-five all traces of the religious impulse had disappeared, as we see in his essay 'Contribution to the Critique of Hegel's *Philosophy of Law*' published in a review he co-edited called *German-French Yearbooks*. Marx drew on the thinking of Hegel (1770-1831) and Ludwig Feuerbach (1804-1872), and the location of Marx in this German Idealist tradition is crucial. While the thought of both Hegel and Feuerbach can be understood again as a debate within religion (largely from a *jnani* impulse), Marx retained not a shred of sympathy towards it.

In 1848 he published with Friedrich Engels *The Communist Manifesto* which was to radically change the world, and which made it clear that history was no longer the domain of religion, but of class struggle. However it is from the earlier essay that nearly all of Marx's famous quotes on religion are to be found, including the dismissal of religion as the 'opiate of the masses'. The essay is an attack on the German state which Marx sees as so backward that its recent proud developments were in fact nothing more than the historical dregs of progress already overcome by means of the French Revolution and the more leisurely changes in Britain. The essay opens with these words:

> For Germany, the criticism of religion has been essentially completed, and the criticism of religion is the prerequisite of all criticism.
>
> The profane existence of error is compromised as soon as its heavenly *oratio pro aris et focis* ['speech for the altars and hearths'] has been refuted. Man, who has found only the reflection of himself in the fantastic reality of heaven, where he sought a superman, will no longer feel disposed to find the mere appearance of himself, the non-man ['Unmensch'], where he seeks and must seek his true reality.
>
> The foundation of irreligious criticism is: Man makes religion, religion does not make man.[171]

Marx is making a secular claim about man, that he is unformed, and needs to develop through *revolution*, a claim that is parallel to the spiritual claim made in most religions that man is unformed and needs to develop through *spiritual practice* (worship in the West, meditation in the East). For Jung the claim is made in psychological terms: man develops through *individuation*. But we note also that Marx is bound to the German terminology of the 'superman', the same term that would guide Nietzsche's more agonised abandonment of 'God'. For Marx the issue of religion's obsolescence is so obvious that he needs to devote only a few more paragraphs to it in the opening of the essay, the most celebrated of which are reproduced here:

> Religious suffering is, at one and the same time, the expression of real suffering and a protest against real suffering. Religion is the sigh of the oppressed creature, the heart of a heartless world, and the soul of soulless conditions. It is the opium of the people.
>
> The abolition of religion as the *illusory* happiness of the people is the demand for their *real* happiness. To call on them to give up their illusions about their condition is to call on them to give up a condition that requires illusions. The criticism of religion is, therefore, in embryo, the criticism of that vale of tears of which religion is the halo.
>
> Criticism has plucked the imaginary flowers on the chain not in order that man shall continue to bear that chain without fantasy or consolation, but so that he shall throw off the chain and pluck the living flower. The criticism of religion disillusions man, so that he will think, act, and fashion his reality like a man who has discarded his illusions and regained his senses, so that he will move around himself as his own true Sun. Religion is only the illusory Sun which revolves around man as long as he does not revolve around himself.

With these words, written in 1843, Marx has done with religion. We note also that his literary gifts as a writer, his capacity to speak in highly memorable images – such as the chain with the living flower and man as his own true Sun – must have greatly contributed to his appeal. It was not of course as a philosophy that his position on religion came to impact on such a huge swathe of humanity, but because his thinking was central to the communist revolutions of the early twentieth century. In Russia, as in Europe, pre-revolutionary intellectuals were grappling with the possibilities of atheism, but we only have to look at the example of Dostoyevsky (born three years after Marx) to see how agonised and tentative this process was. The Russian people were mainly Greek Orthodox in religion at the time of the Revolution, but almost overnight found themselves in a Soviet state predicated on atheism, the atheism of Marx and Engels. As communist revolutions swept across the globe, the religions of those peoples were forbidden as epitomising the oppressive class structures of the old order. In China Taoism was suppressed, in Mongolia the shamanist traditions were suppressed, in all the countries of southern Asia which embraced revolution Buddhism was suppressed – and Islam likewise wherever it was the pre-revolutionary religion. But in each case, where communism after decades eventually gave way to the market economy and there was a relaxation of the ban on religion, the old spiritual traditions have revived. Orthodox churches, Islamic mosques, and Buddhist temples were rebuilt, while the herdsmen of Mongolia returned to the mixture of Buddhism and shamanism practised by their forbears. Marx was wrong: religion is not an illusory happiness, but a real one, as religious thinkers through the ages have shown. (No social or political science that ignores this can properly describe the post 9/11 world.)

But Marx believed that his thinking was 'scientific'. He was not alone amongst non-scientists to make this claim: others include Goethe, Hegel, Steiner, Freud and Jung. But his 'science' bore no relation to the revolution kick-started by Newton, rather, as with so many others, Marx's thought merely attempted to borrow its cachet. We saw for example that Marx believed that the equality of labour value in the production of commodities had been 'scientifically' demonstrated (in the opening chapter of *Capital*),[172] but in reality he had taken a moral *a priori* and read it into the existing situation. His philosophy predicted the inevitable outcome of certain social processes, but there was nothing scientific in this, rather Marx was the victim of the prevailing historicism in German thought at that time. ('Historicism' is the idea that there is an inevitable outcome to social processes that can be predicted with certainty.) Any exposure to the real working methods of science dispel the vapours of historicism like mist in the sun, but Marx was instinctively and viscerally a *philosopher* in the modern sense, and therefore at heart secretly contemptuous of science. It is in Marx that we find the archetypal modern demonstration of the divergence between philosopher, scientist and theologian. He is certain that religion must be tried in the court of philosophy not the other way round. (Engels agreed with him when he attacked Feuerbach for suggesting the opposite, that philosophy must be absorbed into religion.[173]) Philosophy must have sovereign authority over all other fields and Marx quotes Hume to show how absurd it is for religion to question philosophy: 'This puts one in mind of a king arraigned for high treason against his subjects.'[174] Marx is also sure that his philosophy rules over *science*, and this leads him to make 'scientific' predictions which are eventually shown entirely false by the course of events.

Marx held an absolute certainty that religion would be swept away, just as capitalism would. As secularism increased its grip in all corners of the world after WW2, the Institute of Marxism-Leninism in Moscow compiled writings of Marx and Engels on religion (1955). In the introduction it states:

> The correctness of the Marxist-Leninist concept of the ways of overcoming religious views has been fully corroborated by the experience of the Soviet socialist state . . . the social and ideological roots of religion have been torn out for ever; the majority of the population has long been free from religious prejudices. . . . That is why the Party considers it its direct obligation to wage a systematic ideological struggle against the unscientific religious outlook which clouds man's consciousness, dooms him to passivity and fetters his creative energy and initiative.[175]

These views arose directly from Marx's historicism; *history* appeared to have other ideas however. Marx's influence has nevertheless been enormous, but not just through the instrument of revolution. His equation of religion with oppression, the identity of dogma with the structures of power, has influenced thinkers in religion, philosophy and politics ever since. Yet, as is usual in human affairs, we find places where Marxist thought and the religion most obviously its target, Roman Catholicism, create an uneasy synthesis – in Liberation Theology. This movement originated in the poverty of South American states, and represented an impatience with the slow progress of religion in changing the conditions of ordinary people. Naturally the Roman Church is nervous of this development, and in 1984 the *Sacred Congregation for the Doctrine of the Faith* issued a statement called *Instruction on Certain Aspects of 'Theology of Liberation'* in which it warned of the incompatibilities of Marxism and Catholic doctrine. It understands and deplores the poverty in South America and elsewhere which has created the movement, but cannot accept its Marxist basis.

At the same time the Church must recognise the inherent generosity of spirit of Marxism: which is 'scandalised' by poverty and suffering and is not prepared to wait for a top-down solution. In both systems 'love your neighbour as yourself' is effectively a defining principle, but in both systems dogmatic historicisms of different types lead to the perversion of the original generosity of impulse into *control*. Tragically, in both cases their extreme exercise of control became sadism.

Darwin

Where Marxism as a philosophy is directly the product of Marx, Darwinism as a philosophy is not directly the product of Darwin. Charles Darwin (1809-1882) was neither a philosopher nor a theologian, but a scientist in typical Victorian mould, and the implications of his discoveries left him as bewildered as many others in Victorian society. Where Newton was sure that his discoveries were to the greater glory of 'God' and lost not a single night's sleep over their possible danger to theology, Darwin was left with many questions. Neither man could predict the eventual impact of their scientific discoveries however, nor, as we like to put it, the *cultural reception* of their science. In 1860, nine years after the publication of the *Origin of Species*, Darwin wrote:

> With respect to the theological view of the question. This is always painful to me. I am bewildered. I had no intention to write atheistically. But I own that I cannot see as plainly as others do, and as I should wish to do, evidence of design and beneficence on all sides of us. There seems too much misery in the world. . . . On the other hand, I cannot anyhow be contented to view this wonderful universe, and especially the nature of man, and to conclude that everything is the result of brute force. I am inclined to look at everything as resulting from designed laws, with the details, whether good or bad, left to the working out of what we may call chance. Not that this notion at all satisfies me. I feel most deeply that the whole subject is too profound for the human intellect. A dog might as well speculate on Newton.[176]

Where Baron d'Holbach and Richard Dawkins are content to conclude that 'everything is the result of brute force' – though they do not use that form of words – Darwin is not. On the other hand, the idea of 'designed laws' does not completely satisfy Darwin either. This means that the claim Dawkins makes for Darwinism, that it would have given Hume an intellectually satisfied atheism, is clearly not the case for Darwin himself. The above passage is cited in a thoughtful book on science and religion by palaeontologist Stephen Jay Gould (1942-2002) in which he puts forward the idea of Non-Overlapping Magisteria (NOMA). Quite simply his thesis is that science and religion deal with different aspects of life and represent different domains or magisteria of human enquiry and experience. Dawkins argued with Gould over their different interpretations of evolutionary theory, which is not of relevance here, but Dawkins totally dismisses NOMA.

Nietzsche

The legacy of Marx and Darwin, while problematic, has always been incontrovertible. Nietzsche's legacy is less clear however, as he has the Protean quality to be different things to different thinkers. For some he is simply the author of the unforgettable phrase 'the death of God'. For others he is the key nineteenth century thinker who opened the doors to a criticism of the entire Western intellectual tradition, a prefiguring of the postmodern agenda. For others he is the last great Romantic. For some he provided the Nazis with the intellectual underpinning of their fascist adventure, and from whom they took both the idea of the 'Superman' and the 'Underman' (*untermensch*): the latter had to be destroyed for the former to come into being – and could conveniently be defined as the Jew, the Slav, or the communist.

The darker sides of Nietzsche have no defence at all, in particular his monumental contempt for ordinary people. To him they are the 'botched and the bungled', 'the many-too-many', the 'flies of the marketplace',[177] – not for Nietzsche is the dictum 'love your neighbour as yourself.' The Kierkegaard Corrective was not for Nietzsche, nor in fact was anything Christian: although his father had been a Lutheran pastor, Nietzsche lost his faith after reading D.W. Strauss's *Leben Jesu,* and he became by instinct a pagan (as is vividly present in *Twilight of the Idols*). But the secular worldview is predicated on the idea that ordinary people are the basis for society, that the Romantic dream of tragic genius and heroism is elitist twaddle, and, if it is riddled with the kind of contempt that Nietzsche shows for the ordinary lived life, then all the more reason to reject it.

The *Encyclopedia Britannica* tells us that 'two books were standard issue for the rucksacks of German soldiers during World War I, *Thus Spoke Zarathustra* and *The Gospel According to St. John*. It is difficult to say which author was more compromised by this gesture.' Apologists for Nietzsche think that he was compromised by the use of *Zarathustra* as the literary underpinning of militarism, and there is a strong case that Nazism in practice would have horrified him. But what of Nietzsche's views on 'God'? His is no easy, casual atheism, as in the case of Marx. In this sense he is closer to the postmodernists with their ambivalences and shifting sands of discourse. Like Marx he wanted to replace the gods with man, but not man as the ordinary worker – as we have seen Nietzsche has only contempt for this idea – but man who can become a god. He does not want to destroy religious values but to 'revalue' them from a human perspective – again like Marx in its impulse but expressed in late-Romantic terms. Man is a 'going-over' an anguished transitional creature, reminiscent of Marx's view of man as unformed until revolution, religion's view of man as unformed until salvation, and Jung's view of man as unformed until 'individuation'. In fact Nietzsche's journey is both a revolt against religion but at the same time a journey of spiritual growth. Unfortunately the path he chooses leads to madness, literally. His great epic *Zarathustra* can be seen as a regressive descent into madness, the hero unable to resolve anything or offer the great vision he promises to the 'higher men', retreating instead into an absurd and childish gesture at the end of the book. *Zarathustra* reads as a diary of a descent into madness, but for the secular mind it is a quaint expression of the nineteenth century imagination, one that shuts the door on religion not so much because we empathise with Zarathustra when says 'God is dead', but because it is a preposterous flight of Romantic fancy. The secular world may suffer from overt materialism, but it is at least more grounded in the body with its uncomplicated and obvious needs, and social egalitarianism.

Freud

Marx's atheism, well ahead of Nietzsche, was so complete as to make the term 'atheism' itself redundant. Engels thought so: 'It can even be said of the German Social-Democratic workers that atheism has already outlived itself with them: this purely negative word no longer has any application as far as they are concerned inasmuch as their opposition to faith in God is no longer one of theory but one of practice; they have *purely and simply finished with God*, they live and think in the world of reality and are therefore materialists.'[178]

However, in case there were people left with any lingering dualism, for whom the pure world of extended stuff might yet have a counterpoint in thinking stuff – the derided world of spirit – Freud neatly heads them off at the pass. If the materialists and atheists since La Mettrie and Holbach had built the grand structure of secularism, Freud provided the roof that prevented escape: a reductionist philosophy of mind. While neither Marx nor Freud could entertain the slightest possibility of any reality underpinning the visible structures of religion, Freud's analysis of its apparent lure went much further than Marx's. For Marx, religion is an 'illusory sun' around which man had revolved, instead of around himself; it is the instrument of oppression . . . end of story. However, we saw that Freud took the visible elements of religion and subjected them, not to a straight rejection, but to a psychological debunking in terms of a new mechanism: the regressive and the Oedipal. The human mind – the 'thinking stuff' of Descartes – was returned to modern man as just more machinery. Where Buddhist psychology is a psychology of the spirit, a *jnani* framework of the religious mind, the Freudian psychology of modern man is exclusively a *psychopathology*. 'Man a Machine' said La Mettrie in the eighteenth century, 'Man a Broken Machine' said Freud in the twentieth century.

Anyone now either clinging to old religion, or turning to the New Age and new spiritual movements, was irrevocably shut out of the progressive trajectory of Western thought by Darwin, Marx and Freud. Extended stuff (whether in the domain of science or politics) and thinking stuff (whether reluctantly admitted to at all) are equally devoid of 'God' for the politicised, psychologised intellectual.

Who Put the Case for 'God'?

But it doesn't add up. Newton and Voltaire, as scientist and thinker respectively in the seventeenth and eighteenth century, didn't want 'God' to die. Darwin and Marx, as scientist and thinker respectively in the nineteenth century, ensured that he did. What happened in the intervening period? There were plenty of people putting the case for 'God', but how come they did it so badly that Marx could triumph and Darwin, despite himself, cemented that triumph, leaving Freud to cap it with a reductionism of the mind? Stephen Jay Gould's quiet articulation of NOMA is erudite, intellectually credible, and obvious, but is largely ignored, and its earlier homologues disappeared without trace on the intellectual horizon. Voltaire fought superstition and argued for revolution; Marx did the same. Yet Voltaire had not the slightest doubt that 'God' should live and Marx not the slightest doubt that he was an odious fiction. Secularism is made much more in Marx's image than Voltaire's: why should that be?

There were good reasons in fact why 'God' was so hard to defend: we look at them now.

3.3 The Spiritual Wounds of the West

3.3.1 The Inquisition

Voltaire championed the cause of several victims of religious persecution in the 1760s, including Jean Calas of Toulouse, mentioned above. The treatment of Calas would be described today as judicial torture and murder, and it was routine in Roman Catholic France for centuries until Voltaire and the Revolution ended it. Judicial torture and murder are universal features of earlier times across the globe, probably arising wherever city states and militarism took hold. Britain did not outlaw torture until 1640, France not until the 1789 Revolution. Jean Calas had the misfortune to be a Huguenot; he was charged with a crime in the days leading up to the annual celebration of the Massacre of St. Bartholomew's Day (the massacre of Huguenots on 24 August 1572). The citizens of Toulouse pressed for the prosecution of Calas on religious grounds, as Christian mobs have down the centuries. Protestants were no less likely to torture and murder civilians in the name of religion: the 1692 Salem witch trial in Massachusetts hanged 18 innocent people, pressed one man to death with rocks, and brought about the death in prison of at least four more accused of witchcraft, possibly as many as seventeen. Protestants and Catholics alike inherited in the eighteenth century of Voltaire a tradition of religious violence that became symbolised by the single dreaded name of the 'Inquisition'.

The historian Edward Peters makes the point that the 'Inquisition' as a monolithic entity is a cultural construction, a myth created by writers and artists from the eighteenth century on, and continued in twentieth-century film making.[179] The reality was no less appalling however, but in practice it was a sporadic emergence of religious persecution in response to political and cultural crises. In the time of Voltaire writers helped construct the image of the Inquisition as monolithic partly because of its use by the Papacy to resist the advance of Protestant movements. Voltaire's *Candide* satirised the Inquisition (as well as the optimism of Leibniz) and became the model for literary, artistic, filmic and televisual satires to this day. Peters defines an Inquisition as any systematic application of Roman law to religious repression, rather than as a single entity. For our purposes we can understand the Inquisition as the application of Roman law to the sphere of religion, or more specifically as the phenomenon of *Roman cruelty within absolutist monotheism*. Judicial torture and killing was central to Roman law, including the notion that evidence from slaves was not legally admissible *unless* extracted by torture. What Western historians often fail to emphasise however, is that the application of the cruelties of Roman law – prior to Christianity – was almost never for *heresy*; neither was this a feature of most other despotic empires of the period or before. As David Hume pointed out in section nine of *The Natural History of Religion*, polytheism is naturally tolerant, and the Romans extended this toleration to the deities of their conquered subjects. Hume is no friend of polytheism, sharing the ignorant prejudice of his time that it was mere superstition, so his observations on the tolerance of Roman polytheism carry all the more weight. The cruelty of Roman law was only applied to heresy when it became the natural tool of the absolutist monotheism of Christianity.

Although major formal Inquisitions start in 1184 and finish in 1834, the punishment of heretics in Christianity goes back to its earliest days, and the punishment of religious heretics of all kinds is recorded in the Old Testament. Hence it has been a feature of religious life in the West and Middle East for millennia. The nominal purpose of the Inquisition was to preserve orthodoxy from the 'false' doctrines of the heretic, though in practice the major persecutions were pursued for political ends. The spirit of tolerance was not universally absent in antiquity however, the persecution of heretics only intensifying with particular periods and political developments. Neither was the persecution of heretics always imposed from above in the religious hierarchy – there are cases for example of bishops attempting to save an accused from the mob. It would be simplistic to see the phenomenon of the Inquisition as merely imposed by the Roman Catholic Church on its people – there was an instinct in the European mind that led ordinary people to actively support it. Effectively, while the persecution of heretics might be regretted at all levels in society, *it was accepted as the price for social cohesion*.

Here lay a problem for the Christian world: killing was against the moral code of the Bible, so how could the ecclesiastical authorities actually carry out the sentence? The Inquisitions began to provide a centralised control over trials and sentencing (although this was a too-and-fro struggle with local authorities), but could not carry out the sentence on moral grounds. The solution was simple: the convicted heretic was handed over to the *secular* authorities for execution, meaning the court officials of the sovereign ruler.

The initiation by Pope Gregory IX of the Inquisition in 1231 can be seen as an attempt to systematise the persecution of heretics, to keep central control over the dogmas of Christianity, and to prevent counter-hereticism at grassroots level. It was also an effort to control the process by moderating its worst excesses. The recently formed Franciscans (founded in 1209) and Dominicans (founded in 1220) provided the bulk of the early Inquisitors, who were given a peripatetic and proactive role in their inquiries. The medieval Inquisition was in fact a response to the so-called heresies of the Cathari and Waldenses, which were large-scale movements in the eleventh and twelfth centuries. They were Christian sects influenced by Manichaeism, brought from Bulgaria in about the year 1000, and which had spread through Italy, Spain, France and Germany. King Robert the Pious burnt thirteen 'distinguished citizens' in Orléans for adherence to such views, which was followed by sporadic executions by local authorities across Europe, prior to the first efforts to control the process by Gregory IX. Again it must be stressed that kings and bishops were not just imposing their authority on heretical groups, but that the population at large were restive for such persecutions. In 1114 townspeople in France even stormed the prison in the city of Soissons, while its Bishop was seeking advice from his seniors in Beauvais, in order to burn the incarcerated heretics at the stake.[180]

The Spanish Inquisition

In 1478 Pope Sixtus IV authorised the Spanish Inquisition as an independent office devolved to the *secular* direction of sovereigns Ferdinand and Isabella. The Spanish Inquisition was renowned for its cruelty, and has been the subject matter for fictional and satirical treatments by Edgar Allen Poe, Dostoyevsky, Monty Python and many others. From novel to comic book, to feature film, the Spanish Inquisition is indelibly printed on the Western psyche, and is therefore hard to investigate without prejudice. For us however the salient point is that it exemplifies religious persecution as an instrument of state, even of state-forming. Ferdinand and Isabella wished to unite Spain under their rule and faced the competing allegiances of the Jews and Moors to their religions and cultures. As non-Christians the Jews and Moors were exempt from the Inquisition, but that was not the point – they were given the choice to convert to the Roman Church or to be expelled. Those that converted were then subject to the Inquisition which became the test of their orthodoxy. 'Conversos', as they were known, were regarded with suspicion and were the frequent subject of denunciation, not least because those convicted of heresy had their wealth and property confiscated. Situations were engineered that seem almost farcical to us, such as testing converted Jews or Muslims with meals containing bacon or pork sausage, or inviting them to participate in activities on the Sabbath. The slightest hesitation could lead to arrest, torture and burning at the stake.

It is perhaps Dostoyevsky's *Brothers Karamazov* (written in 1879-80) that conveyed to the modern mind the full absurdity of the Spanish Inquisition, a contradiction that the secular mind places at the heart of its rejection of religion. The middle brother Ivan, the intellectual, writes a poem through which Dostoyevsky explores the implications of the Inquisition. The story tells of Christ's return to Earth at the time of the Spanish Inquisition, only to be recognised and arrested as the 'worst of heretics' by the chief inquisitor. Dostoyevsky's aim was thoughtfully religious, exploring issues raised by the contradictions of church authority in the subsequent trial of Jesus. The power of this image has stayed in the contemporary mind, so that a hundred years later the Monty Python team were to create satirical sketches of the Spanish Inquisition, but with a purely *secular* intention.

The Roman Inquisition, Protestantism, and the Index

Pope Paul III established the Roman Inquisition in 1542 to combat Protestantism. The Roman Inquisition was accompanied by a growth in censorship, leading to the Congregation of the Index in 1571. The Index was the Roman Catholic Church's list of banned books, and the Congregation of the Index became the office that systematised the centuries-old practise of censorship. Pope Innocent I had first published a list of forbidden books in 405, and at the end of that century issued the first Index; 1559 saw the Pauline Index and 1564 the Tridentine Index. The last edition of the Index was that of 1948; it was abolished in 1966. The mindset that gave rise to the various inquisitions is visible in the *Catholic Encyclopedia*, completed in 1914, and which predates the Second Vatican Council, responsible for introducing significant changes in Catholic practice. The *Catholic Encyclopedia* is hence neither authoritative in terms of contemporary scholarship, nor in terms of contemporary Catholicism, but it is highly instructive as to the best and worst of Catholic thinking in historical terms. Of particular interest is the way in which it deals with figures such as Dionysius, Eckhart and Erigena, and the challenge posed by their original and profound religious thinking. The *Catholic Encyclopedia* also states this on censorship:

> In general, censorship of books is a supervision of the press in order to prevent any abuse of it. In this sense, every lawful authority, whose duty it is to protect its subjects from the ravages of a pernicious press, has the right of exercising censorship of books.[181]

A secular educated population simply does not accept that it needs protecting from the 'ravages of a pernicious press' – it is in fact an extraordinary admission by the Roman Catholic Church that it lives in the past. So too does the American religious right: since the first publication of the Harry Potter series of books, it has made three thousand attempts to ban them from public libraries.[182] The grounds? That its central theme, wizardry, is unchristian.

The upsurge in censorship in the Roman Inquisition was not a simple case of the Roman Catholic Church resisting the nascent science of its day, but was partly due to the nature of the Reformation and Counter-Reformation of the sixteenth century. The Reformation was initiated in 1517 by Martin Luther when he posted his 'Ninety-five Theses' on the door of the Castle Church, Wittenberg. He attacked the system of indulgences of the period, which allowed absolution of sins by means of payment, and insisted on a return to Scripture. Luther, and then Calvin, were central to the break from medieval Christianity and created a movement in the North of Europe that became, rather inaccurately, known as Protestantism. The Counter-Reformation, the reaction from the Papacy, was initiated by the Council of Trent which met from 1545 to 1563, and from which derived the Tridentine Index already mentioned. Roman Catholicism in the south of Europe produced a more mystical direction in St John of the Cross and Teresa of Avila, while Spain became the bastion of anti-Protestant Catholicism.

The legacy of the Reformation and Counter-Reformation was not a pluralism and more open debate, but tighter censorship. Protestant emphasis on Scripture created a literalist reading of the Bible that had not been known before, producing dogmatism of new heights. Calvin's Switzerland saw the burning of heretics by *Protestant* decree and Catholicism was also forced into a literalist retreat, resulting in the increased banning of books. Given the medieval Church's acceptance of the works of the Scholastics, although a little reluctantly, it might well have permitted Copernicus, Kepler and Galileo to publish their works if it hadn't been for the Reformation and its consequences.

The Witch Hunts of the Sixteenth and Seventeenth Century

With the rapid spread and total domination of Europe by Christianity heresy was largely a question of rival Christian doctrines, rather than of rival *religions*. As we saw, the expulsion of Jews and Muslims from Spain left the main issue as how Christian the 'Conversos' actually were. Once so-called pagans and the Jews and Muslims were eradicated or marginalised there remained only the 'narcissism of minor difference' between the rival Christian doctrines, though this was enough for torture and murder to continue. However, there was one other dynamic in all of this, the very different issue of witchcraft. It was not possible to accuse a so-called witch of adhering to non-Christian faiths – because they were clearly not Jews or Muslims, and neither was 'paganism' a faith – so it remained to accuse them of Satanism. The judicial arrest, torture, trial and murder of 'witches' – both men and women – was a practice carried out by the secular authorities whether or not an Inquisition was formally involved.

Hence the sixteenth century saw the culmination of the witch-hunts in a ferocity that tortured and killed not so much the *doctrinal* heretic as the men or women of rural villages who practised certain arts. In his *Religion and the Decline of Magic* Keith Thomas points out that 'Many other observers testified to the deep-rooted appeal held for contemporaries by the traditional dispenser of magical remedies – the village wizard, or "wise man". . . . ' He adds:

> During the sixteenth and seventeenth centuries these popular magicians went under a variety of names – 'cunning men', 'wise women', 'charmers', 'blessers', 'conjurors', 'sorcerers', 'witches' – and they offered a variety of services, which ranged from healing the sick and finding lost goods to fortune-telling and divination of all kinds.[183]

The Inquisition provide details of the trials of such individuals and hence a comprehensive record of the age, but not of the practices of the accused as expressed in any neutral language. But we have such a language, even though it may be unfamiliar in this context – the language of shamanism. While the Old Testament is mainly concerned with the heresy of idolatry – the worship of idols or entities such as the sun and moon – it expresses this more in polytheist rather than shamanistic terms. We have previously suggested that shamanism is older than polytheism and is quite probably the ur-religion of mankind, prior to the agrarian revolutions which created the cities and city-states conducive to polytheism. In the rural areas of France and Spain, which even today cling to the smallholder's lifestyle so foreign to Britain and the US, it is not hard to imagine that rural populations existed with traditions reaching right back to the mists of shamanism. (We earlier suggested that this was the case in Rudolf Steiner's rural Austria as well.) The witch-hunts in the rural areas of mainland Europe, and also, thanks to Matthew Hopkins – self-designated 'Witch-Finder General' – in England, can be thus understood in a new way. How could the Church own up that its remit ran thin in remote rural areas? Rural areas of Europe that are still relatively remote today? And why should not the ancient, visceral bond with the land that ran from agricultural communities back through the pre-agricultural Iron, Bronze and Stone Age communities not preserve that sense of Nature imbued with spirit?

The targets so often of the sixteenth century witch-hunts were women. They might be forced under torture to denounce other women or might even do so out of spite, but generally this pogrom was a horror carried out by *men* against women. Feminist scholar of religion Starhawk suggests that during the European and

American witch hunts nine million witches were executed in total, of which eighty percent were women.[184] Whatever the true figures (probably far less) the image is deeply entrenched in the Western psyche. In the film adaptation of John Updike's novel *The Witches of Eastwick*, three single women conjure up between them a 'tall, dark stranger, travelling under a curse' – Darryl van Horn (played by Jack Nicholson). He invites them to his mansion which he tells them was a place where witches had been brought. In a style suited to a modern American audience he presents a possible explanation of the witch hunts: 'The medical profession started the witch scare in the fourteenth century. Wanted them out of the birthing profession. They were mostly midwives. It's true. They're scared. Their dicks get limp when confronted by a woman of obvious power. And what did they do about it? Call them witches. Burn them, torture them, until every woman is afraid. Afraid of themselves, afraid of men. And all for what? Fear of losing their hardon.'[185] What Darryl van Horn speaks of in a populist Hollywood fiction, Starhawk is serious about in the re-establishment of the 'Craft':

> to reclaim the word *Witch* is to reclaim our right, as women, to be powerful; as men, to know the feminine within as divine. To be a Witch is to identify with nine million victims of bigotry and hatred and to take responsibility for shaping a world in which prejudice claims no more victims.[186]

Exporting the Inquisition

The concept and practice of Inquisition was exported by the Europeans to all corners of the globe during the centuries of colonisation. The Spanish took the Inquisition to the Americas for example, while the Portuguese took it to India and Japan. The Inquisition was established by the Portuguese in Goa in 1560 and it only officially ended in 1774, though remnants of its practices survived until the British occupation in 1812. Records show that 121 men and women were burned at the stake in the Goan Inquisition, though only 54 were alive at the time of burning. The attempt to introduce Christianity and its Inquisition by the Portuguese in Japan backfired as the Japanese instituted an Inquisition of their own directed at the eradication of Christianity and forcible re-conversion to Buddhism. The entry in Kodansha's *Encyclopedia of Japan* records that the Buddhist church became an instrument of the seventeenth-century Tokugawa regime in this affair. The goal was not religious at all: it was merely to end the rising threat of European influence in Japan. It is one of the very few cases where Buddhism is in any way implicated in violence, in this case leaving some 3,000 Christians judicially murdered for their religion. We can sum up: *Outside lands subject to Western monotheism, the violence of judicial torture and murder was almost never carried out in the name of religion.*

Socrates is a rare exception, but only if we believe that the charge he faced of impiety was the reason for his execution. In India, no less violent in ancient times than any other land, there appears to have been little judicial or mob violence in the name of religion until the monotheist tradition of Islam arrived in the eighth to the tenth centuries. Even then, the so-called 'communal' violence between Hindus and Muslims did not escalate until after Partition in 1947 (though some believe that the British had earlier encouraged inter-religious hostility to further a divide-and-rule policy).

Because of the colonialist dominance of Europe in the world, the European mind has rarely understood other cultures in their own terms. Hence the secular mind when it makes the equation 'religion = God' is unaware that its perception of global religion is framed by European missionary zeal, which reads monotheism into all the world's religions. When the secular mind – titillated by the gothic image of the Inquisition in popular culture – makes the further equation 'religion = Inquisition', it is falling into the same trap.

3.3.2 Suppression of Thought from the Fifteenth to the Eighteenth Centuries

From the adoption of Christianity by Constantine in the fourth century to the time of Voltaire in the eighteenth century, the habit of the cruel Roman law in the sphere of religion had lived fourteen centuries in the European mind. It was so deeply entrenched that a revolution of thought was required to end it, including of course the French Revolution of 1789 itself. Voltaire took enormous risks as we saw in publishing his works, most of which were banned, and placed on the Index. But *Candide,* written in 1759, lampooning the Inquisition, represented a turning of the tide. We saw that Voltaire was greatly impressed by the tolerance of England during his exile there, and in particular by the sentiments expressed in Locke's *Essay on Toleration.* Yet Locke himself had written it while in exile in Holland, just prior to his return to England in 1689 (accompanying the Princess of Orange). It was originally published anonymously, though with the changes that rapidly took place in Britain after the coronation of the Princess as Queen Mary II of England, it became widely known as his work. The *Essay of Toleration* is a rebuttal of Hobbes's view that uniformity of religion was essential for the stability of society. We see in this how painfully slow it was for the European mind to move from one pole of the argument – that uniformity of religion guaranteed stability, to the opposite – that toleration created peaceful society. For those with the mindset of Hobbes, the Inquisition or its equivalents were a necessary evil; for those who sided with Locke it had to be swept away. The argument that Locke used was not available before the Reformation however: he pointed to the increasing number of Christian sects, and the impossibility of determining – on any grounds whatever – which was the 'true' faith. As he says at the beginning of his *Letter*, whatever the boast of any one Church of the orthodoxy of their faith 'everyone is orthodox to himself'.

If Hobbes in the middle of the seventeenth century still adhered to the idea of stability through religious uniformity, then how much more was this the prevailing idea during the fifteenth-century Renaissance, despite a flowering of many fields: the arts, science and . . . religion. In Italy the Florentine Academy, led by Marsilio Ficino (1433-1499), became a centre for new thought based on Neoplatonist, Hermetic and Kabbalistic ideas, and led the young Pico Della Mirandola (1463-1494) to put forward some of the most radical religious ideas of the time. He collected 900 religious ideas or *theses* from the spiritual traditions that he studied, and proposed a debate in which each one would be the subject of scholarly dispute. The centrepiece was his own extraordinary *Oration on the Dignity of Man*, written in 1486.

But Mirandola made his proposition only 56 years before the establishment of the Roman Inquisition, in a time of only *relative* openness, relative that is to the closing down of doctrine in the Reformation and Counter Reformation. Pope Innocent VIII was made to believe that at least thirteen of Mirandola's theses were heretical. The proposed disputation was prohibited and the theses were placed on the Index of Forbidden Books, despite an *Apology* from Mirandola. He fled Italy for France, and it was not until 1493 that Mirandola was finally absolved by a brief of Pope Alexander the Sixth. Mirandola died a year later at the age of 31. (We look at the Renaissance and the spiritual thinking of Mirandola in more detail in section 4.2.4.)

The most visible atrocity committed by the Inquisition against early humanist thinkers was against Giordano Bruno, who was burned at the stake in 1600. This date is significant as the dawn of the seventeenth century, the century more than any other set to overthrow the modes of thought belonging to the past. Bruno's burning at the beginning of this century must have cast a long shadow over its thinkers. But the emerging prestige of Newton and the reputation of the Royal Society led to similar organisations being set up in most leading European countries, and also in Russia. The accelerated pace of scientific discovery and the appearance of new technologies led to the widespread interest in and acceptance of science, so that the thought and books of the scientists became less liable to censorship. We have seen that most scientists were rarely as radical in their religious thought as Newton – who anyway hid his real religious interests – and certainly none of them were atheists. It was amongst the philosophers that we find the truly interesting religious and spiritual thinking of the seventeenth and eighteenth centuries – and they knew it, and hid it. It was the Continental rationalists – Descartes, Spinoza, and Leibniz – who had the most radical spiritual ideas of the age, and who were at the same time forced into the circumlocution we have earlier described.

3.3.3 Early Christianity vs. Early Buddhism

We have traced the intolerance of the Christian church through the various manifestations of the Inquisition, and have seen its long reach even in the most liberal of European countries in the seventeenth century. Having defined Inquisitions as effectively the application of the cruelty of Roman law to the sphere of religion, we need to go further back in time to see how this came about – after all, such an appalling history did not seem to have arisen in the Far East. Clearly a factor was the adoption of Christianity as the state religion of the Roman Empire. However, despite the key role that the Emperor Constantine (280-337) played in this, Christianity did not dominate the Empire until the end of the fourth century, and its experience up to that point was crucial in establishing its character. Christianity prior to Constantine was not the monolith that emerged to create the Inquisition, but consisted rather of a variety of groups that absorbed the few supposed facts and beliefs around Jesus into other traditions, such as Gnosticism, Neoplatonism, Hermeticism and Manichaeism. Above all else it was initially a Jewish sect, persecuted by the Jewish mainstream. One of those vigorously assisting in this persecution was Saul of Tarsus, later to be known as St Paul, perhaps the principal key founder of the new religion.

As Christianity spread through the Empire, due largely to the efforts of the converted Paul, it encountered both popular and official resistance. For the averagely pious polytheists of the Roman

Empire, who were not attracted to the new movement, Christianity was an 'atheist' religion as it denied the gods. Popular prejudice against it – including the conviction that Christians practiced incest and cannibalism – was widespread.[187] However the Roman authorities were tolerant of other religions. We have mentioned that David Hume was intrigued by this – here is what he said: 'idolatry is attended with this evident advantage, that, by limiting the powers and functions of its deities, it naturally admits the gods of other sects and nations to a share of divinity, and renders all the various deities, as well as rites, ceremonies, or traditions, compatible with each other.'[188] He added: 'The Romans commonly adopted the gods of the conquered people; and never disputed the attributes of those local and national deities, in whose territories they resided.' This being so, it highlights the question as to why Christianity, founded on love and charity, should systematically apply the cruellest aspects of the Roman judicial system to the sphere of religion, when the Romans themselves did not.

The Romans were capable of breathtaking brutality (it was the so-called 'barbarians' of the North who finally ended the traditional Roman games on the basis that they were uncivilised). But the Romans did not undertake religious persecution as long as the citizens of the Empire included the Emperor amongst their prayers. This worked fairly well in a world that was largely polytheistic, and, no doubt in many areas, still inclined to shamanism. (Whether the great Goddess religions still lingered within the Empire is not clear.) The one exception was Israel with its bizarre – to the Romans at least – insistence on a single 'God'. However with the good pragmatism required to hold any large empire together, the Romans made an exception for the odd nature of the Jews.

The early Christians were few in number and were only occasionally persecuted, allowing their ideas to spread throughout the empire. The appeal of Christianity lay partly in what was effectively a welfare system: this arose as a direct result of St Paul's emphasis on charity, and became a force for cultural stability. Even one of Christianity's later enemies, the Emperor Julian the Apostate, had to admit that its strong point was charity: 'No Jew is ever seen begging, and the impious Galileans support not only their own poor but ours as well.'[189] But Jews and the 'Galileans' (Christians) shared the oddity of monotheism, and this eventually became the source of serious difficulty for both monotheist religions, despite Roman religious tolerance. By the middle of the third century the Christians were numerous across the Empire, well organised, and multicultural

– the first transnational organisation that could seemingly pose a threat to the Empire. In 250 CE Emperor Decius – at a time of instability due to Northern invasion – decided that he had to deal with this potential threat and issued an edict that all citizens of the empire must sacrifice to the Roman gods and must be able to produce a certificate to this effect signed by a Roman official. A large number of Christians defied this edict and they and many of their bishops lost their lives as a result. Decius died only a year later, with persecutions having largely died down, but the Emperor Diocletian revived the persecutions in 303-304. One interpretation might be that the Christians were suffering religious persecution – and from the Christian side it must have seemed so – but the more informed view must be that Decius and Diocletian had purely political motives.

Christians refused to participate in the normal polytheistic rituals, both religious and civic, which bind society together, and did so on grounds of the prohibition against idolatry. Yet the prohibition against idolatry is not specific to Jesus's teachings – he never mentions it – but derives instead from the Judaic tradition. Their refusal to worship the Emperor had been downplayed or gone unnoticed until Decius and Diocletian forced the issue with certification, and only at this stage did the persecution of Christians take a religious light. The Empire was under attack in the third century, both from without, and from the growing decadence within its administrative structures. The intensity of Decius's persecution was partly a reaction to these forces, but for the Christians it cemented a nascent tradition of martyrdom which had taken place sporadically for a long time. The example of Jesus was naturally a factor in the Christian willingness to die for their beliefs. In Henry Chadwick's history of the early Christian Church he comments that 'Origen always writes as a member of a martyr church. . . .' [190] Judaism, out of which the early Christianity so intimately sprang, had also become a martyr religion when the Empire finally lost patience with the troublesome peoples of Palestine. After the destruction of the second Temple in 70 CE the last stand of the Jews (in terms of a powerful cultural image at least) was in the rock fortress of Masada a few years later. A Roman army of 15,000 besieged the fortress for two years to subdue the 1,000 Jews, a number which included women and children. Even the Romans were horrified by what happened then: all but two women and five children committed suicide. Suicide was forbidden to the Christians, but their many early martyrs and perhaps the image of Masada contributed to the atmosphere of religious persecution. A similar image but closer to home was represented by the Catacombs:

underground caves for the burial of the dead that provided sanctuary for persecuted Christians.

How the Roman Empire would have dealt with the growing Christian 'problem' is unknowable, but the rule of Constantine changed everything, almost overnight. He claimed that his personal adherence to the Christian 'God' had secured him victory in battle, and issued the decree lifting all persecution of Christians. He donated land and property to bishops and so the earliest official Roman church structures were founded. But the establishment of Christianity as the religion of state was not to be plain sailing. The polytheistic traditions of the Roman Empire took time to give way to the new religion. Constantine's conversion can be dated to 314, but the full adoption of Christianity was not complete until the 390s, and even then polytheistic remnants survived long after. Its biggest setback took place under the emperor Julian from 361 to 363, who attempted a root and branch return to the old gods, supplemented by new polytheistic catechisms, Neoplatonist ideas, and an attempt to 'catch up' with Christian charity. The Arian bishop of Alexandria was murdered by a polytheist mob in 361, a violence mirrored by Christian mobs in the city in later years. But the uptake of Christianity – often first by the women in families – proceeded irrevocably, and the emperors after Julian ensured its eventual dominance. After incorporation of Christianity into the State there was a rash of the cult of the martyr, with shrines appearing in honour of the martyrs and their relics. This recognised the suffering and uncompromising faith of the early believers, surely as a call to move into the light of a new era, free from persecution. That was not to be.

Councils

Constantine found that Christianity, despite its radical monotheism, had a great deal of difficulty articulating the exact role and status of Jesus. To sort this out he convened what was to be the first of a series of 'Councils' to establish a coherent doctrine. These became known as *ecumenical* Councils using a word that originally meant something like community, but took on the specific Christian meaning of universality. Constantine's own adherence to Christianity was flexible, as he continued to hold to the divine status of Emperor and his former sun-worship, but his aim in setting up these Councils was to create a doctrine that all the Christian groups would agree upon. In 325 CE in ancient Nicea (part of modern-day Turkey) arguments of stultifying complexity were pursued regarding the so-called Arian heresy, which denied the identity of Jesus with 'God'. This was resolved against Bishop Arias and resulted in his exile, though the heresy persisted throughout Christian history, Newton being a secret adherent for example, as we saw earlier. Other matters were settled, in particular the disagreement about what gospels would be included in the official New Testament. The Council of Nicea was therefore crucial in shaping the emerging Christian religion, while the gospels that were left out of the canon document the pre-Constantine diversity of the various Christian sects. The list of these texts forms the first 'Index' of forbidden books, and for many of them even their existence became a secret. Hence the discovery of the Dead Sea Scrolls in 1947 aroused considerable interest in the Christian community because they largely expanded on or reinforced the Old Testament, whereas the Nag Hammadi discoveries in 1945 have gone largely unnoticed because they unearthed many of the missing early Gospels, dismissed as *Gnostic*. These, in particular the *Gospel of Thomas,* show Jesus in a very different light to orthodox Church doctrine. There was a natural tendency to only include those texts that supported the power structures of the new religion, particularly those that suggested an apostolic succession (which legitimised the power of Rome).

We have earlier pointed out that uniquely in world history the Roman Church is a religion with a central authority that controls the canon of its literature. Polytheism by its nature was without any central authority, and it was only after the first ecumenical Councils that Julian the Apostate attempted a comparable centralisation of religious formalities for polytheism – a short-lived venture.

Church Fathers and Heresiology

In addition to the Councils, Christian belief was forged by the thinking of the so-called Church Fathers, also known as the early heresiologists of Christianity. They determined both what Christianity believed in, and what it emphatically rejected. Irenaeus of Lyons was the first to leave us extensive anti-heretical documents. His dates of birth and death are not know, but in the year 177 or 178 he succeeded the bishop of Lyons after his martyrdom. Irenaeus's major work, the *Exposure and Refutation of the Falsely So Called Works of Gnosis,* is typical of subsequent works that attacked the doctrinal rivals of Christianity, principally Gnosticism, but later on also Manichaeism. The prevailing polytheism of the old empire was not the target – that was dispelled by the authorities as successive Emperors made Christianity the official religion – but rather rival streams of thought that had the potential to absorb the Christian doctrine and take it in different directions. Christianity made great efforts to distance itself from Hellenic spirituality in particular. Amongst the later heresiologists that include Hyppolytus, Tertullian and Origen, it was Tertullian who summed it up: 'What then has Athens to do with Jerusalem, the Academy to do with the Church, the heretics to do with the Christians?'[191]

The emerging concept of 'heretic' in the Christian Roman empire, coupled with the existing judicial machinery, made it possible for heretics to be tortured and murdered by the state . . . but not inevitable. The nearly four centuries prior to the firm establishment of Christianity as religion of state had included many periods and acts of oppression, symbolised by the image of the *Christian* martyr – whether perishing in the flames, eaten by lions, or victim of other cruelties dreamed up by the barbaric imagination of the period – and perhaps more generally by the images of the Catacombs and Masada. The idea that the oppressed inevitably become the oppressors is one possible explanation of later Christian cruelty against heretics, but first we take a look at another faith that was adopted by an emperor as its State religion: Buddhism.

The Spread of Buddhism

Buddhism spread as religion of empire in an interesting parallel with the spread of Christianity. At first glance this is a story of two emperors, Ashoka in the East and Constantine in the West. The Emperor Ashoka reigned in approximately the years 265-238 BCE, some two to three hundred years after the Buddha. Ashoka was an emperor of the Mauryan dynasty, founded by his grandfather Chandragupta in the wake of Alexander the Great's death, controlling all of India except the south; and which spread west to Persia and east to Laos and Cambodia. Chandragupta had adopted the Jain religion, and Bindusara, the father of Ashoka, had adopted the Ajivika religion. Ashoka adopted Buddhism and, more than any of his dynasty, went on to spread the precepts of his chosen religion across his empire.

What we know of Ashoka comes mainly from rock edicts scattered around the empire, which include those addressed directly to Buddhists and the greater number addressed to the general population. These are known as the Minor Rock Inscriptions, the Major Rock Inscriptions, and the two Separate Edicts at Kalinga.[192] The rock edicts record that Ashoka conquered the Kalinga country (modern Orissa state) in the eighth year of his reign, and that the sufferings that the war inflicted on the defeated people moved him to remorse. According to myth he underwent a conversion experience, but more recent scholarship suggests that his commitment to Buddhism simply deepened over a period of time. Under its influence and the promptings of his own conscience he radically modified his approach to conquest and empire. The rock edicts scattered throughout the region bear the messages of Ashoka to the populations of his empire, and are remarkable for their spirit of peace, tolerance, compassion and even vegetarianism. Ashoka's Buddhism was largely an ethical one, and a far cry from the evangelism of most Western religions. While a practising Buddhist himself, he did not want to make direct converts, but encouraged the spread of a Buddhist outlook that respected and absorbed local religions. His general edicts did not give much detail of Buddhist beliefs or practices, but promoted two principle virtues: 'ahimsa' or non-violence, and 'maitri' or active benevolence, principles that the other religions and sects of that time shared. It is this neutral characteristic of Ashoka's reign that led more than anything else to the great variety within the strands of Buddhism in existence today, many of which are much more removed from early Buddhism than Christian variations are from early Christianity. To the Western mind this is an unfamiliar proposition within religion, that it can be so flexible, and prompted the ever-reliable *Catholic Encyclopedia* to make this comment:

> Northern Buddhism became the very opposite of what Buddha taught to men, and in spreading to foreign lands accommodated itself to the degrading superstitions of the peoples it sought to win. It is only the Southern Buddhists of Ceylon, Burma, and Siam who deserve to be identified with the order founded by Buddha.[193]

As usual the *Catholic Encyclopedia* is both perceptive in its comments and also revealing of its prejudices. The entire article on Buddhism contains useful insights mixed with the transparent prejudice of a tradition attempting to discredit what it obviously sees as a serious rival. The objection raised in the above extract is unfounded however: the 'degrading superstitions' of Ashoka's expanded territories are no such thing, but rather a mixture of the shamanic and polytheistic traditions known the world over until Christianity so completely discredited them. The *Catholic Encyclopedia* is correct however to identify what it calls as 'Southern Buddhists' to be closer to the Buddha's teachings – these are regions where the early, Therevadan tradition is strong. It is also correct to say that 'Northern' Buddhism, i.e. Tibetan Buddhism, is a radical departure from the Buddha's teachings, as it is formed by an extraordinary meeting of the incoming Buddhism and the highly advanced indigenous shamanic religion of the *Bon Po*.

Without Ashoka it is doubtful whether Buddhism would have become a world religion. It was initially a smaller rival to the Jain religion, which in modern times retains a tiny number of adherents almost exclusively in India. Ashoka, in seeking a religious teaching to become the religion of empire, could have chosen from a number of rivals that we mentioned before, including Brahmanism, Jainism, and the religion of the Ajivikas. Jainism taught a non-violence and respect for life that went further than Buddhism, so a constructive encounter with Jainism might have had the same revelatory impact on him. The Emperor Constantine, in the fourth century after Christ, had likewise a number of competing sects from which to select as religion of empire: through sheer chance Constantine chose Christianity. It became the religion of empire at a time when Rome was embattled against the 'barbarian' hordes of the north and east, and, as we saw, was profoundly influenced by the militarism of state. Ashoka's empire, by contrast, was ruled by a man who had acquired a genuine abhorrence of violence.

The Mauryan Empire had retained large swathes of Alexander the Great's territory. When judged by the contemporary mind, Alexander seems a man of barbaric cruelty, though in using the term 'barbaric' there is an irony: it was the Greeks who coined the term for the hostile races on their borders, making a word out of the 'bar bar' noises they contemptuously heard in the foreign tongues. The Mauryan Empire was no doubt built on the same cruelty that Alexander or the Romans exhibited in creating their empires, and the Mauryan dynasty disappeared not long after Ashoka's spread of pacifist Buddhism. Kings, such as the father of the Buddha, were feudal warlords whose fiefdoms ran so long as their patron had the vigour, aggression, intelligence, and good luck to hold out against their neighbours. The Sakyamuni clan of Siddhartha Gautama was over-run not many years after the Buddha's visit to his family (some seven years after his departure), and legend has it that the entire family perished.

While the parallels with Constantine's Christian empire are historically remarkable – both emperors experienced conversions to one emerging religion out of any number of competing faiths; both instigated it as religion of state; and both set up Councils for its clarification – the question remains why Christianity retained the barbarous cruelty of Roman law, where Buddhism was relatively untouched by the (presumed) corresponding violence of Mauryan empire formation. This is not to say that the peoples and rulers of Buddhist lands took any less part in warfare than their counterparts in Christendom, merely that doctrinal deviation within Buddhism or adoption of another religion were rarely the excuse for violence. To pin it down further: nothing comparable to the Roman Catholic tradition of systematic Inquisitions ever arose in India or other Buddhist lands (until it was exported there by the West).

Mahatma Gandhi pursued the Buddhist doctrine of non-violence – *ahimsa* – in the twentieth century, leading up to independence from Britain. When he translated the *Bhagavad Gita* into Gujarati he found it most difficult to deal with its central narrative device: Krishna is persuading Arjuna to go to war.[194] The *Gita* is part of a vastly longer epic drama called the *Mahabharata,* which tells in mythic form the war that ensues. When Voltaire surveyed what Indian religious texts were available to him, he was impressed by the lack of violence in them. Whether he was aware of the *Mahabharata* or not, there is still a profound difference between the violence recorded there and that in the Old Testament: the motivation for the great Indian war had not a shred of religious basis.

3.3.4 The Intolerant 'God'

In the building of all the ancient empires slaughter was inevitable; torture its companion. Whether the British can claim – as they may like – to have built their empire with the least amount of slaughter and torture is open to debate. Our question is not why Roman judicial murder and torture existed, but why it was applied to the sphere of religion by the Christians where it had not occurred to the Romans to do so. That an oppressed group on coming to power then become oppressors is a possible explanation: Christianity as a martyr religion under Diocletian had suffered hideously, and Nero for example is supposed to have used Christians as human torches to light his garden parties, but its founding principles were love and charity – from where did it learn intolerance? In examining the emergence of Christianity we see that whatever might be new in it, represented by the Gospels, there was much that was old in it, represented by the Old Testament. In fact, long before the New Testament was available and eventually stabilised as a canon, the Greek-speaking Christians of the Roman Empire read the Old Testament in the form of the Septuagint (a Greek compilation). Vidal calls the Old Testament a 'barbaric Bronze Age text' and

Voltaire extensively criticised the violence in it. Could the religious intolerance of the Septuagint have been the source of the later Christian religious intolerance?

Where it is almost impossible to find an exhortation to religious intolerance in the Hindu, Buddhist and Taoist scriptures – as Voltaire noted – the Old Testament is the opposite: it is hard to find even a small section which does not speak of intolerance, eternal damnation or punishment. Here is an example from Deuteronomy 17:2-5:

> If there is found among you, within any of your towns which the Lord your God gives you, a man or woman who does what is evil in the sight of the Lord your God, in transgressing his covenant, and has gone and served other gods and worshipped them, or the sun or the moon or any of the host of heaven, which I have forbidden, and it is told you and you hear of it; then you shall inquire diligently and if it is true and certain that such an abominable thing has been done in Israel, then you shall bring forth to your gates that man or woman who has done this evil thing, and you shall stone that man or woman to death with stones.

The dreadful punishment of stoning to death was not for any civil crime, that is an act which harms another human being or the environment, but for the crime of the wrong kind of worship, in other words heresy. Thomas Jefferson in framing the American Constitution – which placed a 'wall of separation between church and state' – was well aware of the difference between civil crimes and religious heresy: 'The legitimate powers of government extend only to such acts as are injurious to others. But it does me no injury for my neighbour to say there are twenty gods, or no God. It neither picks my pocket nor breaks my leg.'[195]

Where the Buddha is rather stern with any of his monks who practices or understands the dharma in the wrong way, the worst that the culprit faces is disapproval. The Hindu scriptures don't even suggest that bad karma might follow wrong religious beliefs or practices, and the Taoists at worst deploy humour. But in the three monotheist traditions, and in these alone, we find that 'wrong' religious beliefs or practices are to be punished as if actual harm has been done. What is quite incredible – to us since the time of Voltaire – is that the punishment should be death. While *judicial* murder is carried out against the real or imagined enemies of state, *ecclesiastical* murder is carried out against heretics.

Judaism

For a balanced analysis we need to examine the oppressive aspects of monotheistic religion without flinching, but also to expose its varied genius and how it relates to specific forms of the spiritual impulse. The particular genius of the Christian tradition is examined in section 4.1 *Piety and the West,* and the spirituality of the Greek mind is explored in section 4.2 Jnani *and the Hellenic Influence,* but in this section we look at the unique spiritual genius of Judaism.

The fact is that Judaism is a difficult religion to look into from the outside, partly because it has never sought converts like most other religions, and therefore never needed to explain itself to the outside world. From the outside it is a religion of texts and observances (Karen Armstrong suggests that orthopraxy is in fact more important than orthodoxy in Judaism),[196] but it is also a religion of profound and beautiful *ambiguity*. Where differences of spiritual orientation are clearly articulated and celebrated in Hinduism, the Judaic impulse is towards obscurity, the elision of difference. This means that Judaism holds as one of its core principals the *mystery* of its 'God', and, at the heart of its efforts for thousands of years, there has been the drive to counter the natural and populist instinct for anthropomorphism.

The classic work by Jewish scholar Gershom Scholem, *Major Trends in Jewish Mysticism,* is a complete historical sweep through all the stages of Judaism, its evolution being as complex as that of Christianity, but over a longer period of time. The stages of Jewish mysticism are expounded in detail: the early Merkabah or 'throne' mysticism lasting from the first century BCE to the tenth century CE; the medieval German Hasidism; the early Spanish Kabbalism of Abulafia of the thirteenth century; the Kabbalah of the Zohar; the Lurianic Kabbalah after the Spanish exodus of the 1490s; the impact of Sabbati Zevi's millenarian mysticism in 1666 (which even Newton succumbed to); and finally the late phase of Hasidism. Scholem explores a great variety in the mysticism of these periods, and also how central they were to the development of mainstream Judaism. We suggested in Chapter Two that 'God' could come into being as the object of the devotional vector, thus making an association with theism and spiritual love, or the *bhakti* orientation. Scholem shows us that this is too sweeping an assumption, as the 'God' of the Jewish mystics is rarely the object of religious love, except in the phase of medieval German Hasidism. He says of the Merkabah period that, 'Not only is there for the mystic no divine immanence, there is almost no love of God. What there is of love in the relationship between the Jewish mystic and his God belongs to a much later period. . . .'[197]

This later period is the medieval one, where the Jewish mystics were influenced by the profoundly devotional world of Christianity at that time. Even so we learn from Scholem that the Hassidic mystics are drawn to the elements of altruism, asceticism and self-abnegation as well as to the love of 'God'.[198] At the same time the esoteric and occult elements that weave their way into all religion are present in the different periods, receiving at different times a less or more favourable reception. If the 'God' of the Jewish mystics is not the one of the devotional impulse, neither is it quite the 'God of the philosophers': this makes Judaism unique.

One of the defining features of Jewish mystics, as explained by Scholem in his chapter on their general characteristics, is their conception of the hidden 'God' or *Deus Absconditus*, resistant to anthropomorphic attribution. Yet at the same time it is a *relationship* that is central to the approach to the mysterious absolute. Where other religions have stressed that this relationship may lead to union, the Jewish mystic, to use a sexual metaphor, postpones that climax in order to prolong the intercourse, or even denies its possibility. Two twentieth-century Jewish thinkers that inherit the language of relationship with a profoundly mysterious 'other' are Martin Buber and Emmanuel Levinas. Buber was a friend of Scholem, and both are important influences on Levinas. Buber's best known work, first published in 1923, is simply titled 'I and Thou', and has remained a widely read masterpiece of twentieth-century Jewish thought. In it Buber says simply that we have only two relationships: the *I-Thou* relationship, which can only be spoken by the whole being, and the *I-It* relationship which can never be spoken by the whole being.[199] The human cannot exist apart from these two relationships: a belief he contrasts with those of the Buddha, and which Levinas goes on to contrast with the 'One' of the Greeks. Buber presents to the world in his little book a distillation of all that is best in the original Western monotheism, in full awareness and respect of other traditions, but firm: 'God' is 'other'; and our approach to him is held in the single word (as he calls it) *I-Thou*. There is a mutuality in this term which disallows humans on their own, or 'God' on his own. Buber says that if the goal of the Buddha is only one of the goals of the spiritual life, then it is not 'ours' (by which he means the monotheists) 'but if it is *the* goal, then it is falsely described'.[200] In Buddhism *nirvana* is attained as the *mahamudra*, the great gesture, a climax that takes place only once and in which the concept of separation ceases. Buber says of this: 'All doctrine of absorption is based on the colossal illusion of the human spirit that is bent back on itself, that spirit exists in man.'

Although Buber is writing about Buddhism, his objection to its core principle is also the Judaic objection to the core principle of Greek *jnani* spirituality: the possibility of identity with all that is, or with emptiness, *shunyata*. In Buber's writings – whether we personally prefer the Buddha's religion of transcendence or the Abrahamic religion of monotheism – we find a deep intelligence at work, one that does not resort to violence or anathematisation but rather one that reflects, puts the arguments, and leaves it to the reader to decide. More than that, it is lyrically spiritual.

Where Buber objects to the 'doctrine of absorption' of the Buddha, Levinas objects to the 'primacy of the same' in Socrates. We do not have here the contradistinction between *bhakti* with *jnani* (as we have seen that Judaism is not fundamentally devotional) but of the theistic impulse with the transcendent impulse. Along the way Buber and Levinas touch on many of our other distinctions: between *via positiva* and *via negativa*, between the social and the solitary, but at no stage do they admit either of the transcendent (in the Buddhist or Neoplatonist sense) or anything of the occult.

We suggest that Levinas's *Totality and Infinity* is one of the great spiritual works of the mid-twentieth century, being both a culmination of Jewish religious thought and a convergence with powerful new ideas in postmodernism. What makes it remarkable is its rejection of what Levinas calls the 'totalising' aspect of religion, and at the same time a rejection of the Greek spirituality (which we characterise as *jnani* transcendent). In *Totality and Infinity* we find a recapitulation of the struggle for the Western soul between the Hebraic and the Hellenic forms of spirituality, where Levinas naturally puts forward the Hebraic, the essence of which is *exteriority*. By this Levinas means the 'other' or 'God' as beyond one's self; a fundamental statement of the theistic position.

Monotheism and the world of Nature

Scholem makes an important point early in *Major Trends in Jewish Mysticism*: 'Religion's supreme function is to destroy the dream-harmony of Man, Universe and God, to isolate man from the other elements of the dream stage of his mythical and primitive consciousness.'[201] Scholem is writing in the 1930s and would have been influenced by ideas on consciousness from Freud and Jung, but even taking that into account we have a remarkable statement of how he conceives of religion. It is surely in this Judaic impulse to religion that the root of the West's rejection of the old nature-religion, the shamanistic ur-religion of humanity, must originate.

The polytheism of the Greeks lived more comfortably with the world of Nature, but the monotheism of the Jews did not, as we see expressed so forcefully by Scholem. He understands the formation of monotheism as a stage of religious development which required an 'abyss' to appear between 'Man and God', and between 'Man and Nature'. He sees the earlier stage of humanity as a form of childhood, a mythical epoch, a world 'as being full of gods whom man encounters at every step and whose presence can be experienced without recourse to ecstatic meditation.' In the second stage 'the great monotheistic religions live and unfold in the ever-present consciousness of this bipolarity, of the existence of an abyss which can never be bridged. To them the scene of religion is no longer Nature, but the moral and religious action of man and the community of men, whose interplay brings about history as, in a sense, the stage on which the drama of man's relation to God unfolds.'[202] For Scholem, the Jewish mystic comes into being in response to the abyss, and their mystical longing is to come into relation with 'God' as ultimately 'other'. There is little sense, as in Christian mysticism, or even more so in Hinduism, of the possibility of union with 'God', or identity with 'God'. And completely impossible would be the idea of Nature itself as the site of transcendent union.

Throughout the Old Testament we find a revulsion against 'idolatry', with the injunction against it repeated over and over again. We are going to suggest that there are two forces at work here: firstly a move to destroy the remnants of shamanism as a Nature religion, and secondly the move to eliminate adherence to the rival gods of polytheism. This is speculative of course, but is based on the ideas emerging from recent anthropology which suggest that the transition from the essentially shamanic hunter-gatherer civilisations to the essentially polytheistic early agrarian societies was traumatic. In the Old Testament books of Kings, we find the most sustained account of the destruction of 'shrines, sacred pillars, sacred poles, and idols.' The Revised English bible uses the term 'sacred poles' whereas the King James edition uses the term 'groves' (e.g. I Kings 14:23 and I Kings 16:33). Rupert Sheldrake considers that, despite the persecution of such practices by, for example, King Hezekiah in 622 BCE, the religion of natural sites persisted in Palestine right into the nineteenth century.[203] In fact Jewish scholars in the ecological movement are sensitive to the accusation that their religion was at the forefront of the denigration Nature, but the Old Testament seems to endorse it.

Evolution of the Judaic 'God'

Karen Armstrong's well-received *History of God* traces the evolution of the Jewish 'God', providing at the same time an instructive case study of the transition of a religion from polytheism to monotheism. The Jewish 'God' started life as Yahweh, a god of war to be worshipped alongside other deities, such as Baal the god of fertility. Baal was a *local* deity, specific to Canaan where the Israelites settled after the Exodus. Armstrong makes the point that during this period of polytheism matters of worship were managed as much by women as by men, particularly in respect of the Goddess religions of Ashera, Ishtar or Anat, but as Yahweh came into the ascendant the guardians of religion were increasingly male. Old Testament sources indicate that it took a period of seven hundred years to completely make the transition to monotheism, the final rejection of the other gods being accomplished after the first exile in Babylon. Yahweh being the god of war was also the god of suffering, and he embodied their experience as captives, removed from their land. At the same time Baal, being linked to the land they had lost, seemed lost as a god, and so the allegiance to him failed, as it did to other localised deities.

On the return of the Israelites from exile, granted by Persian king Cyrus in around 550 BCE, they enforced the monotheism of Yahweh, destroying the temples and icons of the other gods. At the same time the traditions of the mother goddess and the equal role of women in society were lost, as the war-god and his male priesthood began to dominate. The return from Babylon is perhaps the most significant moment in Western monotheism. King Cyrus of Persia had conquered Babylon in 539 BCE, bringing with him the religion of Zoroastrianism, a monotheism founded about 1,400 BCE by the prophet Zarathustra. It is quite possible that the Jews in exile were exposed to the idea of the single invisible 'God' of Zoroastrianism, and that this provided the impetus to their own adoption of monotheism. Another theory holds that the monotheism of the Jews derives from the brief Egyptian monotheism of Akhenaton in the fourteenth century BCE. Whatever the reason, there is no doubt that the real turn towards monotheism in the history of the Jews takes place at the return from exile. Even so, the process of eliminating the other gods was slow.

Nehemiah, a Persian official, was sent to Judea about 444 BCE, and served as governor for 12 years, instigating various religious reforms for the Jews, including the observance of the Sabbath. His most crucial reform, continued by the Persian Ezra, was the prohibition against Jewish men taking non-Jewish wives. Ezra initiated an investigation of all mixed marriages enforcing the 'dismissal' of non-Jewish wives and their children (the suffering of which seems to have aroused no comment). The motivation for this was that intermarriage had encouraged the adoption of 'the abominable practices of the Canaanites, Hittites, Perizzites, Jebusites, Ammonites, Moabites, Egyptians and Amorites' (Ezra 9:2). Nehemiah reflected: 'Was it not because of such women, . . . that King Solomon of Israel sinned?' (Nehemiah 13:26) But just what was King Solomon's sin; just what were the 'abominable practices'? It was to allow his foreign wives to teach his children the foreign gods. Whatever the Jews at the time understood the reason to be, as a cultural dynamic it had immense historical ramifications: to this day the Jewish faith is carried through the line of the mother. Of course fathers are involved in the education of their children to some degree, but in their early childhood in a pre-industrial agrarian society, children would learn the basics of religion from their mothers. By ensuring that the mother was Jewish, the monotheism of the Jews became entrenched, and the foreign gods permitted no entry.

Narcissism of minor difference

We need to make a distinction here between monotheism, which postulates that there is one 'God' for the whole of existence, and monolatry, which means the tribal worship of a single 'God' belonging to that tribe. Prior to the Babylonian exile it is possible that the Jews, to the extent that they did worship a single god rather than many, followed the practice of monolatry, meaning that they accepted that other tribes would worship their own 'God'. Freud's expression puts into perspective any reason for selecting one of these gods over any other: the 'narcissism of minor difference'. By this he means that a tribe exaggerates the difference between their 'God' and those of their neighbours, as part of their identity-creating process. This process did not always favour Yahweh, as in the case of Jezebel, the wife of King Ahab (ruler of Israel in approximately 874- 853 BCE), who persuaded him to introduce the worship of the Tyrian god Baal-Melkart, a nature god. She tried to destroy those who opposed her and most of the prophets of Yahweh were killed at her command. Later the priests of Baal were slaughtered. This is recorded many times however, and it seems a common fate of the priests of Ishera and Baal to be slaughtered on their own alters by the Jews, and their bones ground down along with the idols. How historically accurate any of this might be is unknown however, for example there is no reliable historical evidence for Kings David or Solomon, or for that matter Jesus.[204] Our point is that the Old Testament became a key text for Christians through the Septuagint, and its violence towards 'minor difference' an uncontested element of 'Holy' Scripture. The modern mind since Voltaire sees nothing holy in this.

Like Hume, Armstrong points out that polytheism is naturally tolerant, though unlike Hume she points out that it is also more likely to be inclusive of women than the emerging monotheism. The Judaic 'God' was ushered in over the blood of those who resisted, and the religion that emerged contained within it a profound hatred of 'idolatry'. Montesquieu, writing in 1721, saw it as a kind of madness, 'which the Jews took over from the Egyptians and which passed from them, like a common epidemic disease, to the Mohammedans and Christians.'[205] In his *Letter Concerning Toleration*, written in 1689, John Locke was more careful not to offend the Jewish tradition, by saying that its intolerance was right and proper to a theocracy.[206]

The Western revolutions of monotheism over polytheism in the religions of Judaism, Christianity and Islam have created a religious intolerance that shows itself in different ways in the three religions of the book. For Judaism, the adherence to a single 'God' became the mark of a race, the Jewish identity, and therefore once the murderous transition from polytheism was effected, Jewish monotheism was rarely associated with physical violence. As Jews were not permitted to marry outside their faith, and once they were dispersed, their lone 'God' ceased to b a god of war. Jews have never proselytised, believing instead in the idea of themselves as a 'chosen race'. Christians however inherited the 'God' of the Old Testament, and as soon as Christianity became the religion of empire and its instrument of conformity, the Christian 'God' was again a god of war. Christians became 'soldiers' for Christ and adopted an evangelical fervour that resulted in their religion dominating the globe, and in the atrocities of the Crusades. For the Arabs of Islam their 'God' became the god of war through creators of empire from the seventh century onwards (including under the generalship of Mohammed). Islam spread to the East and West creating cultural empires that outstripped Christendom until the Renaissance.

3.3.5 The Unexamined Collective Trauma of Western Religion

We saw that the historian Edward Peters made the distinction between the myth of 'The Inquisition' as a monolithic entity, and the historically verified occurrences of inquisitions (in the plural) instigated in different times and places. He suggests that the Romantic and Gothic novelists of the eighteenth century assisted in the myth-making, preparing the ground for twentieth-century portrayals of the Inquisition on film. More than that, he says that the films of director Carl Dreyer has brought the Inquisition to life 'in a way that no polemic or novel ever had, with the possible exception of Verdi's and Dostoyevsky's work.'[207] He mentions Dreyer's 1919 *Leaves from Satan's Book* and the 1928 *The Passion of Joan of Arc*, but we will use Dreyer's 1943 film *Day of Wrath* to illustrate the horrors of the Inquisition as it plays out in the contemporary mind.

The Day of Wrath

The Danish filmmaker Carl Dreyer wrote and directed *Day of Wrath* to create perhaps the most shocking indictment ever made of religious cruelty perpetrated by men against women. It is made in black and white and is widely regarded as a filmic masterpiece, mostly circulating outside of the mainstream as an arthouse film. It is set in seventeenth-century Denmark, at roughly the time of the Salem witch trials (which are the subject of Arthur Miller's play and screen adaptation *The Crucible*). Dreyer's film is in two parts: the first deals with the flight, capture, and burning of an old woman, Herlofs Marte, for witchcraft, while the second deals with Anne Pedersdotter, the pastor's second wife. Herlofs Marte is thoughtfully and properly tortured – that is according to a clearly defined procedure – by the religious officials concerned only to 'save her soul'. The torture itself is barely shown, rather it is the plain, desperate old woman – who has no allies to speak for her – sitting half-naked in sullen, bewildered defiance, that is unspeakably pitiable. In one shot she is naked from the waist up, her chains being refitted after her session on the rack, while the formally dressed and sombre men speak her native language with words that are remote and deadly: she can neither understand nor answer them in kind. The inevitability of her burning is spelt out in the words recorded by the tribunal: 'Having been tortured to the glory of God she voluntarily confessed. . . . On the day which was exceptionally fine Herlofs Marte was burned, happily.' She is dragged protesting, and tied to a tall ladder; it is propped into the upright position, and then solemnly tipped over onto the raging flames. The ecclesiastical and secular officials watch from the stands, along with the rest of the town.

The pastor, Absalon Pedersson, is asked by his mother 'did she denounce?' to which he replies 'not a living soul'. His mother is making a spiteful enquiry about Anne, the second wife of Absalon, and far his junior, who she suspects of 'powers', perhaps known to Herlofs Marte. Absalon's son Martin returns to the family home and

Anne, deprived of her youth and happiness by the forced marriage to the old pastor, falls in love with him. They have an affair behind the old man's back in which Anne for the first time feels that love and future is possible; Martin is head over heels for her. But when Anne finally confesses the affair to her aging husband, he has a heart attack and dies. She is now the one accused of witchcraft, by Absalon's vengeful mother. The son's evidence is crucial: he promises to defend her. In the final moments of the trial, the young woman, so brought to life by the hope of running away with a man of her own age, looks to Martin. His testimony is clear: 'My father's death is not to be laid at any person's door.' Anne's salvation is apparently secure. But his grandmother dramatically intervenes, declaring to the court that Anne had made him fall in love 'with the help of the evil one', and had murdered his father. Martin falters at the mention of the 'evil one', looking at his grandmother, as the court looks to him. After an agonised struggle within his mind to resolve his conflicting loyalties he finally capitulates, walks solemnly to his grandmother and sits down. It is all over. However, the magistrate gives Anne one last chance: to swear on the dead man's body which is present in court. Slowly she gets up and walks to it. But without Martin's support she is doomed, and so haltingly she confesses to the killing, 'I lured your son with my power, with the help of the evil one.' Anne herself has come to believe in her own power: the miracle of Martin's love must have been her own evil doing; the removal of the obstacle to their love must also have been her work. She has internalised the whole machinery of suspicion and superstition, and accepts her dreadful fate.

We don't see the burning of Anne as we saw the burning of Herlofs Marta; instead the film ends with a passage from Romans 2:5 describing the 'day of wrath', which is the Christian Day of Judgement. Anne Pedersdotter, a woman subjugated in a man's world, had in her naivety dreamed of happiness and the possibility of a woman's 'power' to change the course of events; she had looked at her aged husband and wished for his death. In the time it took for her lover to cross the courtroom, Anne had in her own mind exchanged the possibility of freedom and marital happiness for burning and the certainty of her own guilt. Henry Miller's Salem women made the same mistake: by believing to some degree in their own occult power, they easily fall in with the inquisition that follows. They had danced in the woods as part of a 'demonic' plot, according to their accusers, but what was its goal exactly? Just to attract a husband. Both films make parallels: Dreyer is inviting the audience to make a judgement on the Nazi occupation of Denmark, while Miller invites the audience of *The Crucible* to make a judgement on the American McCarthyism of the 1940s and 1950s. But *The Day of Wrath*'s real indictment is of the Inquisition, or rather its Calvinist homologues in Denmark. It speaks powerfully to a collective trauma, or myth as Peters would prefer, that is present in the Western mind.

For the atheists the Inquisition became another shibboleth of secularism ('religion = Inquisition'), while no substantial effort has been made by Christianity to atone or apologise for the historic wrong it represents. Instead Christians may be glad to see it become the product of titillatory Gothic fantasy or Monty Python humour, because such strategies make the subject more bearable. But Peters shows us that films like Dreyer's come closest to forcing us to confront the reality of an Inquisition: a person firstly betrayed through denunciation and then by the religion of 'love and charity'. Which of the dictums of Jesus were followed in the ecclesiastical torture and murder of those like Herlofs Marte? None. It was rather at the behest of countless exhortations in the Old Testament like that in Exodus 22:18 'Thou shalt not suffer a witch to live.' The victims are ordinary people – epitomised by a fat frightened old woman and a faithless young wife. Only occasionally are they a Hypatia, a Bruno, a Joan of Arc.

Masada and the Catacombs

The Inquisition is the cultural vessel for the Christian practice of cruelty. But the cultural vessels for Christian *suffering* are many . . . and include a filmic trope that draws on what we can call the Masada-catacombs complex. It is quite staggering how many films – often science fiction – draw on a stereotype of oppression that must have its cultural origins in the siege of Jewish Masada and the claustrophobic tunnels of the Christian catacombs. Perhaps the most recent is the *Matrix Revolutions* (2003); others include *Soldier* (1998), *Extreme Measures* (1996), *Demolition Man* (1993), and all the *Terminator* films. The common factors are a community under siege including men, women and children; a badly clothed and badly fed population (in *Demolition Man* they eat ratburgers); a badly armed but brave collective non-hierarchy; an underground environment of city tunnels or mountain caves; and an enemy that is relentless and like a machine. The Roman army of 73 CE is transmogrified in *Matrix Revolutions* into actual machines (and the human city is even called 'Zion'). In the *Terminator* series the enemy are again machines; in *Demolition Man* and *Extreme Measures* they are the police. Also required is a cute unsmiling kid in shabby brown clothes who watches wide-eyed and learns from the hero or heroine conveniently parachuted into the situation. In *Aliens* (1986), the unsmiling urchin is a little girl, the sole survivor of her community: a direct parallel with the surviving children of Masada. These films after all are made to glorify the heroic-hubristic warrior culture of monotheism, though they are thoroughly secularised. It is the specificity of these situations that suggest an unbroken cultural transmission from nearly 2,000 years ago: other major world events manifest themselves in quite different filmic tropes.

The Unconscious Collective Trauma of Religion

Collective traumas live on in culture, and are worked and re-worked for each generation in ways that are not entirely clear. But for individuals Freud had suggested a mechanism: the unconscious, and this idea was extended to the community in Jung's idea of the *collective* unconscious (Freud sniffily responded to this innovation by saying that *of course* the unconscious included the collective). Freuds's radical insight, for which he deserves the credit, is that traumatic events which are emotionally overwhelming are repressed in memory, and become an unconscious site of cathexis. These fester and produce irrational behaviour known as neuroses; Jung merely extended this concept to whole societies. Here we are pointing to two related but very different religious traumas in the Western collective unconscious: firstly religious persecution *of* the Christians, which we can call the 'Masada-catacombs complex', and secondly religious persecution *by* the Christians, which we can call the 'Inquisition complex'.

If the Masada-catacombs complex lives as an unconscious site of cathexis in the Jewish and Christian mind, then the irrational behaviour it generates will be a paranoia about religious criticism, an over-reaction to attacks on one's *belief* as an attack on one's *life*. Any attempt to engage in critical debate over the beliefs of monotheism may trigger an unconscious anxiety that what follows will be persecution, torture and death. For the really devout it may even trigger a martyrdom complex which actually relishes dying for one's beliefs. Neurotic behaviour of one kind or another is inevitable – according to the principles of psychoanalysis – unless the original trauma is revisited by the adult mind in a safe environment, and its hidden emotional cathexis discharged.

In the detailed study of sexual abuse carried out by psychologists since the time of Freud, it has become abundantly clear that the abused, unless he or she is the recipient of effective psychotherapy, goes on to abuse. Collectively the real abuse suffered by Christians, symbolised in the crucifixion of Jesus, was internalised and played back as the Inquisition. The martyr religion went on to martyr those who opposed it, or were imagined to oppose it. Hence the second half of this equation runs: if the Inquisition complex also lives as an unconscious site of cathexis in the Western mind, then the irrational behaviour that it generates may become a rejection of religion in its entirety, a vituperative atheism. In Victorian Britain the emerging secular societies became debating houses over the 'truth' or otherwise of the Bible for working class men and women learning to read, and beginning to agitate for political change. Secularism was forged in this debate which had the potential to bring energy and intelligence into religion; instead energy and intelligence largely deserted it and joined socialist-type movements. The Christians saw religious criticism (which for Marx was the beginning of all criticism) as deeply threatening: were they reacting out of the cultural cathexis that originated in Christianity as a martyr religion, out of the Masada-catacombs complex? Conversely the emerging secularists saw religious authority as deeply threatening: were they reacting out of the cultural cathexis that originated with the Inquisition?

3.3.6 Religious Cruelty and the Enlightenment Thinkers

The legacy of Western religious cruelty may, in the collective unconscious, be the wellspring of twentieth-century vituperative atheism. But for the key Enlightenment period, the eighteenth century, the question remains: did this legacy push the anticlerical majority to become an atheist majority? What proportion of Enlightenment thinkers, very conscious of the horrible legacy of religious cruelty, and who wrote stinging indictments of it, made Diderot's journey from theism, to Deism to atheism (if, in truth, he did)? Surely, this powerful combination, of science and rationalism on the one hand, and the legacy of religious cruelty on the other hand, must have persuaded the Enlightenment thinkers in droves to demand the end of religion?

No, is the answer. The collective revulsion against religious cruelty found sufficient expression in anticlericalism, which can be understood as the determination to reform religion, not destroy it. The vast movement known as the Reformation was the model: a revolt against existing clerical structures with the sole purpose of improving religion. Where a small consideration of religious cruelty might turn a person's mind to anticlericalism, did a deeper consideration of this dreadful history turn a person to vituperative atheism? Apparently not. It seemed that the legacy of religious cruelty suggested to only a few that religion should end: it suggested to most that toleration should begin.

3.3.7 It Doesn't Add Up

The secular mind assumes that religion or religiosity is confined to church and church-going. Hence when the Enlightenment thinkers argued against the church and its powers, the secular mind assumes they wanted the end of religion. The secular mind assumes that religion is equated with a single conception of 'God', so when the Enlightenment thinkers reframed that conception in a variety of ways, the secular mind assumes they wanted the end of religion. The secular mind assumes that religion necessarily requires an Inquisition of some kind, so when the Enlightenment thinkers raised a chorus of protest against religious cruelty, the secular mind assumes they wanted the end of religion. None of these secular shibboleths is true. It doesn't add up.

We have introduced the idea that the Enlightenment thinkers were mostly contributing to religion – when more broadly understood than through the shibboleths of secularism – so why did not this habit of engagement between intellectuals and religion continue into the modern world? Why did the secular era become polarised between faith tradition on the one hand and non-religious cultural innovators on the other? It is clear that the double whammy of the rise of science and the legacy of religious cruelty did not persuade the bulk of Enlightenment thinkers to argue for the end of religion, so why is it that the legacy of the Enlightenment is felt only through the exceptions, such as La Mettrie, Diderot, and Holbach? We call this the enduring puzzle of Enlightenment philosophy or the conundrum of the Enlightenment. Perhaps the clue to its solution is not in the rare atheism of the Enlightenment thinkers, but in their widespread religiosity: *was it of the wrong kind?* To understand this better we need to go more deeply into the issue of devotional and non-devotional spirituality, into the issue of *bhakti* and *jnani*.

Four: *Bhakti* and *Jnani* in Western Development

In this chapter we revisit the idea that Western development took place through the creative tension between the Hebraic and Hellenic religious impulses, using the concepts of *bhakti* and *jnani*. By examining the lives and writings of key exponents of these two orientations, we can help make tangible the very different sensibilities of these spiritual polarities and how they have shaped Western history. In particular, it is crucial to gain a deeper insight into the *jnani* temperament, as this so deeply characterises many of the Enlightenment thinkers. By also looking at the *jnani* sensibility as it arises in the East, we can discover its universality, and at the same time bring home the point that the great religions arise without recourse to the notion of a creator 'God'.

4.1 Piety and the West

4.1.1 The Language of Devotion

We mentioned earlier that neither 'God' nor 'religion' were words that translate into Chinese. In Jordan Paper's excellent survey of Chinese spirituality in *The Spirits are Drunk* he suggests that the Chinese (prior to secular Communism) never enquired of each other 'what is your religion?' – though Westerners assumed that this could be answered with one of Taoism, Confucianism or Buddhism. Rather, the question that people would ask was: 'to whom (or what) do you offer sacrifice?'[208] What is common to the shamanic and polytheistic worlds is the common act of propitiation. To offer sacrifice or to propitiate springs from the same spiritual impulse, and can be directed to spirits of all kinds, to the ancestors, to the more abstract deities of polytheism, or to 'God'.

We saw that Gershom Scholem held that Western religion, i.e. monotheism, was intrinsically an alienation from the shamanic world of nature and spirits; an essential gulf was created between the human and the divine. A 'God' emerged to whom the proper relationship was one of wholesome and humbling fear. Our account of the intolerant 'God' implied only the fearful qualities of a deity, the primacy of the term 'God-fearing'. There is no doubt that fear was an element in the relationship with the multiple deities as they emerged from their origins as wood-, water- and mountain-spirits. However the secular world has made of 'fear' only a debilitating limitation, rather than a constructive element in a relationship with unpredictable beings and forces of Nature. But sacrifice, offering, and propitiation are more than supplicatory; they are more than motivated by fear of what is bad and desire for what is good. They are part of the profound *connectedness* that is the root of the spiritual life. In shamanism that profound connectedness is to Nature, mediated by the spirits. In polytheism the focus shifts to a more human-centred world, mirrored by the anthropomorphic deities and their very human struggles: the profound connection is to a dramatic universe via its deities. In monotheism, as Scholem insists, the relationship to Nature is of no importance, rather it is to each human being, mediated through the one 'God' who is the wellspring of human history, and which becomes the stage on which his will is enacted. Each of these modalities of the spirit allows for a unique form of spiritual connectedness to be explored.

In Christianity the Hebraic form of monotheism evolved in a new way, with new conceptions of 'God' and a corresponding change in the central practice of prayer: it became an expression of spiritual love, or of the *bhakti* impulse. As Scholem, Buber, and Levinas would insist, prayer is a spiritual act of *relationship*, though in Judaism it is not primarily devotional. In the prayer of supplication and propitiation there might be a reaching out to the divinity simply to ameliorate suffering or improve conditions, for oneself, or for others, but in the prayer of devotion that longing is transformed into the language of lovers, the metaphors of the erotic. Supplication in this context becomes the supplication of all lovers, the passionate desire that their beloved return their love. The poetry and prose of the devotional mystics of Europe, Persia and India create a remarkable canon of the *bhakti* impulse, both drawing on secular romantic literature and feeding into it. Don Cupitt is very aware of these literary traditions, but as an obviously *jnani* thinker is puzzled by them, saying:

> It is not necessary here to dwell at length upon the familiar eroticism of wounds, of thraldom, bondage, and ravishment, for everyone knows that both in India and in the West spiritual writing makes extensive use of erotic imagery. What has not hitherto been sufficiently noticed or explained is the extent to which celibate male, as well as celibate female, writers unhesitatingly adopt the point of view and metaphoric of specifically female sexuality for devotional writings.[209]

Because Cupitt mistakes the writings for the mysticism, he cannot see that the metaphors are used because they are appropriate to an innate spiritual impulse of devotion, while the 'God' he is so worried about (is he real, isn't he real) is the metaphor for the objectless object of devotion. Cupitt knows the term *bhakti* (though not apparently *jnani*) and is again puzzled: 'It is a shock to discover that most Hindu piety has long been of the *bhakti* kind which is theistic and devotional.'[210] What Cupitt misses is that when cultural phenomena arise independently, then it cannot originate in language: it must be innate. In Persia the devotional mysticism of Jelalludin Rumi for example had neither the Christian nor the Hindu languages or traditions behind it: it arose from the innate presence of that spiritual impulse which arises in the heart, that core of the human being which is neither male nor female.

If Western philosophy since La Mettrie says that man is a machine, and continues its reductionist journey via the linguistic turn (adhered to by Cupitt), then the 'heart' has certainly disappeared as a philosophical term. For the reductionists the word stands for a biological pump in the ribcage, nothing more. Thinkers like Augustine, Pascal and Kierkegaard knew of the heart as the locus of Christian religiosity, but they became the exception. What on earth can we mean by 'heart' if it is not a pump? La Mettrie is delighted to tell us that a frog's heart moves for an hour or so after being removed from the body; if it stops then stimulation restores movement to 'that hollow muscle'. The head / heart division has been secularised: the Universities develop the head, while popular culture speaks of the heart in sexual romantic terms, but the heart is just a metaphor for emotion. Our discourse of spiritual love – *bhakti* – is doubly removed from the scientist, who dismisses the popular notion of the heart as 'Mills and Boon', and who, unlike Cupitt, has not even heard of devotional piety.

But for the devotional mystics it is the reverse. Their first reality is the heart in which their devotional impulse plays out the drama of spiritual love; they see romantic love as a doomed and shallow adumbration of that spiritual love, as does Kierkegaard. And the so-called 'physical' heart, the 'hollow muscle' that pumps blood around the body . . . *that* is the metaphor, a mere biological image which has no experiential correlate other than the pulse. The mystic *experiences* the spiritual heart directly, only learns of the pump in a biology class; and finds that is just a flimsy ephemeral idea around which no meaningful imperative for living can revolve. The knowledge of science is dry and empty, remote and irrelevant, compared to the lived experience of devotion, a lived experience that, in most religions, then flows into charitable work for the community.

The more atavistic component of the devotional impulse will always include the impulse to appeasement and supplication, a request for the intervention of the deity (whether for crops, children, freedom from oppression, or even to grant more 'spiritual' gifts), inherited from the polytheistic and shamanistic tradition. Prayer for intervention on the behalf of others is a leap forward in the spiritual life, and it is this generosity of prayer that might be called the middle ground of the spiritual life, recognisable in its ethical and moral dimension, but higher still is a prayer that wells up as a deep love and thankfulness, as an at-oneness, and eventually reaching the mystical heights that become problematic for the theistic religions.

The secular world often believes that the only relationship to 'God' is one of fear: even Einstein characterised Judaism in those terms. But fear lives on an emotional spectrum which moves from fear to awe, and from awe to wonder, and from wonder to thankfulness. Hobbes suggests that supplication is the asking of a gift from 'God', and that thankfulness then follows the granting of that gift, both of which he understands to be the proper way to honour 'God'.[211] But thankfulness, awe and wonder can arise with no immediate gift, or even in the context of loss, being rather a profound perception of the grandeur of life (an emotion that seems to continuously overwhelm Richard Dawkins, for example). This emotion *cannot* come from a reductionist mechanical science, because it arises from a response to the whole instead of through a division of matter into smaller and smaller parts. It can be understood as the true origins of prayer, offerings, sacrifice, and ritual.

4.1.2 Exemplars of the Devotional

We have seen that Christianity developed as a religion of simplicity and charity, and was scornful of science and learning. In the period of the thirteenth to the sixteenth centuries there culminated a flowering of Christian mysticism and piety, a unique contribution to the life of the spirit. In this section we will look at some representative devotional mystics of this period, in whom the Christian experiment reaches a peak, and then view one of the iconic figures of devotional mysticism, St Teresa, through the modern eyes of Jean-Paul Sartre.

The religious characteristic of this period is of self-abnegation, withdrawal from the world, and the formation of spiritual communities dedicated to the personal quest for 'God', the encouragement and support for others in that quest, and to helping the wider community. The practice of anchorage became common, meaning the attachment of a single monk or nun to a particular place, usually a small dwelling by a church. Only men were hermits, and wandered, while anchorites could be either men or women. After their 'service of enclosure', they did not leave their anchorage again. Monastic communities experimented during this period with a wide variety of regulations concerning the balance between the solitary and the social (as we have defined these terms). The broader community supported religious orders and anchorages on the basis of a consensual cultural piety. As Gershom Scholem reminds us, the Jewish mystics of this period were also caught up in a similar impulse to piety, including asceticism, altruism, self-abnegation and penitence. For the secular mind it is the twin qualities of self-abnegation and penitence that are the most foreign. We admire many figures for their selflessness and their contribution to public life, but, particularly since Freud, we see the concept of penitence as a 'guilt-trip' with its roots probably in sexual repression. To consider the medieval Christian period in an empathetic way we need to read the texts from the time alert to our contemporary prejudices. We now briefly introduce some of those devotional mystics and their writings: Richard Rolle (c1300-1349), Julian of Norwich (1342-1416), Catherine of Siena (1347-1380), St John of the Cross (1542-1591), and St Teresa of Avila (1515-1582).

Richard Rolle and the *Fire of Love*

Richard Rolle was born some forty years after the great German mystic Meister Eckhart. He studied at Oxford, but returned to his native Yorkshire without graduating, after which he ran away from home to become a hermit. He asked his sister to bring two of her cloaks and her father's rainhood to the woods, where he cut them up to make some semblance of a monk's habit. She thought him mad, but before she could raise the alarm, he ran off and was found later in a church praying. He was taken to the home of the Constable of Pickering Castle, and, despite his reputation for ruthless worldliness, the squire recognised in the boy his spiritual vocation.[212] He clothed

Rolle more suitably and provided him with food and shelter to begin his life as a hermit. The community, mostly involved in its business of prosperity and generation, nevertheless actively supported those whose spiritual vocation meant turning away from such pursuits. Even the ruthless held a worldview imbued with this piety.

It is only late in Rolle's life that his hermitship received official sanction, perhaps due to his outspokenness and unorthodox behaviour: his literary works seem to have been a form of automatic writing, and he was never canonised. Towards the end of his life he moved to a cell close to Doncaster, and became spiritual director to an anchoress named Margaret Kirkby. Here is a passage from his best known work, *The Fire of Love*:

> Nothing is better than mutual love, nothing sweeter than holy charity. To love and be loved is the delightful purpose of all human life; the delight of angels and of God, and the reward of blessedness. If then you want to be loved, love! Love gets love in return. No one has ever lost through loving good, if he has persisted in love to the end. On the other hand he does not know what it is to rejoice who has not know what it is to burn with love. So no one is ever more blessed than the man who is transported out of himself by the vehemence of his love, and who through the greatness of God's love experiences for himself the sweet song of everlasting praise. But this does not happen to a man overnight unless he has been converted to God, and has made remarkable efforts and has rejected every desire and worldly vanity.[213]

The essential medieval devotion is presented in these few lines: that love of 'God' brings the reward of love, and that its precondition is to renounce the world. Rolle was a solitary hermit, against action in the world of any kind, a position that later became the heresy known as 'Quietism' (associated with the Spanish priest Miguel de Molinos in the seventeenth century). Hermits of the late medieval period might perform all kinds of actions for the community, for example the repair of roads and bridges, but Rolle was not of this type. His 'charity' was *caritas*, the love of 'God'.

Julian of Norwich

Mother Julian of Norwich was an anchoress whose visions made her famous, and which she recounts in her *Revelations of Divine Love*. On 13 May, 1373, Julian was healed of a serious illness after experiencing a series of visions of Christ's suffering and of the Blessed Virgin, about which she wrote two accounts; the second, longer version was composed twenty or thirty years after the first. Her calling is therefore the result of a specific and locatable religious experience, which we do not find in Rolle for example. Early in her spiritual life she had asked for three favours of 'God': that she should have 'mind of his Passion'; that she should suffer an illness at the age of thirty; and that she should suffer 'three wounds'. These three are: contrition, compassion and wilful longing for 'God'. Long after these requests were forgotten they were granted during the period of her life-threatening illness. So severe was her condition that the priest had already read the last rites, and left her to look on the face of Christ on a crucifix he gave her. She then received sixteen visions or revelations, which are recorded in her book, and recovered fully from her illness.

The secular mind is quick to pounce on delirium (experienced in serious illness) as the explanation for religious experience, and Julian's visions may have had some connection with physical brain states brought on by fever. But nothing can account for the transformative effects on such individuals: a person may recover from fever unchanged, but not from the impact of such a profound religious experience. This was only first recognised as a legitimate argument in the West by Abraham Maslow, in his theorising of 'peak experiences'. However, his account, located within the tradition of transpersonal psychology, does not fully recognise that the devotional temperament, such as Julian's, makes such experiences a matter of spiritual love. Here is an extract from one of Julian's revelations:

> And he showed me more, a little thing, the size of a hazelnut, on the palm of my hand, round like a ball. I looked at it thoughtfully and wondered, 'What is it?' And the answer came, 'It is all that is made.' I marvelled that it continued to exist and did not suddenly disintegrate; it was so small. And again my mind supplied the answer, 'It exists, both now and for ever, because God loves it.' In short, everything owes its existence to the love of God.[214]

The secular mind regards this as delusional, the medieval mind as inspirational. The imagination must, of course, play a role in the specific imagery of Julian's revelations, but at the same time the poetic image here of the world as a hazelnut, and its continuing existence ensured by love, speaks eloquently to the religious mind. The medieval world that financially supported Julian in her anchorage, along with her salaried servants, was one that valued her inspiration as a way of maintaining the connection of the entire community with 'God'.

Catherine of Siena

As a child of six, Catherine of Siena experienced a vision of Jesus, described thus: 'The grace of this gift was so efficacious, that Catherine, beside herself, and transformed into Him upon whom she gazed with such love, forgetting not only the road she was on, but also herself, although naturally a timid child, stood still for a space with lifted and immovable eyes in the public road, where men and beasts were continually passing, and would certainly have continued to stand there as long as the vision lasted, had she not been violently diverted by others.'[215] Although remote from our time and secular worldview, the circumstances of Catherine's vision were not that different to the young Ramakrishna in nineteenth century India where he likewise succumbed to a roadside vision and had to be forcibly shaken out of it. Both lived to be amongst the world's best known devotional mystics. There are also accounts of Socrates 'beside himself' and unmoving in public places: in the Indian yoga tradition he could plausibly be described as being in the state of *samadhi* (more on this later).

Catherine became a tertiary (taking simple vows) of the Dominican order in 1363, joining the Sisters of Penitence of St Dominic in Siena. Here is a small section taken from her *Dialogues*:

> God was not wont to conceal, from the eye of her intellect, the love which He had for His servants, but rather to manifest it; and, that among other things, He used to say: 'Open the eye of your intellect, and gaze into Me, and you shall see the beauty of My rational creature. And look at those creatures who, among the beauties which I have given to the soul, creating her in My image and similitude, are clothed with the nuptial garment (that is, the garment of love), adorned with many virtues, by which they are united with Me through love. And yet I tell you, if you should ask Me, who these are, I should reply' (said the sweet and amorous Word of God) 'they are another Myself, inasmuch as they have lost and denied their own will, and are clothed with Mine, are united to Mine, are conformed to Mine.' It is therefore true, indeed, that the soul unites herself with God by the affection of love.[216]

The essence of this is that the souls of the mystics not only reach to 'God' and are united with 'God' through love, but they become 'another Myself'. To the extent that they have lost and denied their own will they become 'God', or a likeness of him. If we stand back from this, and bracket out not only the anthropomorphism of the religious language used by Catherine, but also deny any objective independent reality to 'God', then how can we understand the inner world of her experience? Scholem, Buber and Levinas insist on the 'other' and the relationship with that 'other', but the mystic goes beyond this, abandoning all separation: no relationship is now possible but identity. The modern mind will want to know who or what is this 'other' that is transcended here, if not the anthropomorphic 'God' imagined by the mystic? Has not Freud dismissed this as 'oceanic' and Jung dismissed this as 'inflation'? The only answer to these objections is: love. Once the specifics of the passage are allowed for as partly cultural and imaginative products, then the focus must be on the religious purpose they serve, to express the experience of spiritual love experienced at its most intense. Reason, whether secular and psychological, or spiritual and *jnani*, has no purchase here, despite the references in the above passage to 'intellect'.

St John of the Cross

It is in the mystical writings of St John of the Cross that we find the erotic imagery that so baffled Don Cupitt: what was a man doing writing this stuff? Yet male Hindu and Sufi mystics took this language much further, and the explanation is simple: for those who are so deeply committed to the spiritual life, their identity is not constructed from or against biological gender, as the modern mind insists on. Hence for a man to use the *metaphor* of a woman longing for her lover is quite natural, and is widespread in *bhakti* mysticism. It is the metaphor of female sexuality, not male, because the *bhakti* is passive, surrendered, and awaiting the lover, unlike the *jnani* who is active and deploys the will. This may not be politically correct in modern sexual terms, but is accurate in terms of the metaphoric structure.

St. John's classic work *The Dark Night of the Soul* begins with a poem called 'The Song of the Soul', just eight stanzas of five lines each. Here is one verse:

> On my blossoming breast,
> Alone for him entire was kept,
> He fell asleep,
> Whilst I caressed,
> And fanned him with the cedar fan.[217]

It is imagery like this that has led the secular mind to confuse devotional theism in male monks with homosexuality: both the Persian Jellaludin Rumi and the Hindu Ramakrishna have been misunderstood in this way. But St John's 'dark night' entered the language not so much for its devotionality, but for his account of the struggles and despair of those attempting to reach such heights of spiritual love.

Teresa of Avila v St Genet

St John of the Cross is associated with Teresa of Avila, who was assisted by him in the reform of the Carmelite order. She was born to pious parents, and from an early age had only one ambition: to be a nun, or even a martyr. When she played with other little girls, they built imaginary convents, as we learn from her autobiography, *The Life of the Mother Teresa of Jesus*. Her life of poor health, prayer, ecstasies and self-mortification set a standard for the period, epitomised in her statement that she had 'trampled the world beneath her feet'. Perhaps her best known work is *The Interior Castle*, in which she likens to the soul to a castle made of a single diamond with seven interior mansions representing seven clear stages of the soul in its approach to union with 'God'. (It is this structured approach of Teresa's as a devotional mystic that makes Rawlinson's use of the term 'structured' awkward in his scheme, forcing her to share the same quadrant as Rudolf Steiner, with whom she has nothing at all in common.)

Simone de Beauvoir told us that Sartre 'read the mystics' without saying which; it is clear that Sartre had studied at least Teresa of Avila in some depth. In his book *Saint Genet*, Sartre takes Genet's obsession with the word 'saint' and its application to his life, and sets it against the life of Teresa. There is a modern compassion at work here that is curious about the 'saint' but finds intolerable the idea that the 'sinner' is thus created as its antithesis, as the one who must suffer eternal damnation. St Catherine, St John and St Teresa are exemplars who are impossible to emulate, as Karen Armstrong found, and which realisation precipitated her departure from convent life. But Genet and Sartre propose that the degradation of the thief and that of the mystic saint have something in common, if it is *willed* at least. Sartre's analysis is instructive because it shows how the social secularism of Marx produces the most absurd of outcomes from its premises.

Jean Genet (1910-1986) was a French author who started as a criminal and social outcast, ultimately saved from a life sentence in 1948 by French intellectuals, including Sartre. Sartre's analysis of Genet's life pinpoints the moment of crisis at the age of around ten, when the orphan boy's peasant foster-parents caught him stealing. Up to that point he was a paragon of virtue, with a quite religious bent of mind, but in a certain sense the boy 'died' in that moment, and from then on he lived the eternal recurrence of that moment of death. The subsequent experience of a notorious reform school – that became the material for his novel *Miracle of the Rose* – and his life of thievery and male prostitution played out his obsession with the idea of 'saint'. Genet called this word the most beautiful in the French language.[218] But Sartre approaches the issue from his Existentialist-Marxist background, saying: 'The phenomenon of saintliness appears chiefly in societies of consumers.'[219] He explains this briefly in standard Marxist terms, effectively reproducing a shibboleth of left-wing intellectualism: that commodities are arbitrary objects of an alienated mode of production. In this light, Genet as thief 'appropriates' commodities whose ethical status in the first place is compromised by their location in bourgeois society. But it is the continued practice of theft as the willed ground of his degradation and rejection that Sartre understands to have common ground with the saint.

In the chapter in *Saint Genet* where Sartre considers sainthood in general and St Teresa in particular, he chooses a quote from St John of the Cross as its heading: 'To succeed in being all, strive to be nothing in anything.' This goes to the core of medieval Christian mysticism, a statement of the *via negativa* that would be at home in any Buddhist tradition, in particular the Mahayana which had a precise term for this 'nothing': *shunyata*. But for Sartre the parallels are clear: the saint, like the thief, consumes without producing. The saint like the thief is 'nothing' – beyond respectability, and beyond utility to society. And both have something of the numinous about them: society is fascinated by them. However, this does not apply to the ordinary thief, but only to a man like Genet who *wills* his degradation, just as the saint wills poverty, hunger and censure.

The mystic certainly courts censure from society, even torture and death. St John of the Cross was imprisoned, and although released and made head of a religious order, he was later discharged from it and humiliated. Sartre understands this as the 'concerted practice of humility',[220] but regards it as a sham with the saints, and genuine with Genet. In Teresa's case he points out that whatever setbacks and criticism she suffered, 'the starting point is *general esteem*.' Sartre is right in a sense because the medieval period, however violent its Inquisitions were to the heretic, was built on the consensus of piety. Teresa had the specific esteem of her religious superiors and the general esteem of her age, even if she shocked them with the extremity of her devotional self-abnegation.

In Sartre's time however there is no *general esteem* for piety, so

Sartre can proceed with the construction of an inversion of the whole concept of sainthood, much as *Nausea* represents mysticism as a pathological inversion of its original impulse. At the same time he powerfully conveys the modern scepticism in its full moral strength: what of the *ordinary* person? The practice of humility for Sartre is a sham in a society of consumers, whereas theft is a courageous act which wills the opposite of esteem, which makes the thief a true renunciate. While Teresa suffered the privations that even the poorest of peasants were unlikely to experience, she luxuriated in the esteem of her peers and the approval of her age. Genet, in the act of consuming the appropriated commodity, makes the true sacrifice: he places himself beyond humanity itself. Teresa and all the saints occupy the moral high ground, smug, self-absorbed and impossible to emulate: as exemplars they humiliate not elevate.

This at least is Sartre's analysis. He is keen, as we are, to understand Teresa in the historical context, but his secular Marxism allows only for an understanding of history in terms of the modes of production: feudalist then, capitalist now (both apparently equally 'societies of consumers'). These are useful as far as they go, and predicated on a compassion for ordinary people and a loathing of privilege, but are rigorously secularised. Teresa might be smug to us now, Catherine reprehensible in her call to Crusade (which shows that in respect of prejudice against Islam she certainly did not rise above her age), Julian delusional, and Rolle merely useless as a hermit who neither built nor maintained bridges that would prevent the common folk from drowning. But Genet (remarkable and honourable as he is in some ways) and Teresa cannot be confused other than through a category mistake: not the categories of saint and sinner, but in the categories of *impulse*. Despite Genet's preoccupation with sainthood, he shows no spiritual impulse, which we have defined as that which seeks a deeper connectedness. In the various modalities of the spirit that we explore in this book, this connectedness plays out in different ways; for the devotional mystic its core is love. Rolle says: 'If you want to be loved, love!' He recognises that the volitional aspect of the mystic requires that the first step comes from the self – then 'God' takes ten steps towards one. But Sartre demonstrates that Genet starts with the certainty that he cannot be loved, and so his volitional step is to betray each gesture of friendship: starting with the peasant couple who gave him a home and sustenance. Each horrified reaction from society then validates in him who he is: a thief.

The impulse of Teresa, which is love, leads to an abandonment of self and the profound connectedness in 'God'. The impulse of Genet, which is self-hatred, leads to the definition of self over and against society and the profound alienation from it as an extension of self-alienation. He had no impulse of connectedness, quite the reverse, in a culture that had lost the religious language of connection and replaced it with a Marxist language of economics or a Freudian language of pyschology.

4.1.3 Augustine and Aquinas

Devotion is sometimes seen as the religious impulse of the uneducated and of women (Hume was not the least embarrassed to express this prejudice, as we saw). Yet it was one of Christianity's greatest early male intellectuals, St Augustine, who set the devotional tone for its development. If Augustine stands at the beginning of Christianity as a major world religion of a largely devotional nature, then St Thomas Aquinas stands at its first great moment of transition, where the intellectual, or *jnani* impulse, begins to re-assert itself. Putting it another way, Augustine represents the continuity, through St Paul, with the Hebraic, while Aquinas represents the opening up again to the Hellenic. Hence St Augustine (354-430) and St Thomas Aquinas (1224/25-1274) represent key stages in the development of Christianity. Augustine is often credited with bringing the Platonic stream of thought into Christianity, but the fact is that he quite rejected the 'Platonists'. Aquinas is writing some 850 years after Augustine, towards the end of the period termed the 'Dark Ages' by Renaissance thinkers; in contrast to Augustine he brings Aristotle into Christianity.

Augustine

We can see in historical terms that Augustine was born only 51 years after Constantine adopted Christianity, but his writings as a Christian show confidence of its centrality to the world he lived in. The New Testament was still being moulded by the various Councils into its modern shape, but Augustine is principally influenced by St Paul, and credits his conversion to Paul. As we saw earlier Augustine was for nearly ten years a Manichaean, and in his later writings attacking the doctrines of Mani, he acknowledges its continued influence on him. Augustine is best known for his *Confessions*, and even secular thinkers find the philosophy in it of interest (his thinking on time is highly influential), and his frankness in autobiography commendable. Rousseau modelled his own famous *Confessions* on Augustine. Postmodern writers have also drawn on him: Derrida in his *Circumfessions* and Lyotard in the posthumously published

Confessions of Augustine. Wittgenstein takes Augustine's brief account of language in the *Confessions* as the starting point for his *Philosophical Investigations*. However a more important work in terms of his influence on his time and probably the whole of the Middle Ages was his *City of God*, a polemic against those who claimed that the sack of Rome in 410 was the result of abandoning the pagan gods.

Augustine is very concerned about matters of Christian doctrine, and he also writes some good philosophy, according to Bertrand Russell.[221] Russell makes the point that Augustine's work is to reconcile Platonic ideas, particularly on cosmology, with the Old Testament and with St Paul, though Russell naturally praises the more purely philosophical thinking. However, of interest to us is how even the most philosophical of sections are broken up with devotional passages. Book XI of *Confessions*, the section that enquires into the nature of time, is partly the development of an argument and partly devotional, regularly breaking off from the sequence of thought to praise the Lord and to confess Augustine's inadequacy in confronting the problem. 'My mind is burning to solve this intricate puzzle', he says half way through,[222] and then continues:

> O Lord my God, good Father, it is a problem at once so familiar and so mysterious. I long to find the answer. Through Christ I beseech you do not keep it hidden away but make it clear to me. Let your mercy give me light. To whom am I to put my questions? To whom can I confess my ignorance with greater profit than you? For my burning desire to study your Scriptures is not displeasing to you. Grant me what I love, for it was your gift that I should love it. Grant me this, Father, for *you know well enough how to give your children what is good for them.* Grant me this, for *I have set myself to read the riddle, but it will prove a hard search* until you reveal it to me. Through Christ I beseech you, in the name of him who is the Holy of Holies, let no man stand in my way. This is my hope and for this I live, that I may *gaze at the beauty of the Lord. You have measured my years with a brief span;* they pass away; how they pass, I do not know.

Augustine then returns to his inquiry into the nature of time. In the above passage we can detect an innocence of enquiry – he is 'burning' to solve the problem of time, but actually his piety comes first. He is in a state of profound *trust* in his 'God', such that one cannot help but see the image of a child, likewise burning to know the answer to a question, but deep down much more taken up with the pleasure of his or her father's company. The prime motivation is not philosophy but communion, a being-with that rests on the mutually enjoyable occasion of a question put in good faith. Should it prove that he is not up to the task he has set himself, or that 'God' should withhold the grace of an answer, he will not be set back, his faith will grow whatever the outcome. As a child ready for sleep at the end of the day, what counts was the *occasion*, not the answer.

But Augustine *does* also have the mind of an intellectual and a scientist. In Book V of the *Confessions* he details his disillusionment with Manichaeism and his casting around for a true religion (as mentioned in Chapter Three). Augustine had studied the scientific accounts of astronomy and knew that certain events such as eclipses could be predicted with some accuracy, yet he found the teachings of Manes on these subjects 'extremely incoherent'.[223] For nearly the whole of the nine years that Augustine was a Manichaean he eagerly awaited the arrival of leading Manichaean bishop Faustus to resolve this quandary. But Faustus was barely less confused than Manes on astronomy, and his inability to settle the scientific question marked the turning point for Augustine. Once he was convinced that Manes was ignorant on points of science, he concluded that he must also be wrong on religion. As Augustine puts it: 'But who asked that any Manichee should write about science as well as religion, when we can learn our duty to God without a knowledge of these things?' Although Augustine ranks science way below religion, Manes had made the mistake of attempting to speak authoritatively on science, when he could have remained silent on the subject.

Augustine then cast around for a teaching that he could commit himself to, and in this period he read 'some of the books of the Platonists'. Whether he read Plotinus, the key Platonist of the previous century, is unclear, but Augustine is adamant that what he found did not ultimately satisfy him, eventually equating their teachings with idolatry.[224] However the Platonists gave him a glimpse of 'God', which Augustine explains with a quote from Corinthians 11:19 'I *caught sight of your invisible nature, as it is known through your creatures*.'[225] Whether this was the Plotinian 'One' we do not know, but Augustine, looking back on this period considers that the knowledge he gained from the Platonists 'only bred self-conceit'. He says: 'I used to talk glibly as though I knew the meaning of it all, but unless I had looked for the way which leads to you in Christ our Saviour, instead of finding knowledge I should have found my end.' This is a very clear statement that for Augustine religion is not about knowledge. He then makes it clear what religion *is* about: 'But how could I expect the Platonist books would ever teach me charity?'[226]

The key word for Augustine is *charity* which here has the meaning of religious love, or *agape*, or in our own terminology the love at the core of the *bhakti* impulse. This was missing both in the language of the Manicheans and in the language of the Platonists; in other words they had too strong a *jnani* emphasis. But Augustine is not as extreme as the exemplars of the devotional that we just explored, so we can usefully apply our *bhakti* checklist to see where he lies on the spectrum. He fills the first requirement of the emphasis on **love** and the **devotional impulse**, as is seen throughout the *Confessions*, perhaps best pictured in the image of Augustine as devoted child to 'God' as his loving father. His **devotional behaviour** is not however of the demonstrative type; he does not show visible signs such as swooning or ecstasy, rather his writings can be seen as a devotional act. His **tendency to anthropomorphism** in his understanding of 'God' is everywhere visible in the manner in which he addresses his 'God'. His **emphasis on submission and surrender** is also everywhere present in the *Confessions*, though again it does not take the extreme form of the later devotional mystics. He does not often use **sexual metaphors** (apart from the Church as bride); his activity lies not so much in **preparation for the 'beloved'** as in administrative zeal; and he is not **drawn to the esoteric** (as his distaste for astrology shows). On nearly all counts then, we should consider Augustine as a *bhakti*, though not in the extreme. His one *jnani* trait, his inclination to learning, is always subordinate to his devotional.

Europe and the North African regions that formed the core of the Roman Empire at the time of Augustine were still a melting-pot of religious ideas in which Christianity was yet to totally dominate. It is obvious that Augustine could choose freely between Manichaeism, Neoplatonism and Christianity. What our analysis additionally offers is an understanding of *why* he should have chosen Christianity: because his personal impulse lies more to the *bhakti* end of the spectrum than the *jnani*. But the tragedy of Augustine is a legacy of intolerance. We pointed out that the word heresy comes from a Greek word meaning to *choose*. Where Augustine could freely choose his religion (based on his own personal impulse), his writings and activities as a Christian bishop ensured that heresy – the act of choosing – would become punishable for others. It is in *City of God* and *Christian Doctrine* that Augustine lays out the blueprint for the development of Christianity, and in neither of these books does Augustine ever contemplate the extension of his own freedom of choice to others. The dreadful doctrine is endorsed, that Christianity is the only *true* religion.

Thomas Aquinas

Thomas Aquinas studied under the leading Aristotelian scholar Albertus Magnus (scientist, philosopher and alchemist) in Paris, and joined the Dominicans whose members were suspected of the 'Averroist heresy'. While studying in Paris Aquinas was concerned to undo the harm done to Aristotle's standing in the Christian world, and succeeded in persuading the Church that Aristotle's was in fact a better system than Plato's, and that the Averroists had misunderstood Aristotle. Averroës (1126-1198), or to give him his Arabic name, Ibn Rushd, was a scholar who wrote extensive commentaries on Aristotle and which were of considerable influence in the Christian world. Although Plato had lived through the Middle Ages as dimly refracted through the Church Fathers and Augustine, Aristotle only reappeared through the Arab world and in particular through Averroës.

But the revolutions in Christian thought that would arise with the new interest in the Greek leaning towards the empirical and the rational were resisted at first. Despite Aquinas's seeming success in integrating Christian piety and Aristotelian rationality, his work attracted censure after his death in 1274. In 1277 the Church included twelve of Aquinas's innovations in thought as heretical, but the mood of Europe's thinkers had been shifted towards the rational and could not be stopped. Still supreme in its confidence in the fourteenth century, the Catholic Church gradually moved to absorb the new ideas and in 1323 Aquinas was canonised as a saint. By the nineteenth century the Roman Catholic Church had made Thomist philosophy supreme after the Platonist revival in the Renaissance period, leading Russell to complain that while Roman Catholics interested in philosophy in the seventeenth century were free to follow Plato, 'such freedoms are things of the past.' To criticise Aristotle in matters of philosophy is 'almost impious' for twentieth-century Roman.[227] However, we saw earlier on that it was just as impious for Galileo to do so in the seventeenth century, whatever Russell asserts.

To understand Aquinas in terms of Aristotle may have value, but to understand Augustine in terms of Plato is misleading – in terms of religion at least. Augustine owes all to St Paul, not to Plato. But Aquinas *can* be associated with the Greek mind of Aristotle, because Aquinas rejects Tertullian's plaintive dictum that 'Christianity is of Jerusalem and the Church, not of Athens and the Academy' (to paraphrase him). It is the word 'Academy' that is crucial here: everything that was rejected in the Greek mind, and symbolised by the closing of the Platonic Academy, throughout the 'Dark Ages', was re-emerging in the time of Aquinas. It is no surprise at all that the Church resisted it.

Matthew Fox and Aquinas

We can look at the mind of Aquinas, and his spiritual impulses, through the work of contemporary radical theologian Matthew Fox. This will help us not only articulate the *bhakti* / *jnani* distinction between Augustine and Aquinas, but, just as crucially, the *via negativa* / *via positiva* distinction. Matthew Fox, founder of Creation Spirituality, was born in Madison, Wisconsin in 1940 and joined the Dominicans in adulthood. In 1977 he founded the Institute of Culture and Creation Spirituality based on the ideas in his published books, but, in the mid 1980s the Vatican investigated his institute and ordered him to silence from the end of 1988 through most of 1989. During this period Fox went to South America and visited the Leonardo Boff, a Brazilian proponent of liberation theology. Subsequent books by Fox led to him being expelled from the Dominican order in 1993. (In fact, scores of other leading theologians and priests besides Fox were silenced or expelled around this time by Cardinal Ratzinger in the purging of dissent while he was Head of the Congregation of the Doctrine of the Faith, including Leonardo Boff and Hans Kung, a German theologian.) In 1999 Fox joined the Episcopal Church.

Fox's key work on Aquinas is called *Sheer Joy – Conversations with Thomas Aquinas on Creation Spirituality*. The introduction is titled 'Thomas Aquinas's Spirituality and the Emerging Renaissance' which is a helpful pointer to what Fox considers to be central to Aquinas's thought – a re-awakening of engagement with creation via arts and science. The rest of the book is structured around what Fox calls the 'Four Paths of creation spirituality', devoting a section to each path, comprising imaginary question and answer sessions with Aquinas. Fox poses questions that relate to one of the four paths, and selects suitable responses from the vast writings of Aquinas. The resulting dialogues naturally reveal as much of Fox's thought as it does of Aquinas's. Fox's four paths are of interest to us, because the first two are named the Via Positiva and the Via Negativa; the remaining two are named the Via Creativa and the Via Transformativa.

Just as we have done in this book, Fox has taken the terms *via positiva* and *via negativa* in theology, which are normally synonymous with cataphatic and apophatic respectively, and has slightly redefined them. His short definition of the Via Positiva is 'experiencing the God of delight and joy', and the Via Negativa is 'experiencing the God of darkness, letting go, and suffering.'[228] It is telling that elsewhere he accidentally capitalises 'Cataphatic', while failing to do so with 'apophatic', in the same sentence.[229] Fox is thoroughly pre-secular in his religiosity, but thoroughly modern in his passion for the manifest world, hence the apophatic or negative theology for him is problematic. We could say that for contemporary theists like Fox it is natural to understand modernity as ushered in at the time of Aquinas, while for contemporary secularists it is natural to understand modernity as ushered in at the time of Voltaire. But Fox has quite instinctively grasped the revolution in Christian thought that Aquinas heralded. Using our terminology and definition, it was a shift from *via negativa* to *via positiva*. The use here is somewhat different to Fox's however. From his imaginary conversations with Aquinas, it is clear that Fox is nervous of what he defines as the Via Negativa, and falls perhaps into a linguistic trap of associating negative emotions with it. The apophatic, as so beautifully expressed by St John of the Cross, has nothing to do with negative emotions, but to do with stripping away the created world, until in the 'Dark Night' (or the 'Cloud of Unknowing' to use the metaphor of another medieval mystic), he is utterly alone with 'God', the object of his spiritual love. His emotions of the *via negativa* are entirely positive, and of a spiritual kind, whereas his negative emotions derive only from the *struggle* to attain to the 'divine darkness'. From the perspective of those truly on the *via negativa*, suffering is born precisely out of the manifest world: the Buddha gives an entire lecture for example on the theme: 'suffering is born from those we hold dear.'[230] Hence for the Buddha and for Catherine of Siena for example the negative emotions arise on the *via positiva*, not the *via negativa*. Or at the very least they could not ascribe them solely to the *via negativa* as Fox does.

But Fox's work only helps to highlight how radical the thinking of Aquinas was: he was the first mainstream Church writer to argue that the manifest world was in itself a path to 'God' rather than the obstacle that the ascetics, recluses, anchorites and mystics of the period believed it to be. The key dictum of Aquinas on this says it all: 'Revelation comes in two volumes – that of nature and that of the Bible.'[231] What this means for Fox is the return of cosmology as a valid religious topic, one that Augustine had considered to be incidental, and a failing of the Manicheans. Both Augustine and Aquinas as good Christians devote much of their writings to the demonstration of how wrong other religions or creeds are, but Aquinas is beginning to show a conciliatory nature that marks the nascent spiritual pluralism of the later Renaissance. Fox demonstrates this with a selection of quotes from the work of Aquinas:

> Every truth without exception – and whoever may utter it – is from the Holy Spirit.
> The old pagan virtues were from God.
> Revelation has been made to many pagans.[232]

The Hellenic is 'pagan' and cosmology is 'pagan', but these re-asserted themselves with Aquinas. That Aquinas and Fox have a profoundly *via positiva* instinct, in the way that we define it, is easy to demonstrate, though Fox's is modern: it includes Goddess spirituality, sexuality, and ecology. Not quite so obvious in both men is their impulse to *jnani* rather than *bhakti* because the simplicity of the pure *jnani* expression is submerged in erudition. What is obvious by its absence in the writings of both men is any expression typical of the *bhakti* orientation that we have seen so vividly in the great devotional mystics living around the time of Aquinas.

While neither Augustine is a highly developed *bhakti* like for example St John of the Cross, or Aquinas a highly developed *jnani* like, for example, his near contemporary Meister Eckhart, their opposing orientations mark out an epochal difference. To direct again religious attention to the material world, arts, and science, was a sea-change in attitude. Augustine, whatever his interests in astronomy and science might have been, makes it abundantly clear that these must be subordinated to piety, or even best forgotten. He says:

> Whenever I hear a brother Christian talk in such a way as to show that he is ignorant of these scientific matters and confuses one thing with another, I listen with patience to his theories and think it no harm to him that he does not know the true facts about material things, provided that he holds no beliefs unworthy of you, O Lord, who are the Creator of them all.[233]

This idea still holds sway amongst Christians today, and is what Richard Dawkins finds impossible to understand: it does a person no harm to be ignorant of scientific facts. One can digest a sandwich without knowing the enzymes involved, and one can drive a car without knowing the chemistry of internal combustion. Similarly, why should the love of 'God' be contingent upon knowledge of the date of the creation, or any other cosmological issue? But Fox demonstrates very well how the impulse to awe and wonder at the natural world is indeed a spiritual one, even though it is foreign to the earlier part of Christian history. Fox's work is also particularly valuable where he makes clear the nature of religious happiness, drawing on Eckhart as well as Aquinas. His work is a significant attempt to correct the secular impression that Christianity dwells on original sin, and that religion is necessarily gloomy and world-denying: the very title of one of his books, *Original Blessing*, indicates this radical stance.

4.1.4 Christianity as a *Bhakti* Religion

We are now in a position to assemble further evidence that Christianity became a religion with its centre of gravity in a *bhakti, via negativa,* anti-esoteric cluster of impulses. An Eastern perspective on this can be useful. Dion Fortune (1890-1946), an esotericist with connections to the Theosophists and a wide understanding of Eastern ideas, clearly understood the *bhakti* element within Christianity:

> The Bhakti Yoga of the Catholic Church is only suited for those whose temperament is naturally devotional and who find their readiest expression in loving self-sacrifice. But it is not everybody who is of this type, and Christianity is unfortunate in not having any choice of systems to offer its aspirants. The East, being tolerant, is wise, and has developed various Yoga methods, each of which is pursued by its adherents to the exclusion of the others, and yet none would deny that the other methods are also paths to God for those to whom they are suited.[234]

Fortune uses the word 'yoga' to mean a spiritual practise (as Ken Wilber does), and points out the pluralism of the East, lacking in Christianity. However, any major religion has to have some flexibility, and as we saw with Aquinas, Catholicism, though initially hesitant, began to show some tolerance towards a more *jnani, via positiva* approach leading into the Renaissance.

Karen Armstrong says (in the context of Indian religions): 'It is not difficult to see that this *bhakti* (devotion) to the Buddhas and the *bodhisattvas* was similar to the Christian devotion to Jesus. It also made the faith accessible to more people, rather as Paul had wished to make Judaism available to the *goyim*.'[235] Paul's role in making love central to the new religion is also put forward by Mircia Eliade in his *History of Religious Ideas*, citing a well-known passage from I Corinthians 13:

> Though I speak with the tongues of man and of angels, and have not charity, I am become *as* sounding brass or a tinkling cymbal. And though I have *the gift of* prophecy, and understand all mysteries, and all knowledge; and though I have all faith,

> so that I could remove mountains, and have not charity, I am nothing. And though I bestow all my goods to feed *the poor*, and though I give my body to be burned, and have not charity, it profiteth me nothing.

Eliade considers this to be a 'summit of Pauline thought'.[236] Note that this is quoted from the King James Bible of 1611, and that the passage refers to charity. As we suggested earlier, the word 'charity' in 1611 held a different meaning than today, as is obvious when Paul talks about bestowing goods to feed the poor, yet not having charity. Hence the Revised English Bible uses the word 'love' instead, which is clearly more appropriate for a modern audience. Armstrong's point was also that a devotional religion had more appeal to the masses, as was the case with certain developments in Buddhism. In fact Paul's concern for the welfare of the growing Christian community is evident in his letters, and his continual exhortations for the community to support each other with the material necessities of life, as a practical acting out of their faith, became central to the new religion. A *welfare system* as central to church activity was deeply appealing, both to the needy and to the generous of heart.

The religious scholar Huston Smith, in his classic *The Religions of Man* says: 'All the basic principles of *bhakti* are richly exemplified in Christianity. Indeed from the Hindu point of view Christianity is one great brilliantly lit *bhakti* highway toward God, other paths being not neglected but not as clearly marked.'[237] Smith is one of the few Western scholars to not only understand and use the term *bhakti* in this sense, but also to refer to its opposite spiritual impulse, *jnani*, making clear that the conception of 'God' in these two paths will be quite different. Don Cupitt, an approximate contemporary of Smith is patently not so able to view Christianity from the outside, but nevertheless does understand *bhakti* as a term applicable to Christianity, saying that 'It seems that Saivite love-mysticism [*bhakti*] and Christian devotion could reach much the same religious experience from very different doctrinal and psychological starting points.'[238] We are saying that, while the cultural location and doctrine may represent different starting points, *bhakti* represents a universal impulse, and to the extent that we can use the secularised concept of 'psychology' at all in this context, the Hindu *bhakti* and the Christian devotional mystic *do* start from the same inner or psychological stance. Where Smith's understanding goes much deeper than Cupitt's is that he accepts this idea, and further accepts that to properly understand *bhakti*, we need to understand its counterpoint *jnani*.

It is in the rejection of the Platonists that Augustine sets the *bhakti* direction for Christianity, as he makes clear in the musings he directs to 'God': 'So I seized eagerly upon the venerable writings inspired by your Holy Spirit, especially those of the apostle Paul. . . . None of this is contained in the Platonist's books. Their pages have not the mien of the true love of God.'[239] Augustine's words are poignant: he is listening to the *feeling tone* of the Platonist works, and finds them lacking in spiritual love.

Augustine's joyous balance of spiritual impulses, though markedly in favour of *bhakti*, extends to the issue of *via positiva / via negativa*. The struggle against Gnosticism in the writings of the early heresiologists is irrevocably tainted for the modern mind by the invective and even hatred expressed towards their doctrinal enemies, and of course this translated into imprisonment, exile, torture and killing. But behind ignorant bigotry there was a genuine attempt to find a balance between the extremes of world-denying renunciation on the one hand, and world-indulging hedonism on the other. St Paul calls the latter 'carnally minded' and is extreme in his views on this (Romans 8:6-7): 'For to be carnally minded *is* death, but to be spiritually minded *is* life and peace. Because the carnal mind *is* enmity against God: for it is not subject to the law of God, neither indeed can it be.' Paul's position here is Gnostic, *via negativa*, and went very deep into the Christian mind.

4.2 Jnani, *the East, and Hellenic Influence*

4.2.1 The Spiritual Language of the Non-Devotional

For the secular mind, the *bhakti* devotional impulse may be utterly foreign, but its description here is probably recognisable to some degree within a conventional secular account of religion: the image of a monk or nun kneeling in prayer is obviously devotional. The *jnani* non-devotional impulse, which has been placed so centrally in the arguments of this book, is a much harder notion to convey in the Western context. Hence this section is crucial to the argument that follows. Without some active empathy for the *jnani* impulse, it is not really possible to understand the basis from which the bulk of the Enlightenment thinkers – whose revolution in thought was made possible by the radical turn of Aquinas – understood religion.

We have noted that the *jnani* will fiercely resist the anthropomorphism that creeps in or even dominates the *bhakti* conception of 'God' – more than this: the *jnani* may do without 'God' altogether. To obtain a clearer picture of this we need to draw on Eastern traditions such as Yoga, Buddhism, Advaita and Taoism. But we pointed out earlier that the Western religious and spiritual thinkers are ambivalent about the religious and spiritual East, if not downright hostile. Those who consider Christianity to be the true religious heritage of the West (even if they have some knowledge of the East) often reject the East: as well as C. G. Jung and Rudolf Steiner we can also include Matthew Fox here. Those with more of a New Age leaning, such as Ken Wilber, believe that Eastern insights must be integrated into Western spirituality. Here, we are not necessarily seeking integration, but rather a perspective from which to answer some of the puzzling problems of the emergence of the secular mind.

Yoga

Yoga is one of the oldest non-devotional traditions in India, and its chief text is the *Yoga Sutras of Patanjali,* possibly about 2,000 years old. The book is a systematisation of much older knowledge. The Yoga of Patanjali is technically known as Kriya Yoga, to distinguish it from other 'yogas' or spiritual practices, for example the Hatha Yoga familiar in the West as a set of physical exercises. The *Yoga Sutras* are written in Sankskrit in a highly condensed or aphoristic form, making translation difficult, but it clearly describes a practice of disciplining the mind in order to attain *samadhi,* a form of ecstatic absorption that in other Hindu traditions are variously assumed to be or lead to liberation, *moksha, nirvana* or enlightenment. Its methods are inward, mental and non-devotional, though typically for the Hindu instinct for pluralism and tolerance it acknowledges that the same result can be achieved by devotion to the guru or to the 'Lord' (Ishvara).[240] However, while *bhakti* is acknowledged as a legitimate path, it is only mentioned so as to dispose of the question: the text is entirely *jnani*. Without going into details of the Kriya Yoga system we can apply our *jnani* checklist. Firstly, **compassion** is not emphasised; but there is strongly **an impulse towards knowing, 'seeing,' knowledge, wisdom, and learning** (though not learning in any academic sense). **Enquiry, possibly doubt, as core behaviour** is not that obvious, but there is enormous **emphasis on the will**. It is **instinctively anti-anthropomorphic**, but the issue does not really arise except in the passing mention of Ishvara (where the Hindu conception of 'the Lord' is made problematic by the assumptions of Western monotheism). **'Infinite' and 'eternal' are key words** which do not appear that much, perhaps because the system is presented more as an instruction for practice than a philosophy, but the text is greatly **drawn to esoteric knowledge and practices**.

In this light, Patanjali's *Yoga Sutras* can be understood as an extreme *jnani* system, with a leaning towards the esoteric that sets a crucial early example of the relationship between *jnani* and the *esoteric,* one that greatly influenced the esotericists of the twentieth century (the Theosophist Alice Bailey wrote an extensive esoteric commentary on the *Yoga Sutras* for example[241]). The first two chapters of the *Yoga Sutras* set out the method for attaining the goal of ultimate *samadhi* (known as 'unseeded', i.e. in which there are no seeds of disturbance that could cause a reversal of the exalted state). It is only in the third chapter that esoteric 'powers' are listed as arising from the clarity and power of the mind that has successfully undergone the training, and the text does not suggest that these are part of the *path.* On the other hand, neither does it suggest, as other *jnani* traditions do, that exercise of these powers might be problematic, or a temptation for the unwary: the text is as neutral as a science manual.

Buddhism

The original teachings of the Buddha are not represented in a single text like the *Yoga Sutras,* and he remains, along with Patanjali and Jesus, a historically unrecoverable figure. The extensive texts known collectively as the Pali Canon are *probably* the closest to the teachings of the Buddha, but were not written down until hundreds of years after his death, making them more historically remote from the Buddha than even the gospels are from Jesus. However, none of that concerns us here, rather that the Pali Canon represents a fine example of an exclusively *jnani* teaching. In one of its major texts the *Majjhima Nikaya* (The Middle Length Discourses of the Buddha), the concept of *bhakti* is not mentioned even once, and the only time that the Buddha uses the word 'love' is when he makes it clear that the monks may have faith and love for him, but it is not as devotion towards the guru that would be found in a *bhakti* system.[242] The Buddha's *jnani* teachings appear to owe nothing to Yoga, and he is generally dismissive of the Brahmanic tradition of his day (as he dismisses all rivals). The Buddha's personality appears through this text as immensely serene and courteous on the one hand, but masterful and overweening on the other: a 'lion roaring in the forest'. In *The Shorter Discourse to Saccaka* the rueful Saccaka tells us that he was 'bold and impudent' to debate with the Buddha; a man might as well attack a mad elephant, or a blazing fire, or a terrible poisonous snake, and expect safety.[243]

The central goal of the Buddha's system is *nirvana,* not as in popular Western consciousness a kind of paradise or state of bliss, but etymologically derived from 'to extinguish'. It is the separate egoic self that is extinguished, or in the Buddha's own words it is the 'end of I-making tendencies',[244] which is the same goal as, for example, expressed in the work of Teresa of Avila, only the method varies. We can assess the Buddha's system according to our *jnani* checklist. Firstly, **compassion** is a core feature of the Buddha's teachings; and there is strongly **an impulse towards knowing, 'seeing,' knowledge, wisdom, and learning**, especially 'insight' (encapsulated in the Buddhist term 'vipassana') though emphatically not learning in any academic sense. **Enquiry, possibly doubt, as core behaviour** is evidenced partly by his own journey of discovery, and also in that he calls his monks 'enquirers'. There is some **emphasis on the will** as part of a wholesome discipline, but insight has a higher priority. It is **instinctively anti-anthropomorphic**: the issue simply doesn't arise, and as Martin Buber wryly comments, the Buddha makes the 'gods' his pupils.[245] **'Infinite' and 'eternal' are key words** which appear in homologues like 'imperishable', 'stainless', and so on; his key preoccupation being with the issue of impermanence and its transcending. The Buddha is **drawn to esoteric knowledge and practices** only in a very limited sense that he assumes, like the Yoga Sutras, that attainment brings esoteric powers in its wake. In particular the ability to recall past lives is assumed by the Buddha to be a sign of imminent attainment, a dubious extrapolation from his own experience. However, unlike the *Yoga Sutras*, he makes it plain that the spirit world is occupied by beings with a less favourable chance of enlightenment than those incarnate in a human body. Hence the intercession of spirits or 'gods' is less than useful, and the seeking of esoteric powers or interventions discouraged. It is in the Pali Canon that we find a clear case for the separation of the transcendent and the esoteric by a practitioner who clearly had knowledge of the spirit world, but utterly discounted it. Ken Wilber (himself a practicing Buddhist) acknowledges this: 'only in a human body can enlightenment be attained. Not gods, not animals, not demons, not angels: only with this human body can I awaken to the empty Ground that is equally present in all other sentient beings.'[246]

Advaita

The Advaita Vedanta (to give it its full name) originates in the ancient Hindu scriptures of the Upanishads, and was fully systematised by the sage Shankara in the eight century CE. He agrees with the Buddha on the necessity of human incarnation, listing the three good fortunes of the spiritual life as: human birth, longing for liberation, and discipleship to an illumined teacher.[247] The term 'Advaita' means simply 'not two' and sums up the simplicity and transcendence of its teachings. In more recent times the Advaita is associated with the sage Ramana Maharshi (1879-1950), from whose lineage several contemporary neo-Advaita teachers are drawn, including Andrew Cohen (as mentioned in Chapter Three). Maharshi was born into a middle-class Brahmin family in South India, showed no special aptitude for religion and had no training in spiritual philosophy, but, at the age of seventeen underwent a spontaneous transformation. His experience was so overwhelming that it took him years to properly recover, leaving home and gravitating towards the sacred mountain called Arunachala that would be his home for the rest of his life. He was quickly recognised as a holy man, and he lived as a guru with a small ashram around him which eventually drew thousands from all over the world.

Compassion is a not an overt feature of Maharshi or the Advaita as it is supremely indifferent to the manifest world, yet paradoxically it was clear in every word and act of Maharshi. There is **an impulse towards knowing, 'seeing,' knowledge, wisdom, and learning**, in a narrow sense of 'seeing' the truth of identity with all that is – the truth of 'not two'. The extreme simplicity of this rules out knowledge in any articulated or systematic form, though the attainment of liberation is in itself a form of wisdom. **Enquiry, possibly doubt, as core behaviour** is evidenced in the question Maharshi gave as the key practice to his followers: 'who am I?'

There is little **emphasis on the will** other than as part of the pursuit of this question. It is **instinctively anti-anthropomorphic**: the 'gods' are not part of the discourse. **'Infinite' and 'eternal' are key words** that appear readily. The Advaita and Maharshi in particular are not at all **drawn to esoteric knowledge and practices**, and Maharshi is one of the most vociferous of modern teachers in warning aspirants away from such *siddhis* as they are called in India. When his disciples asked him about attaining such powers he said: 'To abide firm in the Reality which is eternal is the true *siddhi*. Other attainments are all such as are possessed in dreams.'[248]

Taoism

Where Yoga, the Buddha, and Advaita are deeply products of the Indian mind, Taoism is uniquely Chinese. The historical origins with the legendary sage Lao Tse are unrecoverable, but the three principle Taoist texts, the *Tao Te Ching*, the *Chuang Tzu* and the *Leih Tzu*, present a coherent and deeply non-devotional spiritual path. Where Patanjali, the Buddha (of the Pali Canon) and Maharshi speak of the ending of identification of the self with body, emotion, possessions and family, and follow an extreme of renunciation of the 'world', the Taoist texts hold to a balance: their keyword is harmony. The Taoist sage may be married, a smallholder, administrator or ruler, but the practice of the 'way' is a continuous movement to restore balance which the human penchant for drama is always disturbing. Taoism, as presented in the vivid language and cameos of the Taoist texts, avoids the vainglorious, either in the pursuit of material wealth, or in the pursuit of spiritual attainment. Its model, source, inspiration, and destination is Nature.

At the core of Patanjali's system is a cessation of superfluous mental activity, known as 'modifications of the mind'; at the heart of the Buddha's system is 'extinguishing of I-making tendencies', and at the heart of the Advaita is the abandonment of the separate sense of self in the 'not two'. The world is illusion or 'maya', and the arduous task of the aspirant is to penetrate its veils to see the truth hidden behind the apparent play of manifest forces. The Taoist goal is to rest serene *within* the play of manifest forces, aware of the Source, the Tao, and its manifestations. Their knowing of the unmanifest Source is *through* its manifest harmony in Nature; the one cannot be cut off from the other.

In Taoism, **compassion** takes the form of interest rather than action. There is a **strong impulse towards knowing, 'seeing,' and wisdom** but not to **knowledge and learning** (Taoism is anti-intellectual and anti-technology). **Enquiry, possibly doubt, as core behaviour** is not that obvious, and there is no **emphasis on the will,** rather a 'let go'. It is **instinctively anti-anthropomorphic**, accepting the existence of spirits and 'gods' but does not ascribe human qualities to its core principle of Tao. **'Infinite' and 'eternal' are key words** that appear only through other metaphors, and the texts are **not in the least drawn to esoteric knowledge and practices**. However there are many esoteric accretions to Taoism, particularly a populist quest for immortality through alchemical and other practices. Taoism, with its emphasis on Nature, clearly has its roots in the shamanic past of China, but, like the Buddha, regards the denizens of the spirit world as unimportant to its goals. The *Chuang Tzu* even contains a humorous telling of the encounter between a Taoist sage and a local shaman; the sage confounds the shaman.[249] Where the Indian systems discussed represent an extreme of the *via negativa,* Taoism represents a *via positiva* of moderation.

Via Positiva / Via Negativa

Edward Carpenter, Victorian mystic and social reformer, visited the American poet Walt Whitman and drew this interesting comparison:

> I have two portraits — photographs — which I am fond of comparing with each other. One is of Whitman, taken in 1890; the other, taken about the same time and at the same age (seventy years), is of an Indian Gnani or seer. Both are faces of the highest interest and import; but how different! That of Whitman deeply lined, bearing the marks of life-long passion and emotion; aggressive and determined, yet wistful and tender, full of suffering and full of love, indicating serenity, yet markedly turbid and clouded, ample in brow and frame and flowing hair, as of one touching and mingling with humanity at all points — withal of a wonderful majesty and grandeur, as of the great rock (to return always to that simile) whose summit pierces at last the highest domain.
>
> The other portrait, of a man equally aged, shows scarcely a line on the face; you might think for that and for the lithe, active form that he was not more than forty years old; a brow absolutely calm and unruffled, gracious, expressive lips, well-formed features, and eyes — the dominant characteristic of his countenance — dark and intense, and illuminated by the vision of the seer. In this face you discern command, control, gentleness, and the most absolute inward unity, serenity, and peace; no wandering emotions or passions flit across the crystal mirror of the soul; self-hood in any but the highest sense has vanished — the self has, as it were, returned to its birthplace — leaving behind the most childlike, single-hearted, uncensorious, fearless character imaginable.
>
> Yet just here one seems to miss something in the last character — the touch of human and earthly entanglement. Here is not exactly the great loving heart which goes a few steps on the way with every child of man; here is not the ample-domed brow which tackles each new problem of life and science. Notwithstanding evident signs of culture and experience in the past, notwithstanding vast powers of concentration in any given matter or affair when necessary, the face shows that the heart and intellect have become quiescent, that interest in the actual has passed or is passing away.[250]

This passage is interesting historically because it is located at a time of much greater interest in Hinduism than at present, and because

Walt Whitman was understood by many thinkers of that period to be a great spiritual teacher; more recently Paul Hourihan, an American professor of literature, has presented evidence for this.[251] Despite the Victorian language of Carpenter, the passage is particularly revealing of what we have identified as the difference between *via positiva* and *via negativa*. We don't know which the sage is that Carpenter refers to, but as a composite image of the Yogi, the Buddhist or the Advaita practitioner, it sums up the *jnani, via negativa*. (Although fifty years too early, the description actually conveys an uncanny resemblance to Maharshi). Carpenter has captured the modern anxiety about the *via negativa* however: 'that interest in the actual has passed away.' In the entire *Majjhima Nikaya*, for example, whose verses are not selected by subject matter, only by length, the Buddha expresses no interest in the manifest world whatsoever. His passion and intensity are for teaching, and for the *sangha*, his spiritual community. Carpenter's expression of the Western unease with this apparent indifference is easily matched, as we saw, by Jung when he calls nirvana 'an amputation',[252] or when Steiner calls Christian mysticism an 'impotence'.[253]

Jnani and Philosophy

While great philosophical treatises arise from the three Indian *jnani* systems we have described, we should not imagine that these bear any resemblance to philosophy in the modern Western sense. Instead, despite the lack of 'God', we should understand these writings more as theology, or as exegesis (the attempt to explain, interpret and expand on the original spiritual texts). While all three Indian systems are as determinedly anti-intellectual as Taoism, the human mind inevitably constructs systems of speculative thought out of texts that are initially texts of instruction. Hence in India there are vast libraries of Yogic, Buddhist and Advaita commentaries. The devotional *bhakti* traditions do not suffer from this phenomenon, spiritual love being an active force that does not require exegesis. The dividing line between a typical *jnani* text and the work of mere philosophy is in the requirement of the aspirant in a *jnani* system to undertake a practice. While the reading of spiritual texts can in itself be a practice, it is not sufficient: in Yoga and Buddhism in particular an arduous path of personal development is required; in the Advaita it is more an arduous path of direct insight. Between them we have the argument over 'gradual' and 'sudden' enlightenment, which divides the schools within Buddhism for example.

The Buddha is adamant that mere philosophical speculation is futile, refusing to answer the so-called 'great questions of life', such as whether the world is eternal, or infinite, or what the destination of the soul is after death or after enlightenment. In the *Aggivacchagotta Sutta* ('To Vacchagotta on Fire'), the Buddha patiently counters an inquiry from an increasingly bewildered wandering ascetic, Vacchagotta, who poses questions of the type just mentioned. After the Buddha makes it clear that he holds not a shred of opinion on such matters as to whether the world is eternal or infinite, and that to pursue such speculative thinking represents a 'thicket of views, a wilderness of views, a contortion of views, a vacillation of views and a fetter of views, leading to suffering, vexation, despair and fever',[254] Vacchagotta asks him in astonishment: 'Then does Master Gotama hold any speculative views at all?' The Buddha patiently repeats that he does not, and that furthermore, by the abandonment of speculation and other mental activities that prolong the illusion of 'I-making' and 'mine-making', he is liberated by not clinging. Vacchagotta, missing the point entirely, is captivated by the word 'liberation' and immediately poses a speculative question about it: 'where does the soul of a liberated person go to after death?' The Buddha patiently denies the validity of any possible answer to this question and Vacchagotta is now left utterly bewildered. The Buddha so far has led Vacchagotta to this state of confusion by responding to his questions with paradox, effectively short-circuiting his rational mind. The Buddha now goes on the offensive, pursuing a line of his own questioning: what does fire depend on? He builds in Vacchagotta's mind a compelling analogy of the anxious ego-bound mind as a raging fire, feeding on sticks and grass which represent the material for speculative thought. Once the sticks and grass upon which it feeds run out, the fire ceases to burn, but Vacchagotta cannot tell him whether the fire has gone east, west, north or south once it is extinguished. By co-opting Vacchagotta into the construction of a metaphor directly linked to his own question, the Buddha guides him to the insight that cessation of fruitless mental activity leads directly to liberation. Fire does not go anywhere when extinguished, neither does ego. Vacchagotta is quite overwhelmed by and grateful for the insight. In the tradition of the 'sutras' or verses of the Pali canon, Vacchagotta, as all those defeated in debate by the 'lion roaring in the forest', then becomes a lay follower of the Buddha for life. But the essence of this exchange has nothing to do with speculative philosophy: Vacchagotta encountered an immovable object in the Buddha so compelling and illuminated, a world so more powerfully real than that of philosophical speculation, that he became a disciple on the spot.

The encounter between Vacchagotta and the Buddha reinforces the point made earlier, that the East is a tradition of teachers, while the West is a tradition of texts. And once a tradition of texts dominates, then exegesis and philosophy follow naturally, and the raw human encounter between teacher and aspirant is forgotten.

Because of the British influence in India and on the construction of its universities, Western philosophy provides the concepts and language in which India's own traditions are now examined by its scholars. Nowhere is this more apparent than in the widely regarded work by T.R.V. Murti *The Central Philosophy of Buddhism,* in which he compares the Mahayana or Madhyamika system of Buddhist thought to that of Kant.[255] The analogy works on a number of levels, mainly the common argument that what the religious or spiritual seeker is working towards lies beyond rational thought. However, the book-length arguments that Murti and Kant put forward to support this idea are precisely what the Buddha cuts away at a stroke. Where Murti and Kant leave the reader convinced of the futility of reason, but exhausted in the process of getting to that point through rational means, teachers like the Buddha prevent any such circumlocution in the first place. Where the Buddha and Kant are utterly different is that the Buddha starts where Kant leaves off, requiring no justification for the rejection of reason, and providing both the inspiration and personal example to pursue spiritual *practice* instead of philosophy.

No Creator 'God'

One more essential feature of the Eastern spiritual traditions just described needs to be emphasised: the absence of a creator 'God'. It is not just that a 'God' to resemble the Western monotheistic deity is absent, but that there is no creator of any kind. The West, always seeking a linear derivation of all phenomena, has long and furiously debated the question of creation *ex nihilo,* meaning the creation of the universe by 'God' out of nothing. While the East often agrees that the universe derives out of nothing, the traditions under discussion do not require any action on the part of any divine being: creation arises spontaneously and usually in a cyclical fashion. In Buddhist terminology it arises through 'co-conditioned origination', meaning the action of all things upon other things. In Taoism the creation, or the 'thousand things' arise from yin and yang, which arise out of the Tao – no intervention required. And in the Advaita and Yoga traditions the question of creation does not even arise. The point here is that many *jnani* traditions do not require a creator 'God' – but this does not make them any less profoundly religious.

4.2.2 The Greek *Jnani* and Neoplatonism

When Tertullian distanced Christianity from the Greek world by elevating Jerusalem over Athens and the Church over the Academy, the secular mind may assume that it was philosophy that was being rejected. While this is partly true, we may now consider that it was also the *jnani* religious element in Greek thought that was suspect. Because the Western mind reads its secular assumptions into Greek writings they become reduced to a varied exercise in logical propositions and speculative investigation. However when the Greek texts are laid side by side with those of Yoga, Buddhism, Advaita and Taoism, it can become easier to spot where the writings are primarily *jnani* spiritual texts, and only secondarily interpretable as mere philosophy. In addition to the texts there are also whatever historical accounts are available of the lives and actions of the writers, and which may bear more resemblance to that of a spiritual teacher than a speculative thinker. There is potentially a rigorous methodology available here, but to apply it to the Greek tradition in any systematic way would require a book-length exposition. Hence we can only sum up the most promising direction that such an enquiry would take, bearing in mind that we are only seeking to establish that the *jnani* impulse was present in Greek thought, and that it lived on in the Western mind broadly speaking through Neoplatonism.

Those Greek thinkers who most immediately appear to be *jnani* teachers rather than speculative philosophers include Pythagoras, Heraclitus, and Socrates. Colourfully, we have termed them 'lost buddhas of the West', using 'buddha' as generic term meaning an Enlightened One, following the tradition in India. Or to be more precise: it is used here to represent a person with a highly developed *jnani* impulse.

Pythagoras

Pythagoras (c580-c500 BCE) and Heraclitus (c540-c480 BCE) are approximate contemporaries of the Buddha, although very little is known about them. It is extremely unlikely that the ideas of the Buddha would have reached either of them in their own lifetimes, though Buddhist ideas did spread westward to some degree in later centuries. In the case of Pythagoras his considerable following during his lifetime and the survival of 'Pythagoreans' as a sect and as an influence through the ancient world speaks of a personality with some force. As far as we know from the various references made to Pythagoras by other ancient Greeks he pursued research into mathematics and music alongside a practice of the spiritual life.

The known existing material on Pythagoras has been compiled in *The Pythagorean Sourcebook and Library*, edited and translated by Guthrie and Fideler.[256] Although each fragment is somewhat inconclusive on its own, taken together there is a compelling account of a spiritual teacher of the *jnani* orientation with a strong *via positiva* bias. In his foreword the music scholar and Neoplatonist Joscelyn Godwin writes:

> If Western civilisation has taken, as it seems, not one but several wrong turnings between the time of Pythagoras and the present, it is because it has been unfaithful to him who should by rights have been its tutelary genius. . . . If he failed as the avatar of the passing age, perhaps he is coming into his own as a new one dawns.

The 'wrong turning' that Godwin refers to is, in our analysis, the dynamic within early Christianity that placed its centre of gravity with the *bhakti, via negativa,* anti-esoteric impulses, and the instinct to repress everything else. Pythagoras represents its diametrical opposite: *jnani, via positiva,* and firmly esoteric. Godwin is right that the modern age does tend to the *jnani, via positiva,* and in the New Age at least the esoteric is given free reign. Pythagoras stands as the first 'avatar' of Neoplatonism, as it was to grow and develop into its Renaissance form. The secular world understands the Renaissance as the commingling of art and science, the 'two cultures' of C.P. Snow, but that view omits the third element so well represented in Pythagoras and the Neoplatonist tradition: spirituality. Pythagoras it seems was also the first to articulate that uniquely Renaissance idea: that of man as microcosm, also expressed in the image of 'as above, so below'. This is a very Western expression of the *via positiva,* and it is a deeply Hellenic idea.

That Pythagoras founded a spiritual community or *sangha* is clear. David Fideler tells us: 'The school of Pythagoras in Croton [southern Italy] appears to have been a religious society centred around the Muses, the goddesses of learning and culture, and their leader Apollo. Iamblichus' description of the school gives it something of a monastic flavour, and there was indeed a 'rule' of life, but while the Pythagoreans gathered together at certain times of the day, most of them did not live together.'[257] Pythagoras seems to have had the compelling personality of the Buddha, his broadly *jnani* spirituality, and his instinct for spiritual community bound together by rules conducive to the spiritual life. However the radical difference is obvious: where the Buddha is intensely *via negativa,* Pythagoras is intensely *via positiva,* centring his spiritual path on the arts and sciences: the role of the Muses symbolising this. Pythagoras himself made claims that he knew his past lives, and had once been Euphorbus, son of Panthus, conqueror of Patroclus[258] – in this we find another parallel with the Buddha, who, in a number of passages, gives quite detailed accounts of his past lives.[259]

Heraclitus

Heraclitus is known for his 'Fragments', a collection of aphoristic sayings which can be read as a spiritual *jnani* text on the mysticism of unity, and which show a resonance in temperament with the archetypal Zen master, though of course no actual connection. It is not clear in fact if Heraclitus had much of a following, as he seemed uncompromising and contradictory – his usual appellation is 'the obscure,' a term he even applies to himself. It is also clear that he follows a *via negativa,* interested in the world only so far as to comment on its essential dynamics – as a system regularly kindling itself and then going out – and on the folly of most men. He also specifically criticises Pythagoras for 'doing too much research', an obvious hint of Heraclitus' preference for the *via negativa.* (We saw earlier that Plato also mocked the Pythagoreans for their empirical research on the vibrations of strings.)

Heraclitus is therefore closer in temperament to the Buddha than Pythagoras, but further away in his legacy, as he does not appear to have had the combination of compassion and teaching ability to found a school or *sangha.* But we are representing both Pythagoras and Heraclitus as 'buddhas' because there is sufficient evidence in what we know of them to support the idea that they were highly developed *jnanis.* The *Fragments* of Heraclitus start with:

> It is wise, listening not to me but the Logos,
> to agree that all things are one. *(v.1)*[260]

The unitive principle announced here at the very start of the text is a good basis for assuming a transcendent teaching. Where the *Golden Verses* of Pythagoras emphasise the golden mean, Heraclitus is more interested in the opposites or extremes, and understanding life as the tension between these polar opposites. Heraclitus pursues the idea that all is one by first identifying opposites and then stating that they are one, for example:

The good and the bad are one and the same. *(v. 29)*
The way up and the way down are one and the same. *(v.30)*

Heraclitus has the detachment of the *via negativa*; he is not interested in the idea of a 'good day' or a 'bad day', as he points out in this verse:

Concerning days of ill fortune, whether it is
necessary to reckon some as such or whether
Heraclitus justly rebuked Hesiod for regarding
some as favourable and some of no account, not
recognising the nature of each day as being one
thing; elsewhere it has been a matter of dispute.
One day is equal to all. *(v.31)*

To anyone with a contemporary outlook on philosophy this would be meaningless, but for the *jnani* this is a natural expression of being-at-one with the universe in the particular context of living outside of time. Once the clamour of future-orientation is gone, then each moment is eternal and unjudged as either 'favourable or of no account'. But Heraclitus has altogether a more sanguine temperament than the Buddha concerning strife, as this verse shows:

Heraclitus censures the poet who says 'Would
that strife might depart from gods and men',
for there would be no harmony without high and
low, nor could there be life without male and
female, these being opposites. *(v.40)*
It is not better that many things come about as
men wish. Disease makes health pleasant, evil good,
hunger surfeit, and toil rest. *(v.41)*

This cosmic perspective may have made Heraclitus unpopular in his time. The Buddha's message, his First Noble Truth, that life is suffering, finds many more ears than Heraclitus' 'hidden harmony'. Heraclitus 'sought himself', and he understood, with the Buddha, the problem of desires; and also has a keen awareness of the impermanence of the manifest world:

I sought myself. *(v.100)*
It is difficult to contend with desires: whatsoever a
man desires, he pays for with the soul. *(v.101)*
For it is not possible to step twice into the same
river, according to Heracleitus. Neither is it
possible to hold permanently onto the same mortal

> substance. But by sharp, keen and quick changes it is scattered and brought together again, not in the past or the future, but at once it combines and disperses and comes and goes. (v.33)

Where the Buddha finds impermanence a matter for regret, Heraclitus celebrates the fluidity of our 'mortal substance'. What lies in common is the perception of the shifting nature of experience, and the necessity for identification with the eternal ground from which we know it. Obscurity, contradiction and paradox are the language of the mystic, not in order to obfuscate, but because the teachings of transcendence are difficult.

Socrates

We now turn to the life of Socrates (470-399 BCE) and explore the idea that not only can he be fruitfully be represented as one of the 'lost buddhas' of the West, but that his legacy best illustrates how East and West have diverged from what were comparable origins. The great difficulty in approaching Socrates is that he wrote nothing, but is present in a large proportion of the Platonic canon as a character in the Dialogues. The nature of Plato's writings means that it is difficult to extract a credible portrait of Socrates and his teachings from the canon.

The trial and execution of Socrates has parallels, in religion and mysticism, with that of Jesus and Mansur (a tenth-century Muslim martyred in Baghdad), to give just two examples. It seems that Socrates was indicted on two counts: impiety, and corrupting the morals of the young. Plato devotes four dialogues, *Euthyphro, The Apology, Crito,* and *Phaedo,* to the events leading up to his trial and execution. The charge of impiety is not easily refuted, according to the evidence in Plato: Socrates was not readily inclined to accept the common views held on the gods and their activities, preferring to draw on his own inner resources in moral and religious questions. Plato does however show us that the idea that Socrates had a detrimental effect on the morals of the young was absurd, and essentially a trumped-up charge. There is no doubt that these four dialogues have left a vivid impression on the Western mind of a noble character unjustly imprisoned and executed for holding to the 'truth'. The appellation of 'gadfly' implies that many citizens were not particularly grateful for Socrates' efforts as a spiritual teacher: they were stung by his peculiar method of questioning. Where Pythagoras founded a spiritual community at Croton of which he was the leader, Socrates appeared to treat the whole of Athens as his *sangha*, and clearly alienated the political elite as a result. Other factors may have led to his imprisonment and trial, such as his political views in a time of stress for the citizens of Athens (they were under siege from Sparta).

There must be considerable doubt as to how much of Socrates' thinking is present in the Dialogues, and how much of it is an interpolation of Plato's. Plato himself does not appear in the Dialogues, indeed he absents himself from almost all his writings. Socrates seems to put forward a wide variety of views which are not entirely self-consistent. Hence it is a reasonable objection that some or all of the Dialogues might be a case of Plato putting words into Socrates' mouth. (Thomas Jefferson went so far as to say that Plato 'only used the name of Socrates to cover the whimsies of his own brain'.[261]) But it is possible to reconstruct a different Socrates, particularly with the assistance of Xenophon, the only other significant source of information on him. Xenophon was a practical man, a naval commander and historian, and more inclined to give practical advice (for example on renting out rooms to make up for a loss in income) than indulge in philosophising. Xenophon is probably more reliable than Plato in recording the activities of Socrates as man, and is unlikely to have made the kind of interpolations in Socrates' speech that we suspect Plato of doing, because he was no intellectual. Xenophon is also present in his text, eliminating the suspicion that he wished in some way to manipulate his reader. It is also clear that Xenophon had enormous regard for Socrates as a profoundly *good* man, whereas his temperament would rule out a merely intellectual admiration. Xenophon's portrait of Socrates is consistent with that of a spiritual teacher like the Buddha, operating in a like manner of individual encounter and debate. One of Socrates' best-known aphorisms is: 'one thing only I know, and that is that I know nothing': hardly the culmination of a philosophical enterprise, and quite in keeping with the Buddha's repudiation of speculative views. Socrates is fiercely *via negativa*; for example he calls pleasure and pain the 'rivets' that bind the soul to the body (an echo of Heraclitus),[262] and tells his grieving disciples that he is confident that once he has departed the body he goes to a better place (this earned him censure from Nietzsche).

The case for Socrates as a highly developed *jnani* can only be summarised here. The evidence relates to his serene attitude to execution, his fits of abstraction (which we suggested above is better interpreted as the experience of *samadhi*), his 'voices', a conviction of immortality, and those activities that are open to interpretation as natural to a spiritual Master, guru, or teacher. Of the latter we can mention the well-known metaphor that Socrates presents of himself as 'midwife', interpreted in a secular context as involving the birth of philosophical ideas, but reinterpreted here as the spiritual rebirth of the initiate or disciple.

Plotinus

The confluence of Greek Platonism and other spiritual traditions was given impetus by the work of Plotinus (205-270) of unknown birthplace but who settled in Rome. He is generally regarded as the founder of Neoplatonism as distinct from the Greek schools generally, or the thought of Plato in particular. We can say that the *Enneads* of Plotinus are amongst the finest *jnani* writings from antiquity in the West. We also have some useful information about his later life though the biography written by his disciple Porphry. Plotinus is interested in learning of a kind, as Pythagoras was, but is not so assiduous in 'research'. He has neither the deliberate obscurity of Heraclitus, the rejection of knowledge of Socrates, or the cool aloofness of the Buddha. The *Encyclopedia Britannica* finds a striking similarity between the *Enneads* and the *Yoga Sutras of Patanjali*, but it was unlikely that Plotinus knew of them, being exposed rather to the ideas of Plato and the Gnostic and Manichaean sects. Plotinus went to Alexandria at the age of twenty-eight to study philosophy, which apparently reduced him to a state of depression until he found a teacher called Ammonius Saccas. When Plotinus heard him he said, 'This is the man I was looking for' and stayed with him for eleven years.

Plotinus' conception of the *One* was an image that lived on and had an influence in Christianity. But Plotinus' conception of the One is not theistic, for despite the piety and sublimity of his prose, it is not a devotional work; neither does the 'One' create the world or intervene in it. Neither on the other hand is it a pure *via negativa* teaching like the Buddha's, containing as it does speculative, philosophical, cosmological and occult elements. What we find however are sections that are pure *jnani*, that is they describe a unitive state and a path to that state not in terms of speculation, but in the language of one who has directly experienced it.

Because of the resonance of the concept of the 'One' with the Christian 'God' that later became the universal spiritual language of the West, some scholars have unfortunately translated the 'One' as 'God' or 'Father'. The translation of Elmer O'Brien[263] avoids the

Christian terminology that is found in the translation of Stephen Mackenna,[264] and is hence more true to its non-Christian roots. Here is part of the 'Three Primal Hypostases' in O'Brien's translation:

> Our souls know neither who nor whence they are, because they hold themselves cheap and accord their admiration and honour to everything except themselves. They bestow esteem, love, and sympathy on anything rather than on themselves. They cut themselves off, as much as may be, from the things above. They forget their worth. Ignorance of origin is caused by excessive valuation of sense objects and disdain of self, for to pursue something and hold it dear implies acknowledgement of inferiority to what is pursued. As soon as the soul thinks it is worth less than things subject to birth and death, considers itself least honourable and enduring of all, it can no longer grasp the nature and power of the divinity.
>
> A soul in such condition can be turned about and led back to the world above and the supreme existent, the One and first, by a twofold discipline: by showing it the low value of the things its esteems at present, and by informing – reminding! – it of its nature and worth. (Enneads, V,1,1)

This beautiful passage has many resonances with Buddhist thought, for example where it points out that a preoccupation with 'things subject to birth and death' leads a person away from the One. Also obvious in the above passage is that Plotinus has a teaching or path, equivalent to the Dhamma (Dharma) of the Buddha, and, we note, Plotinus corrects himself that the role of spiritual teacher is not to inform the disciple of the nature and worth of their soul, but to *remind* them – echoes of Socrates.

To some degree Plotinus is following the *via negativa*, but not through a sense of sin or guilt as can be found in the Abrahamic traditions. Plotinus has a warmth and positive approach, and, like the Buddha, he promises liberation or 'divinity' in this lifetime, soon in fact:

> So divine and precious is the Soul, be confident that, by its power, you can attain to divinity. Start your ascent. You will not need to search long. Few are the steps that separate you from your goal. Take as your guide the most divine part of the Soul, that which 'borders' upon the superior realm from which it came. *(Enneads, V,1,3)*

Let us turn for a moment to Porphyry's brief account of Plotinus's life to see what there is in it that might be consistent with that of a 'buddha', or *jnani* spiritual teacher. This single paragraph stands out (note that the extracts from Porphyry's *Life of Plotinus* are from Mackenna's translation):

> Thus he was able to live at once within himself and for others; he never relaxed from his interior attention unless in sleep; and even his sleep was kept light by an abstemiousness that often prevented him taking as much as a piece of bread, and by this unbroken concentration upon his own highest nature. *(Life, 8)*

This is an interesting characterisation of the spiritual teacher: that he lived 'at once within himself and for others'. In the later Buddhist traditions this would be the quality of the Bhodisattva, one who lived for others, or to be more precise, one who saw the buddha-nature in others and strove continuously to help them realise it. Porphyry tells us that several men and women 'of position' were also impressed with Plotinus' practical abilities and honesty, to the extent that they entrusted their orphans into his care along with all their inherited property. Let us look at another extract which indicates something about Plotinus's conception of the 'One':

> The One is absent from nothing and from everything. It is present only to those who are prepared for it and are able to receive it, to enter into harmony with it, to grasp and to touch it by virtue of their likeness to it, by virtue of that inner power similar to and stemming from the One when it is in that state in which it was when it originated from the One. Thus will The One be 'seen' as far as it can become an object of contemplation. *(Enneads, V1,9,4)*

Plotinus praises 'detachment', but his place on the *via positiva / via negativa* spectrum is worth considering in a bit more detail. He devotes a whole treatise called 'Against the Gnostics' to refuting their main arguments, particularly that the world itself is corrupt or fallen. True, Plotinus regards the acquisition of a body by the soul as a 'descent', an idea common in antiquity and taking various forms including the story of the expulsion from Eden, but he does not like the extreme view of the descent as original sin, a fall into evil, and a matter for lamentation, as we see in this passage (translation, Mackenna):

> To those who assert that creation is the work of the Soul after the failing of its wings, [a reference to Plato's *Phaedrus*] we answer that no such disgrace could overtake the Soul of the All. If they

> tell us of its falling, they must tell us also what caused the fall. And when did it take place? If from eternity, then the Soul must be essentially a fallen thing: if at some one moment, why not before that?
>
> We assert its creative act to be a proof not of decline but rather of its steadfast hold. Its decline could consist only in its forgetting the Divine: but if it forgot, how could it create? Whence does it create but from the things it knew in the Divine? If it creates from the memory of that vision, it never fell. Even supposing it to be in some dim intermediate state, it need not be supposed more likely to decline: any inclination would be towards its Prior; in an effort to the clearer vision. If any memory at all remained, what other desire could it have than to retrace the way?
>
> What could it have been planning to gain by world-creating? Glory? That would be absurd – a motive borrowed from the sculptors of our earth.
>
> Finally, if the Soul created by policy and not by sheer need of its nature, by being characteristically the creative power – how explain the making of this universe?
>
> And when will it destroy the work? If it repents of its work, what is it waiting for? If it has not yet repented, then it will never repent: it must be already accustomed to the world, must be growing more tender towards it with the passing of time. *(Enneads, II,9,4)*

This defence of the manifest world is not as complete as in Pythagoras, but it is positive indeed when placed alongside the Buddha's indifference to it. It is even less Gnostic in its retreat from the world than often found in Christianity.

From Porphyry's account we can see that Plotinus created a spiritual community or *sangha* around him as did Pythagoras, but with less emphasis on the arts and science. Buddhism became a major religion of the East, and in Plotinus the West nearly made the source of its major religion a similarly *jnani* teacher. If Julian the Apostate had been successful in reversing the tide of Christianity, his reformed state religion may have been largely based on the teachings of Plotinus.

Neoplatonism as the Alternative Western Religion

Neoplatonism *could* have become the religion of the West, rather than Christianity. It had early founders such as Pythagoras, a key martyr in the form of Socrates, and a brilliant expositor in Plotinus. The *Enneads* alone could from the core text of the religion, with many other writings of the Greeks to complete its 'bible'. It had a *jnani* parallel in Buddhism, which became a world religion through adoption by the emperor Ashoka. What if Constantine had adopted Neoplatonism instead of Christianity? Or if Julian had succeeded in his reforms? Or if Augustine had not decided that Neoplatonism was all ultimately 'idolatry'? All this is mere speculation of course, though it is instructive to consider what Neoplatonism might have lacked: a charismatic proselytiser like St Paul, and a core conception that could have popular appeal. Buddhism, however abstract and aloof its central ideas about impermanence and nirvana, had at its core a profound compassion, the *meta* of loving kindness, while Christianity had its *charitas*, the active principle of love. Perhaps this was missing in Neoplatonism (as Augustine complained of), a core impulse to love and compassion embodied in such a way that it could reach out to ordinary people.

4.2.3 Dionysius, Erigena and Eckhart

At the time of Plotinus the Roman Empire was polytheistic, pluralist and religiously tolerant. Christianity was an obscure Jewish sect, only one of a number of small religious movements competing with Gnosticism, Manichaeism, and other smaller groupings and mystery schools. We now jump into a world where Christianity dominates and shuts out these rivals. Although mainstream Christianity was *via negativa* in its placing of 'God' as infinitely superior to the world, it was also intensely social through Christian charity in the Pauline sense. The example of remote piety in the devotional mystics was counterbalanced by the uniquely Christian corrective that Kierkegaard so skilfully laid bare: 'love thy neighbour as thy self.'

We have suggested that Neoplatonism was the *jnani* vine that grew around the *bhakti* tree of Christianity, and both were shaped by each other. Its Hellenic character appealed to the more intellectual of Christians, and Augustine did indicate that there were *some* acceptable elements in it. Its most visible influence on Christian ideas lies in what was much later called the *negative theology*. We saw in Chapter Two that the *Encyclopedia Britannica* says of the negative theology that 'it could be called an Eastern idea present and effective in the Occident'. This simply betrays the millennia-old prejudice within the Christian world against the Hellenic *jnani* non-theistic tradition. The negative theology is not foreign to the West at all, but is joyously present in the Greek mind.

However, the most celebrated early writer of the negative theology seems to have fooled Christendom into a grudging acceptance, firstly by falsely or accidentally attributing his work to a convert of St Paul, and secondly by forcing Christian conceptions to serve a radical but not easily detectable purpose. He was originally assumed to be Dionysius of the city of Areopagus, mentioned in the New Testament as converted by St Paul at Athens (Acts 17:34). This was much later proved to be an error, so he is now formally known as Dionysius the pseudo-Areopagite, or the pseudo-Dionysius. For brevity we will refer to him simply as Dionysius, in awareness that he was an anonymous fifth-century Syrian monk.

His work received attention from the ninth-century Irish theologian, Johannes Scotus Erigena (also spelt 'Eriugena'), who wrote commentaries on it and other major works of a Neoplatonist flavour. Meister Eckhart of the thirteenth century completes this trilogy of early Christians whose thought is radically *jnani*, somewhat Neoplatonist, and understood as proponents of the negative theology. They are significant as deeply mystical Christians whose spiritual impulse, as revealed in their writings, would be quite at home in the Eastern *jnani* traditions of Yoga, Buddhism and Advaita, rather than in the Abrahamic devotional monotheism of their milieu. They also anticipate the brief re-emergence of the Neoplatonist undercurrent as it flowered in the Renaissance.

Dionysius

The *Catholic Encyclopedia* simply says of the writings of Dionysius that they are 'peculiar', avoiding the term 'error' which is its usual way of flagging up heresy. Whatever the sources of and influences on Dionysius his work is highly original, reading at times like a Plotinus, or more generally Greek, transposed into Christian terminology. Here is an extract from the beginning of *The Divine Names*:

> By no means then is it permitted to speak, or even to think, anything, concerning the superessential and hidden Deity, beyond those things divinely revealed to us in the sacred Oracles. For Agnosia, (supra-knowledge) of its superessentiality above reason and mind and essence . . . to it must we attribute the super-essential science, so far aspiring to the Highest, as the ray of the supremely Divine Oracles imparts itself, whilst we restrain ourselves in our approach to the higher glories by prudence and piety as regards things Divine. . . . For, as things intelligible cannot be comprehended and contemplated by things of sense, and things uncompounded and unformed by things compounded and formed; and the intangible and unshaped formlessness of things without body, by those formed

> according to the shapes of bodies; in accordance with the self-same analogy of the truth, the superessential Illimitability is placed above things essential, and the Unity above mind above the Minds; and the One above conception is inconceivable to all conceptions; and the Good above word is unutterable by word. . . . Unit making one every unit, and superessential essence and mind inconceivable, and Word unutterable, speechlessness and inconception, and namelessness . . . being after the manner of no existing being, and Cause of being to all, but Itself not being, as beyond every essence, and as It may manifest Itself properly and scientifically concerning Itself.

This passage uses the term 'One' as if it had come from Plotinus, but referring, apparently, to the Christian 'God'. We note that Dionysius uses the preposition 'it' to refer to 'God' rather than 'Him' – a characteristic of a profoundly anti-anthropomorphic thinker. Yet the essence of Dionysius' thought is that we cannot *know* 'it', the 'One', 'divinity' or whatever rather un-Christian term he adopts for 'God'. His delicacy about naming the divinity and his anti-anthropomorphic stance derives as much perhaps from Judaic mysticism as it might from Neoplatonism or Plotinus, but he is rather startling in his array of terms for our not-knowing. The most radical of these is 'agnosia' which is simply the negation of gnosis, to know. Thomas Huxley ('Darwin's bulldog') proudly re-invented 'agnostic' in the nineteenth century apparently unaware that he had a precedent in the fifth century. But Dionysius uses the term agnosia in a way familiar to another fiercely *jnani via negativa* tradition . . . Zen (and it is no accident that modern America's leading enthusiast on Zen, Alan Watts, should write on Dionysius). The negative theology in the West and Buddhism in the East, particularly Zen, agree on the ineffable quality of the ultimate. Buddhism manages without any 'God' and it is not at all certain that Dionysius really needs the concept either.

In one of his other well-known works, the *Mystical Theology*, his opening does include a reference to 'God' however:

> TRIAD supernal, both super-God and super-good, Guardian of the Theosophy of Christian men, direct us aright to the super-unknown and super-brilliant and highest summit of the mystic Oracles, where the simple and absolute and changeless mysteries of theology lie hidden within the super-luminous gloom of the silence, revealing hidden things, which in its deepest darkness shines above the most super-brilliant, and in

> the altogether impalpable and invisible, fills to overflowing the eyeless minds with glories of surpassing beauty.

His originality now extends beyond 'agnosia' and includes the 'divine darkness', a metaphor rare in the mystical literature of the world, more usually filled with images of light. This darkness however fills 'eyeless minds with glories of surpassing beauty', in an image that is reminiscent of the *Dark Night of the Soul* of St John of the Cross. They are archetypal images of the *via negativa*, but one comes from a great Christian *jnani*, the other from a great Christian *bhakti*.

Erigena

While Dionysius is 'peculiar' for the Church, Erigena does not get off so lightly. The following extract from the *Catholic Encyclopedia* is instructive on the Church's attitude to both men, and to the unwelcome 'mingling' of Christian ideas with ideas alien to it:

> Although the errors into which Eriugena fell both in theology and in philosophy were many and serious, there can be no doubt that he himself abhorred heresy, was disposed to treat the heretic with no small degree of harshness (as is evident from his strictures on Gotteschalk), and all through his life believed himself an unswervingly loyal son of the Church. Taking for granted the authenticity of the works ascribed to Dionysius the Areopagite, he considered that the doctrines he discovered in them were not only philosophically true, but also theologically acceptable, since they carried with them the authority of the distinguished Athenian convert of St. Paul. He did not for a moment suspect that in those writings he had to deal with a loosely articulated system of thought in which Christian teachings were mingled with the tenets of a subtle but profoundly anti-Christian pantheism. He was accused by his contemporaries of leaning too much towards the Greeks.[265]

We note the horror of the Church towards pantheism, and the nervousness of anything 'Greek'. Erigena is today regarded as a highly original theologian and philosopher, with obvious Neoplatonist leanings in addition to what he owed to Dionysius. His writings display both the extreme *via negativa* of Dionysius, and also an 'emanationist' treatment of the manifest world which is entirely in keeping with the *via positiva* strand of Neoplatonism. The anxiety that the Church felt about this ninth century monk eventually led to his books being placed on the Index in 1681. His work was an influence on the Albigensians, and also on Meister Eckhart, and much later, on the German Idealists.

Eckhart

Johannes Eckhart was born in 1260 in Germany, acquired the title of 'Meister' when the degree of Master of Theology was conferred upon him in 1302, and died in 1327 in France. Towards the end of his life Eckhart was charged with heresy. The bull of Pope John XXII, dated 27 March, 1329, condemned 28 propositions extracted from the writings of Eckhart. He retracted previously identified errors as charged, but the damage was done to his reputation in the period following his death, and he lapsed into obscurity. The struggle between Eckhart and the Inquisition is symbolised by the eventual excommunication by Papal bull, though he died shortly before he received it. The papal Inquisition had been active for almost exactly a hundred years at this point, and the universal atmosphere of suspicion that this created was part of Eckhart's Europe.

Meister Eckhart's legacy is valuable, as we find recent and contemporary philosophers and theologians such as Heidegger, Levinas, Derrida, Caputo, Fox and Cupitt making reference to him. The *Catholic Encyclopedia* tells us: 'Although he frequently quotes from the writings of the Pseudo-Areopagite and of John Scotus Eriugena, in his mysticism he follows more closely the teaching of Hugh of St. Victor'. It also states: 'Although a profound mystic he was also an able man of affairs, admirably manifesting the spirit of his order by uniting throughout his career great activity with contemplation'. The contemporary scholar of religion John D. Caputo, writing on Eckhart in connection with Heidegger, also calls Eckhart 'a man of prodigious activity'[266] Caputo considers the parallels between Eckhart's *On Detachment* and the philosopher Martin Heidegger's *What is Metaphysics*, to be remarkable considering the six centuries of German thought that lay between them: 'It is as though in these two works – as in these two authors as a whole – the beginning and end of German thought touch each other'.[267]

Caputo is writing in a postmodern context, as is Don Cupitt when he makes the absurd assessment: 'Eckhart is the extreme case of the mystic as writer. . . . He merely plays dazzling games with language'.[268] Far from a dilettantish series of language-games, Eckhart's writings represent some of the most profound accounts of the *jnani* spiritual impulse available to the modern mind and this is attested to by the wide range of contemporary interest in him.

Here is an extract from one of his most famous tractates, *On Detachment*:

> I have read many works of both heathen masters and prophets, and books of the Old and New Testaments, and have sought earnestly and with the utmost diligence to find out what is the best and highest virtue, with the aid of which man could be most closely united with God, by which man could become by grace what God is by nature, and by which man would be most like the image of what he was when he was in God, when there was no difference between him and God, before God had created the world.
>
> And when I search the Scriptures thoroughly, as far as my reason can fathom and know, I just find that pure detachment stands above all things, for all virtues pay some regard to the creatures, yet detachment is free from all creatures. Hence it was that our Lord said to Martha: 'One thing is needful', that is to say, he who wished to be untroubled and pure must have one thing, namely detachment.
>
> The teachers praise love most highly, as St Paul does when he says: 'In whatever tribulation I may find myself, if I have not love, then I am nothing'. But I praise detachment more than all love. First, because the best thing about love is that it forces me to love God. On the other hand, detachment forces God to love me. Now it is much nobler that I should force God to myself than that I should force myself to God. And the reason is that God can join Himself to me more closely and unite Himself with me better than I could unite myself with God. That detachment forces God to me I can prove by the fact that everything likes to be in its own natural place. Now God's own and natural place is unity and purity and they come from detachment. Hence God must of necessity give Himself to a detached heart.[269]

With only a few changes, much of this could come from Plotinus, who praises detachment in 'The Good or the One'. With Eckhart we have a clear statement that detachment is above love, in contradiction to Paul, but a sure sign of the pure *jnani* temperament. In fact it contradicts not just St Paul, but the whole development of Christianity throughout the Middle Ages. We are not suggesting however that Eckhart derives his thought from Plotinus, because Eckhart's emphasis on detachment is so far-reaching that it can only come from his inner experience. It is also possible that a full translation or edition of the *Enneads* was not available to Eckhart.

We also note in the above passage that not only is Paul contradicted, who shaped the religion of Christ in the first place, but that Christ is not needed either. In much of Eckhart's writings we could replace all references to 'God' with 'the One', and all uses of 'He', 'His' and "Himself' with "it', "its' and "itself'. If we try it then the difference is a change in the deity from an active principle to something less anthropomorphically engaged with the world. If "God' is changed to "the One' then He, or it, becomes more passive in the exact proportion that the aspirant becomes more active, i.e. we recognise the emergence of the typical *jnani* characteristic, the spiritual *will*. Eckhart himself recognises the active principle in the personal Christian "God' as problematic and often talks instead of *Gottheit* as a higher and less active principle which is usually translated as the "Godhead'. Let Eckhart explain:

> God and the Godhead are as different from each other as heaven and earth. . . . Creatures speak of God – but why do they not mention the Godhead? Because there is only unity in the Godhead and there is nothing to talk about. God acts. The Godhead does not. . . . The difference between God and the Godhead is the difference between action and non-action.

This is alarmingly heretical, bearing close resemblance to the Gnostic distinction between "God' and "Godhead'. Later in his essay *On Detachment* Eckhart even removes the central Christian practice, prayer, though it is interesting that "God' becomes again a more active principle in this passage:

> Hence, if the heart is to find preparedness for the highest of all flights, it must aim at a pure nothing and in this there is the greatest possibility that can exist. For when the detached heart has the higher aim, it must be toward the Nothing. . . . In the same way, when God wishes to write on my heart in the most sublime manner, everything must come out of my heart that can be called 'this' or 'that'; thus it is with the detached heart. Then God can work in the sublimest manner and according to His highest will. Hence the object of the detached heart is neither this nor that.
>
> But now I ask: what is the prayer of the detached heart? I answer that detachment and purity cannot pray. For if anyone prays he asks God that something may be given to him, or asks that God may take something away from him. But the detached

> heart does not ask for anything at all, nor has it anything at all that it would like to be rid of. Therefore it is free from all prayer and its prayer is nothing else than to be uniform with God. On this alone the prayer of detachment rests.[270]

To see prayer as mere supplication or petitioning is a mistake that only a *jnani* would make, and this is what Eckhart does by assuming that prayer must involve asking for a good to be given or an evil to be taken away. His *jnani* instincts create no possibility for understanding the heights of devotional prayer that a *bhakti* like St John of the Cross so ecstatically describes. But while Eckhart introduces the "Nothing' in the beginning of this extract, a term that Buddhism would recognise (though more usually called "emptiness' or *shunyata*) we cannot entirely avoid the *active* nature of Eckhart's "God', in so far as he is obliged to use the term. But is it really any more than a convenient, culturally sanctioned metaphor with Eckhart?

Let us turn to the question of the *via negativa* / *via positiva* distinction in connection with his teachings. As we have seen it is rare to find an extreme example of either path, so it is a question of teasing out where the balance lies. Eckhart's praise of detachment from the "creatures' (a term that was by now well-established to describe the world as a whole) indicates a *via negativa*, but there are *via positiva* elements in his cultural context that he is not immune to. Firstly, like Plotinus, he wrote philosophical treatises, the Latin works that were meant to be read by the scholars and theologians – we can see this as an active engagement in the world of a kind that the Buddha shunned. Secondly, the Christian concept of active love, which means good works of various kind, were endorsed by Eckhart, and his "prodigious activity' was directed selflessly at the good governance and instruction of his Christian districts. Let us leave Eckhart with a last extract, one that shows the pure *jnani* expression of transcendence:

> In that breaking-through, when I come to be free of my own will and of God's will and of all his works and of God himself, then I am above all created things, and I am neither God nor creature, but I am what I was and what I shall remain, now and eternally. . . .
>
> When I stood in my first cause, I then had no 'God', and then I was my own cause. I wanted nothing, I was empty Being and the only truth in which I rejoiced was in the knowledge of my Self. Then it was my Self I wanted and nothing else. What I wanted I was, and what I was I wanted and so I stood empty of God and every thing. (*Sermons*)

We see that the fully-realised *jnani* living in a theistic *bhakti* tradition has no choice: slowly and reluctantly he ultimately even has to dispense with 'God'. It is also entirely perverse for the postmodern Cupitt to assert that a mystic like Eckhart had no mystical 'experience', only the act of writing. In the above passage where Eckhart talks of the 'breaking-through' he does so in such forceful terms, and in a manner so akin to highly developed *jnanis* in other unrelated cultures, that it is inconceivable that this was mere speculation, a mere literary flourish.

Pantheism and Anti-Anthropomorphism

The *Catholic Encyclopedia* was alert to the 'anti-Christian pantheism' of Dionysius. It also says of Eckhart: 'While at Strasbourg, although he had no relations with the Beghards, he was suspected of holding their mystical pantheism.' We pointed out in section 1.4 that pantheism was a negatively charged word for Christians, even for thoughtful scholars like Evelyn Underhill (writing in the early twentieth century) and Keith Ward (writing in the early twenty-first century). It represents a puzzle for the non-Christian. We are now in a position to suggest a solution to this puzzle: it is to do with the question of the anthropomorphism of 'God'. We have seen that the conception of 'God' varies enormously, in particular on a sliding scale of his resemblance to a human being. For a religion that was largely devotional and highly intolerant of deviation, that sliding scale was narrowed to a small part of the spectrum that was *largely* anthropomorphic. Though emphatically not a human being in most respects, 'God' was nevertheless a being who loved, willed, and acted: the 'supreme being'. He was a personal 'God'. This is the natural outcome of two major forces acting on the conceptualising of him: the Judaic insistence on his separate existence or 'exteriority' as Levinas calls it, and the poetic tendency of the devotional impulse to give him human qualities: recall how St John of the Cross had 'God' rest his head on his breast.

For the three great *jnanis*, Dionysius, Erigena and Eckhart, their 'God' has no anthropomorphic trace: they are Christian non-devotional mystics who have abandoned the Judaic insistence on a 'God' as 'out there', or exterior, and who lean to the Greek, Neoplatonist abstraction which finds 'God' more like an 'it' or nameless divinity or the 'One'. Dionysius insists: 'God' is not a being. If 'God' is not a discrete entity, then he is everywhere and nowhere, particularly if, as in the case of Erigena, the manifest world emanates from him, and it becomes possible to equate 'God' with Nature. This equation of 'God' and Nature appears as horrific to the conventional Christian;

it is the entire loss of the personal 'God'. This gulf between the 'God' of the *jnanis* (which appears horribly pantheistic to Christians) and the anthropomorphic 'God' became central to Christianity's pained withdrawal from an emerging modernity, as we will see.

4.2.4 The Renaissance and the Reformation

The *via negativa* inwardness and world-denying stance of the Christian adventure gradually gave way to a guardedly world-curious attitude in favour of art and science, reaching an early peak in the Renaissance. Here we enter a period that the secular mind has some empathy for. Its hero is Leonardo da Vinci, universally recognised as one of the greatest artists of all time, and amongst scholars like Michael White as the first true European scientist.[271] Leonardo makes clear in his notebooks that, while Greek thought may preoccupy his contemporaries, he refuses to 'deck himself out with their purple prose'. Instead he advocates, 150 years before Francis Bacon, the power of observation. He applies this principle in art and in science, creating the first wave of revolt against Aristotle that was essential for the birth of modern science.

But the modern mind brackets out the core feature of the Renaissance without which its renewal of the arts and science cannot be properly understood: its religious dimension, or to be more precise the flowering of Neoplatonism. Where Stephen Jay Gould understands and promotes the idea of science and religion as separate magisteria (his NOMA theory), he fails to spot when Leonardo is speaking from a religious basis: Gould criticises him for pursuing a key tenet of the Renaissance: 'as above so below'. Gould published an attack on the reputation of Leonardo of being vastly ahead of his time, suggesting instead that Leonardo actually held to the 'antiquated' fifteenth century views of his era. It was the parallel that Leonardo drew between the human body as microcosm and the planet as macrocosm that Gould found unscientific.[272] But it was this very idea that galvanised the period, and which is quintessentially Neoplatonist, and derives, as we saw, from Pythagoras. To criticise Leonardo for holding this view is to believe that religion and science could at that period have possibly been so easily separable. It was the hallmark of Neoplatonism, with its roots in Greek thought, that religion and science were one and the same discipline. In our analysis the Renaissance marks the point where the Neoplatonism of Dionysius, Erigena and Eckhart, which were largely *jnani via negativa,* bursts forth as the *jnani via positiva*. It is the original vision of Pythagoras – repressed by the *bhakti via negativa* of Christianity – that flowered, along with an explosion of interest in the occult.

Bertrand Russell comments that the fifteenth-century Italian Renaissance replaced the authority of the Church with that of the 'ancients' – the Greek philosophers, and particularly Plato and Aristotle.[273] Russell adds 'The humanists of the time, however, were too busy acquiring knowledge of antiquity to be able to produce anything original in philosophy.' He goes on: 'The first effect of emancipation from the Church was not to make men think rationally, but to open their minds to every sort of antique nonsense.'[274] (This unfounded criticism is interesting because it parallels the criticism of the New Age and new religious movements of the present time, that with the end of religious authority people now believe in anything rather than nothing.) While Leonardo drew little directly from either Plato or Aristotle, other artists were profoundly drawn to the Greek revival, and many of them through the Florentine Academy of Marsilio Ficino (1443-1499).

Marsilio Ficino and the Florentine Academy

Ficino was born in Florence and barely travelled away from it, but was the leading influence on the intellectual and artistic life of that period. The great Renaissance artists, Botticelli, Michelangelo, Raphael, Titian, Dürer, and many others were directly inspired by Ficino. He founded the Florentine Academy in 1462 as a direct successor to Plato's Academy; it survived his death until the year 1522. Although Ficino modelled himself and the Academy on Plato's thought, the greater emphasis on music and the visual arts meant that the Florentine Academy was more Pythagorean than Platonic, that it ultimately owed more to Croton than to Athens. Plato was sceptical of music and the visual arts, and it was rather the Pythagorean idea of the 'music of the spheres' that entered the ethos of the Italian Renaissance. Ficino's vision was deeply spiritual, and he translated not only the whole of Plato into the contemporary Latin, but also the works of the Hermetic tradition and the *Enneads* of Plotinus. He stands like Pythagoras and Plotinus in the West as founder of a spiritual community or *sangha*, comparable to that of the Buddha's but with one radical difference: in our terminology Ficino's community was not just *jnani* in its spirituality, but profoundly *via positiva* (and also esoteric). The central metaphor of this spiritual teaching was, as mentioned earlier, 'as above, so below', or man as microcosm. It is a world-view incompatible with secular thought, as we saw with Gould, because the cosmology it represents is poetic or imaginal rather than of modern science. (Its contemporary prophet is Richard Tarnas – his book *Cosmos and Psyche* is a re-assertion of the Renaissance vision in semi-contemporary terms.)[275]

The idea that the human being is a mirror of the universe, or that it in some way recapitulates or contains the cosmos, is deeply spiritual because it speaks of a fundamental identity of self with the world, a fundamental connectedness. In the reductionist, materialist worldview, where there is only 'extended stuff', the human body is its own limit and boundary, irrevocably separate from other bodies, the planet and the universe. Hence the Renaissance vision is incommensurable with the modern scientific world view, and Gould's discomfort with Leonardo inevitable. The Neoplatonist or rather Neopythagorean vision of Ficino's Academy (which lives on in the New Age) is a holistic vision prior to the alienation of the modern mind.

In the spiritual teachings of the *via negativa* there is rarely an attempt to account for the origins of man or the cosmos, and certainly in none of the canonical and non-canonical sayings of Jesus do we find any such discussion, for example. This *via negativa* passage from the Gospel of Thomas shows how he typically avoids a discussion of cosmogony, cosmology or teleology:

> The disciples said to Jesus, 'Tell us how our end will be.' Jesus said, 'Have you discovered, then, the beginning, that you look for the end? For where the beginning is, there will the end be. Blessed is he who will take his place in the beginning; he will know the end and will not experience death.'[276]

The Buddha, as we saw, is quite explicit that he holds no views on such matters, as they are irrelevant to the spiritual path that he teaches. But the Judaic tradition is intensely bound up with such descriptions, and even though Jesus, and for that matter Paul, does not dwell on cosmology, it enters Christianity through Augustine and the Hebraic heritage of the Old Testament. Once the Papacy became the supreme arbiter of doctrine, it gradually fell to it to hold views on cosmology and cosmogony, leading to its famous quarrel with Galileo. The cosmogony of the Church is know as Mosaic cosmogony, i.e. of Moses, while the cosmogony of the Neoplatonists was radically different, and ultimately had to be repressed.

The very term 'as above, so below' speaks of an interest in the stars, and the belief that their configurations above influenced the destinies of people below. Hence the Neoplatonist tradition is intimately bound up with astrology, one of the esoteric disciplines that flourished in this period. It was the intense interest in astrology that led to the close observations of the planets, which in turn led to modern astronomy. It was this child of astrology that initiated the scientific revolution that was to deny its heritage and the discrediting of astrology and the entire Neoplatonist worldview.

Pico Della Mirandola

Ficino's Academy attracted possibly the most daring of Renaissance thinkers, Pico Della Mirandola (1463-1494), who, at the age of 23 published his *Oration On the Dignity Of Man*, a pamphlet that has summed up the essence of the Renaissance for generations, possibly its best 'manifesto'. Pico defines the dignity of man in a way that would be strange to us, as a creature with no 'visage' of his own:

> At last, the Supreme Maker decreed that this creature, to whom He could give nothing wholly his own, should have a share in the particular endowment of every other creature. Taking man, therefore, this creature of indeterminate image, He set him in the middle of the world and thus spoke to him:
>
> 'We have given you, O Adam, no visage proper to yourself, nor endowment properly your own, in order that whatever place, whatever form, whatever gifts you may, with premeditation, select, these same you may have and possess through your own judgement and decision. The nature of all other creatures is defined and restricted within laws which We have laid down; you, by contrast, impeded by no such restrictions, may, by your own free will, to whose custody We have assigned you, trace for yourself the lineaments of your own nature. I have placed you at the very centre of the world, so that from that vantage point you may with greater ease glance round about you on all that the world contains. We have made you a creature neither of heaven nor of earth, neither mortal nor immortal, in order that you may, as the free and proud shaper of your own being, fashion yourself in the form you may prefer. It will be in your power to descend to the lower, brutish forms of life; you will be able, through your own decision, to rise again to the superior orders whose life is divine.'[277]

This *mutability* of man is a strange idea to us, but expressive of the medieval view of 'man as microcosm', that in some way the human spirit contained within it as a reflection all of existence. But this mutability was a new optimism that man was unshackled by the existing order, and could achieve things for self or humanity through his efforts. Pico's oration is not humanist in the modern sense, however, because it is *entirely* religious. It was an introduction to his extraordinary call to the Renaissance world to shift its religious thinking radically. Pico is a quintessentially Neoplatonist thinker, but that does no justice to the range of his religious receptivity. As mentioned earlier, he constructed a challenge to debate 900 theses from pretty much all the known religious and spiritual sources of the time, many provided by Ficino's translations. If Pico's dream of open spiritual debate had been realised the Church would have had to undergo not just one, but two revolutions. For the Church to truly open itself to the non-devotional was an impossibility of centuries. Secondly, it would have had to embrace an alien pluralism: in effect it to undo the prohibition of idolatry. Instead, this was the period in which 'syncretism' became a dirty word. Pico's oration contains a proposition repugnant to the Church at the time, that the mutability of man not only allowed for a descent to any level of depravity or brutishness, but also to rise about the angels, to be one with 'God' himself. This is how Pico puts it:

> If vegetative, he will become a plant; if sensual, he will become brutish; if rational, he will reveal himself a heavenly being; if intellectual, he will be an angel and the son of God. And if, dissatisfied with the lot of all creatures, he should recollect himself into the centre of his own unity, he will there become one spirit with God, in the solitary darkness of the Father, Who is set above all things, himself transcending all creatures.

The Buddha himself would have recognised such a statement, particularly in the notion that the human being can rise above the angels or gods. To 'transcend all creatures' is quite central to Buddhism, the 'solitary darkness' recognisable to the Buddhists and in the West as part of the negative theology of Dionysius, but to the Church of Pico's era such a statement is blasphemy. Man's mutability was not an accepted understanding of man either, and may well have been Pico's genuinely original insight for the era. It echoes later in Nietszche and Sartre. In Nietszche's *Zarathustra* the story opens with a tightrope walker attempting to cross between two towers: the image symbolises man stretched between animal nature and godlike destiny. For Pico this was an entirely positive idea, that an individual could make themselves into the likeness of anything; for Nietzsche the choice was an unbearable tension, and led to madness. In Sartre it was the starting point of his Existentialism, that a person was pure potential. But he saw it in a secular context, initially as a personal struggle to become what one has chosen to be; later on, as the Marxist side of his thinking dominated, he saw it in political terms. 'Saint' Genet, in Sartre's view had exercised his dignity in choosing the 'lower' path; in some way Pico had made that a possibility. It is worth adding that the mutability of man has another spiritual progenitor: in the 'shape-shifting' ability of the shaman. The shaman likewise 'has no visage proper to himself', but through contemplation on the manner-of-being-in-the-world of animals takes on their inner characteristics, even donning their masks.

Giordano Bruno

The Renaissance opened up the mind to Greek influence, but also allowed for interest in many traditions now derided as superstitious, e.g. astrology and alchemy. The instinct towards magic, clairvoyance, astrology and alchemy can be subsumed under the term esotericism or occultism as we discussed in Chapter Two. We can find this spiritual orientation in the key Renaissance figure of Giordano Bruno, born in 1548 and burned to death as a heretic in 1600. At first glance Bruno is clearly of a *jnani* orientation, inheritor to the Neoplatonist tradition, and inevitably at odds with the devotional orientation of the Christian world he finds himself in. But, like Pico and many others, his interests are also clearly occult as we have defined them, making Bruno an awkward hero for the contemporary humanist.

In 1590 he took up residence in the Carmelite convent in Frankfurt am Main (Germany), the Prior of which made this famous comment: 'Bruno was a universal man who did not possess a trace of religion and who was chiefly occupied in writing and in the vain and chimerical imagining of novelties.' This comment is the tragic judgement of the Western religious tradition on the *jnani via positiva* temperament. Bruno's contribution is dismissed by the *Catholic Encyclopedia* thus: 'His system of thought is an incoherent materialistic pantheism.' It comments: 'In Germany, whither he went in 1587, he showed the same spirit of insolent self-assertion as at Oxford.' This 'insolent self-assertion' – his independence of thought – is precisely what humanists have revered Bruno for, and led to his becoming an iconic figure of the Renaissance for subsequent freethinkers.

The Renaissance period offers difficulties to the secular mind. It is baffling to find that the 'freethinking' of the time was more concerned to think freely *about* religion rather than *against* it. Neither does any real science come into being, that being reserved for the middle and late seventeenth century with Galileo and Newton. Bruno's and Pico's vision is *learned*, resting on an erudition beyond any but the wealthiest, and ignoring the real spiritual impulse of the time that produced Luther and Calvin, that is a grassroots piety that wanted freedom from a corrupt Church, and had no access to Neoplatonist learning.

Reformation: Iconoclasm and Biblical Literalism

The spiritual pluralism of the Florentine Academy might well have spread if it had not been for a return to Biblical literalism engendered later on by the Reformation and Counter Reformation. The start of the Reformation dates from 1517 when Luther nailed his '95 theses' to the door of the Castle Church, Wittenberg. Martin Luther (1483-1546) and

John Calvin (1509-1564) were the principal figures in the revolt against the medieval Church. Although much useful reform and clearing away of corruption was achieved, the results in terms of *toleration* were negative. The new ideas might have signalled, in the Northern countries at least, a new spiritual pluralism, such as Pico was putting forward. But the deeply conservative nature of Christian religiosity made that impossible. Instead the major dynamic can be characterised as 'the narcissism of minor difference', resulting for example in the judicial torture and murder of Jean Calas, as we saw earlier, motivated by his Huguenot origins. It might have been expected that Luther and Calvin would reject the very principle of Inquisition, given their claims of the right to dissent, but, it turns out, they precluded such rights from others. Calvin viciously persecuted the Spaniard Michael Servetus, who was burnt alive on 27 October, 1553, while Luther wrote: 'Whoever teaches otherwise than I teach, condemns God, and must remain a child of hell.' John Stuart Mill commented: 'Those who first broke the yoke of what called itself the Universal Church [The Roman Catholic Church] were in general as little willing to permit difference of religious opinion as that church itself.' [278]

Apart from an increase in religious violence, the two worst aspects of the Reformation were the retreat to biblical literalism on both sides, and the eruption of iconoclasm on the Protestant side. Where Ficino's Academy provided the new model for a profound relationship between the arts and the spiritual impulse, Protestantism recapitulated the hatred of 'idols' that motivated the early Christian mobs, who for example defaced the statues of the Parthenon in the fifth century. In Britain under Cromwell, his uneducated soldiers were fired to such a hatred of all things 'Papist', that they were given free reign to destroy Church art up and down the country: Britain lost some of its greatest art treasures in that period.

The Suppression of *Jnani*

We can speculate that without the Reformation, the Church, which had initiated astronomical questions in the first place in order to improve its calendar, might have accommodated itself with science, as it did with the early attempts of the Scholastics. It might even have accommodated the radically *jnani via positiva* ideas of the Florentine Academy. But it retrenched in the face of the Reformation schism, and introduced a new Inquisition that brutally suppressed Protestantism where it had sufficient military power. Without the Reformation it is possible that the Church, confident enough not just to reinstate Aquinas, but to canonise him, might have absorbed some of these influences. It is also possible that reforming religions could have opened doors to Neoplatonist impulses, but the retrenchment was complete on both sides, and the Reformers in particular went 'back to basics' – a *via negativa* austerity and Hebraic fundamentalism – the origins of the fundamentalism in the American religious right today.

We can recapitulate the history of Christian religious repression of the Hellenic *jnani* spiritual impulse through some iconic moments. St Paul opposed Greek learning: 'knowledge puffeth up, but charity edifieth.' (I Corinthians 8:1). Tertullian in the third century gave priority to Jerusalem and the Church over Athens and the Academy. Hypatia, a leading early Neoplatonist, was murdered by a Christian mob in 415; the same mob was encouraged to burn down the Alexandrian Library. Christians throughout the Roman world defaced sculptures and other works of art as 'idolatry' (the Taliban in 2001 blew up the Bamiyan Buddhas for the same reason). Erigena's work was suspect in the ninth century and was placed on the Index in 1681, at the height of post-Reformation religious intolerance. Eckhart's work of the twelfth and thirteenth century was subject to the Inquisition, Pico della Mirandola's work placed on the Index and he was forced into exile, and Giordano Bruno was burned at the stake in 1600. Religious repression, including the murderous witch-hunts of the sixteenth and seventeenth centuries, was a feature of Christianity. The incidents listed above have a common pattern: they target the *jnani via positiva* and esoteric spiritual impulses. Rather like the Communist pogroms against 'intellectuals' that forced academics to work and die in labour camps across Russia and Asia, Christian Inquisitions and persecutions (while targeting uneducated village herbalists such as Herlofs Marte) made the life of the mind a dangerous one. Effectively the post-Renaissance West held a huge, but deeply suppressed, collective impulse to the intellectual, the rational, and the *jnani*. When its time finally came, it would burst forth in the Enlightenment as the revenge of rationalism.

There is another significant piece of evidence that Christianity is at heart a devotional religion, nervous of the natural counterpart, which we identify as *jnani*. We have earlier discussed the nature of prayer as a Christian practice, in the context of the devotional mystics, as in its heights being an offering of love to 'God', Jesus, the Virgin Mary or the saints. In the *bhakti* traditions of India this object could be the goddess Kali, in the case of Ramakrishna, or Ram or Krishna, and often just the guru. The term 'meditation' is more generally used for Buddhist-style *jnani* practices, which include *meta*, the meditation on loving-kindness, and the mindfulness of breathing. As a shorthand, 'prayer' tends to signify a *bhakti* practice, while 'meditation' a *jnani* practice. Hence it is significant that the Roman Catholic Church, which became increasingly hostile to Buddhism in the twentieth century, also condemns 'meditation' when understood as an Eastern or Buddhist practice. Cardinal Ratzinger (inaugurated Pope Benedict XVI in 2005) created much offence in Buddhist communities in 1997 when he characterised Buddhism as an 'auto-erotic' spirituality, meaning a narcissistic practice, and the attainment of *nirvana* as complete indifference to the world. Even worse was his statement: 'it seeks transcendence without imposing concrete religious obligations.'[279] This is in complete ignorance of the overwhelming place that the 'rules of the *sangha*' take up in the Pali Canon, being precisely a listing of religious obligations or precepts that are central to all forms of later Buddhism. But it is the particular anxiety about 'meditation' that is significant. In the 'Letter to the Bishops of the Catholic Church on Some Aspects of Christian Meditation' issued by the Vatican's Congregation for the Doctrine of the Faith the Church in 1989, it makes clear that 'Eastern' forms of meditation are unsuitable for Christians for much the same reasons as Buddhism is generally in 'error'.[280]

4.3 A Radical History of Western Development

A conventional account of Western cultural history considers that its development took place within the creative tension between Hebraic and Hellenic models of thought. The secular viewpoint generally understands this as a struggle between a religious, or theological worldview, pitted against Greek rationality, in other words it brackets out Greek spirituality. The account pursued in this book expands the conception of the Hellenic as including both the rational/scientific and a religiosity precisely defined in terms of a particular clustering of spiritual impulses. The Two-Fold Model of Spiritual Difference allows for a quite radical refinement of the Hebraic/Hellenic argument. Where the secular mind has assumed that the inevitable trajectory of Western development must lead to the triumph of a Hellenic rationality, and the eradication of religion as mere superstition, the analysis here shows that the defining creative tension in the Western mind is between two different modalities of the spirit. The concept of *jnani* is crucial to this, and, while it is exemplified in the work of Plotinus, it needs an Eastern context in which to understand it, and to rescue Plotinus – and the whole of Neoplatonism – from the label of 'philosophy'. The East also provides important examples of religious traditions in which there is no creator 'God'. This helps place in perspective a key issue that dogs all of Enlightenment religious debate, as we will show in the next chapter.

Five:
The Undefended Western 'God'

In this chapter we return to the Enlightenment to further probe its role in the origins of the secular mind. We have seen that countless scientists found in their science no contradiction to their religion, no disenchantment or atheism; on the contrary their science contributed to their religiosity because, in the words of Newton, it helped them 'see into the mind of God'.

Neither did most theorists of the period find in their broader rationalism any route to atheism, but rather to a *jnani* religiosity, which can be considered as variants of an emerging Deism. Only Holbach and La Mettrie took scientific rationalism to an atheistic extreme, and perhaps only Holbach in the entire period also argued that religious cruelty was a reason to abandon religion entirely, rather than to initiate tolerance. Amongst the vast majority of Enlightenment thinkers we discover a massive *engagement* with religion, so how is it that later thinkers, such as Marx and Bentham, became the exemplars of the modern world? How is it that the total disengagement of Marx and Bentham became the model for modernity? Religious history is one of continuous change and innovation, the most recent upheaval prior to the Enlightenment being the Reformation. Why was it then that the collective efforts of the majority of Enlightenment thinkers did not create a new spiritual revolution? Why is the legacy of the Enlightenment the opposite: a successful secular revolution and a failed religious revolution? Or, to put it more colourfully, why did the deeply creative religious intellectuals of the Enlightenment fail to prevent the death of 'God' at the hands of a tiny atheist minority?

5.1 The Enlightenment Reconsidered

5.1.1 The Early Enlightenment

Isaac Kramnick, in his introduction to *The Portable Enlightenment Reader*, says: 'Central to the Enlightenment agenda was the assault on religious superstition and its replacement by a rational religion. . . .' This rational religion was to be known as Deism, which Kramnick goes on to describe: 'in which God became no more than the supreme intelligence or craftsman who had set the machine that was the world to run according to its own natural and scientifically predictable laws.'[281] But this is a caricature of Deism, one that both conventional Christianity and modern secularism conspired to create. From the preceding chapters it will be clear that something much richer was brewing as a potential new religion of the West, one that could have been the Enlightenment legacy.

But the Enlightenment is not an easy period to fix with definite dates, as significant precursors go back to the Renaissance; though Kramnick for example suggests the period from the 1680s to the 1790s.[282] It is useful therefore to consider three broad periods: an early Enlightenment, the heyday of the Enlightenment, and the Enlightenment legacy. We could rather arbitrarily include amongst early Enlightenment figures those who died within the seventeenth century, which would include Bacon, Kepler, Mersenne, Hobbes, Descartes, Pascal, and Spinoza, all of whose thought contributed to the foundations of the Enlightenment project. In seeking out the precursors to the secular worldview, it is impossible to find in the thinking of these men *any* hint that they wished to see an end to religion itself. But the Enlightenment period is a time of a rationalism re-asserted, perhaps even the revenge of rationalism – understood as a Hellenic impulse – over the Hebraic. Hence it is instructive to look in depth at the work of the three Continental rationalists.

5.1.2 Descartes, Spinoza and Leibniz

Descartes, Spinoza and Leibniz did not appear to succumb to the double whammy of science-as-atheism and the conscious or unconscious revolt against religious cruelty. They were deeply religious men. We saw in earlier that Descartes' intended epitaph 'He who hid well lived well' stood for all the philosophers of that period, so that these three men were forced to guard their true opinions. But we are not concerned here with what their contribution is to formal philosophy. Instead we want to show that they are at times as spiritually interesting as Dionysius, Erigena and Eckhart, and in some ways their natural inheritors. This is by way of pursuing the question: what exactly was the spiritual revolution of the Enlightenment, and why was it ultimately unsuccessful? We will also show just how little of their spiritual legacy is understood today, by looking at Thomas Damasio's books on Descartes and Spinoza and the question of consciousness.

Proofs of the Existence of 'God'

For any modern secular reader approaching the works of Descartes, Spinoza and Leibniz, the greatest stumbling block is the sheer volume of writing devoted to the subject of 'God', including proofs of his existence. For Bertrand Russell, as a professional philosopher and determined atheist, the proofs are an exercise in philosophical logic, located within the history of such proofs starting with Aristotle and continued by Aquinas (who added four more). Russell considers that 'Leibniz brought into their final form the metaphysical proofs of God's existence', including those of Descartes,[283] though as Russell points out 'modern theologians no longer rely on them.'[284]

Descartes uses the familiar Scholastic justification for venturing to prove the existence of 'God': to convert the infidel. He points out that mere recourse to scripture is circular, because the infidel does not accept Scripture in the first place, and instead suggests, touchingly, that pure reason will do the job. He says 'all which can be known of God may be made manifest by reasons obtained from no other source than the inspection of our own minds.'[285] That such a venture is a complete failure is demonstrated by the entire secular world of philosophy that reads Descartes, considers his proofs in terms of their philosophical virtuosity or lack of it, and remains unmoved. For a *bhakti* the 'God' of their devotional impulse is created in each movement of that impulse, and to offer rational proofs through 'the inspection of our own *minds*' is absurd, as the movement is of the heart. For a *jnani*, like the Buddha, Dionysius or even Descartes himself, the *jnani* impulse is prior to any rational process of thought: 'God', in whatever terminology that ultimate principle is expressed, is 'proved' long before the argument. Hence we can say that the proofs of the existence of 'God' in Descartes, Spinoza and Liebniz are not amongst their most original contributions to the life of the spirit, but are the first evidence of their *jnani* engagement with religion, as it was with the Scholastics.

The Conundrum of Consciousness

But the three Rationalists – Descartes in particular – left a much more interesting legacy: the conundrum of consciousness. It is inevitable that a *jnani* spiritual orientation, with its emphasis on enquiry, should lead to profound questions regarding the nature of consciousness (the *bhakti* on the other hand is not usually inclined to such enquiry or believe that any answers so derived would be at all illuminating). When that *jnani* orientation is coupled with a philosophical inclination, then speculation on 'mind' or consciousness is pressed hard. The later Buddhist traditions were aware of the Buddha's reluctance to engage in speculative enquiry, but despite this unleashed a vast philosophy of mind that is to this day far more sophisticated than that of the West. The principle reason for this is that the East had in the practice of 'meditation' a tool for examining mind which simply did not have a counterpart in the Western devotional practice of 'prayer' (as discussed earlier). In any case, the Enlightenment thinkers may not always have pursued a consistent or appropriate religious practice (yoga) to underpin their religious thinking.

Descartes, Spinoza and Leibniz were confronted with questions about mind or consciousness that were greatly sharpened by the emerging science of the time, and the image – pursued so vividly by La Mettrie – of nature as a machine. If 'extended stuff' – to use Descartes' terminology – was so rapidly becoming reducible to different kinds of energetic exchange between material substances, then 'thinking stuff' became all the more the repository of the sacred. Science, in reducing the domain of the objective to machinery, had made the subjective domain more visible as 'other' posing the never previously necessary question about the relation between two kinds of 'stuff'.

Descartes

Descartes is principally remembered for his mind-body distinction, or, using the Latin terms, his distinction between *res cogitans* and *res extensans*, thinking stuff and extended stuff. His best-known aphorism is: 'I think, therefore I am.' But what exactly did Descartes mean by 'think'? On this rests the true meaning of *res cogitans*, and also our ability to relate Descartes's thinking to Eastern ideas, and to the *jnani* principle in the Western tradition. Hence it is worth quoting Descartes in full on this subject:

> What is a thinking thing? It is a thing that doubts, understands (conceives), affirms, denies, wills, refuses, that imagines also and perceives. (*Meditations*)[286]

> By the word thought, I understand all that which so takes place in us that we of ourselves are immediately conscious of it; and, accordingly, not only to understand (intelligere, entendre), to will (velle), to imagine (imaginari), but even to perceive (sentire, sentir), are here the same as to think (cogitare, penser). (*Principles of Philosophy*)[287]

This is remarkable because of its breadth: it effectively covers the *entire* contents of consciousness. Most commentators have assumed that Descartes's *cogito* means to think as in *ratiocination*, in other words they assume that Descartes is talking about discursive, analytical, language-based thought. From Descartes' own definition, stated unambiguously and in two separate major works, it is clear that it covers nearly every conceivable element that might present itself to consciousness, that can enter into the scope of human awareness. There is one possible omission however: feeling or emotion. Was this a deliberate omission, one that would place feeling or emotion under a category other then *res cogitans*? Clearly not, as he does not mention feeling or emotion when discussing *res extensans*. The conclusion must be that he omitted feeling or emotion because, in Jungian terms, he was himself a 'thinking type', confirmed for example by his dismissal of the heart in his book *The Passions of the Soul*.[288] Would it distort his thesis if we included thinking or feeling into his definition of 'thought'? Not in the slightest. However, the point here is that 'thought' for Descartes can only be properly understood as 'contents of consciousness', or even consciousness itself. It is here that the Buddhist theory of mind is infinitely more sophisticated than the Western tradition, but Descartes had no access to it. What we can look at is the possible analogy between the practice of meditation in Eastern *jnani* traditions, and Descartes own attempt to penetrate the mystery of consciousness. Descartes tells us as at the start of the *Third Meditation*:

> I will now close my eyes, I will stop my ears, I will turn away my senses from their objects, I will even efface from my consciousness all the images of corporeal things; or at least, because this can hardly be accomplished, I will consider them as empty and false; and thus, holding converse only with myself, and closely examining my nature, I will endeavour to obtain by degrees a more intimate and familiar knowledge of myself.[289]

Clearly Descartes is describing the mental preparation for what he termed the 'inspection of our own minds', that inspection which he believed would yield sure knowledge of 'God', and which is reminiscent of Plotinus. Descartes calls these writings 'meditations' and the description in this passage if presented to any Buddhist would strike them as nothing less than preparation for the act of meditation as understood in their tradition. Any meditator attempting to efface from consciousness 'all the images of corporeal things' would agree with Descartes that this 'can hardly be accomplished' (i.e. that it is hard to do), but whether it is a Christian contemplative instructed in the methods of the *Cloud of Unknowing* or whether it is a Mahayana Buddhist instructed to see the emptiness inherent in perception, they would find it a familiar proposition 'to consider them [images of corporeal things] empty and false.'

In saying 'I think, therefore I am', Descartes was not therefore identifying self with discursive thought, but with consciousness.

This was over and against the alternative: to identify oneself with the body as extended stuff. Here we find a resonance with all the *via negativa* thinkers, both *bhakti* and *jnani*, who saw in the finite and mortal nature of the body no resting place for identity. Descartes' practice of doubt was extended to the manifest world, to extended stuff, in the same spirit as the Buddha characterised it as impermanent or perishable, but Descartes did not doubt thinking stuff. His division between mind and body is further made clear when he says 'we cannot conceive body unless as divisible; while on the other hand, mind cannot be conceived unless as indivisible.'[290] (New Agers note: Descartes is holistic in his approach to mind, and reiterates it in *The Passions of the Soul* where he says 'For there is only a single soul within us, and this soul has within itself no diversity of parts.' This is in contradiction to modern psychology which endlessly multiplies components of the psyche.) In the *Meditations* Descartes goes on to say that 'the destruction of the mind does not follow from the corruption of the body'. Such ideas are clearly prior to his attempt at a philosophical and 'scientific' demonstration of them; they belong to the worldview of the mystic, and are articulated in the language of the *jnani*.

It is worth examining how Descartes is drawn to spiritual metaphors around infinity – a fact picked up on much later by the twentieth-century philosopher Emmanuel Levinas. Infinity as a mathematical concept is empty of spiritual significance, but in the history of the spiritual life it is used over and over again to convey an expansion of the soul, and the deep connectedness of existence. It conveys a sense of boundlessness, or indivisibility, Descartes' principle characterisation of mind. If the bounded is loathed by its possessor, as Blake says, then the unbounded inspires the opposite: a soaring elevation of the soul. But we have to ask, given that Descartes is also a brilliant mathematician, how is he speaking of the unbounded, of infinity? The following passage should answer that question:

> And I must not imagine that I do not apprehend the infinite by a true idea, but only by the negation of the finite, in the same way that I comprehend repose and darkness by the negation of motion and light: since, on the contrary, I clearly perceive that there is more reality in the infinite substance than in the finite, and therefore that in some way I possess the perception (notion) of the infinite before that of the finite, that is, the perception of God before that of myself, for how could I know that I doubt, desire, or that something is wanting to me, and that I am not wholly perfect, if I possessed no idea of a being more perfect than myself, by comparison of which I knew the deficiencies of my nature?[291]

This passage is densely packed with spiritual significance. Central to its *jnani* character is his observation that *he possesses the perception of the infinite before that of the finite*. He has made clear that this is a positive perception, rather than one arrived at by the negation of another perception, and it is also clear that his discussion is not one of mathematical principles. If we make the completely plausible translation of the 'infinite' in its Western context to the 'imperishable' in the Indian context, then it is remarkable to note Mahatma Gandhi's assertion that one 'cannot apprehend the imperishable'.[292] Yet the world remembers Gandhi as a spiritual giant, and Descartes as the founder of modern atheist philosophy! Here is evidence that Descartes is much more of a mystic than Gandhi, a more gifted spiritual thinker. We find references to both the infinite and the eternal throughout Descartes' work, but he leans to the metaphor of space more than time. It is no accident therefore, when he was casting about for a term to convey the nature of matter, that he coined *res extensans*. For a thinker inclined to spatial metaphors, the principle characteristic of matter is rather naturally extension.

Descartes, Modern Science and Damasio

Daniel Dennett in his book *Consciousness Explained* reproduces an engraving due to Descartes showing in schematic form how sensations impinge on the optic nerve, travel to the brain and on to the pineal gland, from which all sensations are transmitted to the mind or soul. The soul then acts by willing a bodily response, such as the movement of a hand, and this is transmitted via the pineal gland to the brain, and then to the muscles of the arm.[293] Dennett dismisses this, not because of the choice of organ, but because the whole scheme lies outside of science as we know it. Unless the 'soul', 'mind', or 'consciousness' is a physical thing, there can be no interchange of energy with the body, and hence no interchange of information as perception or instruction. This means that the 'upward causation' involving perception as Descartes describes it has no proper basis, and the same is true for 'downward causation', the acting of the will on our bodies. (Note that Holbach, writing in 1770, already ridiculed this idea.[294])

Antonio Damasio in his book *Descartes' Error* agrees with Dennett that Descartes is simply wrong, but on neurological, not cognitive grounds. Despite Descartes' contributions to the science of his day via his mathematics (the Cartesian coordinates by which extended stuff is mapped out), he did not have the empirical instincts of Newton; his mechanics was plain wrong (though a slight improvement on Aristotle);[295] and his biology of the pineal gland even more mistaken. Science rejects Descartes' division of mind and matter by denying mind, while the New Age dismiss him as the cause of the alienating 'split' between mind and body; religionists ignore him as his religious insights are outside of the bounds of theology. But Descartes' great contribution to the modern world is precisely his Cartesian division: it poses in the starkest possible terms a problem that *has* to be confronted: the conundrum of consciousness.

Damasio believes that Descartes' own solution doesn't work, calling it 'Descartes' Error'. (Damasio is otherwise gracious to him in acknowledging the man's substantial contributions to the thought of the day.) Damasio says: 'This is Descartes' error: the abyssal separation between body and mind, between the sizeable, dimensional, mechanically operated, infinitely divisible body stuff, on the one hand, and the unsizable, undimensioned, un-pushpullable, nondivisible mind stuff; . . .'[296] As a materialist neurologist, Damasio concurs with Dennett when he says that there is only one kind of stuff. Here then is Damasio's solution to the problem posed by Descartes:

> From my perspective, it is just that soul and spirit, with all their dignity and human scale, are now complex and unique states of an organism. . . . And this is of course the difficult job, is it not: to move the spirit from its nowhere pedestal to a somewhere place, while preserving its dignity and importance; to recognise its humble origin and vulnerability, yet still call upon its guidance.[297]

This is a good, though unintended, epitomisation of what we call the 'neurologising of the self': the attempted movement of the 'spirit' from its 'nowhere pedestal' into the brain. But it is absurd, doomed to failure, and ultimately a violence to human subjectivity. Damasio is a fascinating researcher and writer on contemporary neuroscience, but it seems only fitting to call his reduction of spirit to brain 'Damasio's Error' in full cognition that it is held to by all of neuroscience, and is rapidly becoming the latest shibboleth of secularism. His error is pinpointed in the division made by philosopher David Chalmers – amongst the challenges confronting the recently emerging field of consciousness studies – between the 'easy problem' and the 'hard problem'.[298] Without diminishing in any way the extraordinary achievements of brain science, Chalmers sees this kind of empirical science as only able to answer questions of brain functioning – challenging in terms of technology and science, but easy because the materialist basis of the research is clear and established. The hard problem is the question of experience, the subjective qualities of such things as the colour red, or of pain, which are known as 'qualia'. In other words, whatever neuroscience tells us, it cannot tell us about *res cogitans*, 'mind stuff', or the relationship between the brain and subjective experience, or consciousness. The error of all reductionist science of the brain is to ignore Chalmer's 'hard problem': it is this we call Damasio's Error.

Spinoza

We suggested that Descartes' designation of *res extensans* for the material world is consistent with his conception of the infinite as designation for 'God'. Both are spatial metaphors. Spinoza, while just as happy to use the term infinity to describe the deity tends more to the term 'eternal'. From his writings at least, Spinoza seems to be a more fully developed *jnani* than Descartes. The German poet Novalis described Spinoza as a 'God-intoxicated man', a description more appropriate to the *bhakti* type, but which conveys in any case an intensity of spirituality clearly present in Spinoza.[299] Neither Spinoza's *Ethics* nor the *Treatise on the Emendation of the Intellect (TEI)* was published in his own lifetime, while the works that Spinoza did publish were placed on the Roman Catholic Index. *TEI* was not completed, but contains Spinoza's account of his journey of discovery, rather like the account in Descartes' *Discourse on Method,* in terms of his aims and principles. Like many spiritual thinkers throughout history, he saw that he needed to find a source of joy that was beyond the transient, the impermanent, and the conditioned world, the so-called pleasures of which brought only anxiety in their wake. To start with he could see intellectually that by seeking the more certain good he would be happier, but found it hard: no less a saint than Augustine confessed the same difficulty. Spinoza says: 'Then persistent meditation enabled me to see that, if only I could be thoroughly resolute, I should be abandoning certain evils for the sake of a certain good'[300] His crucial insight however was that 'These evils, moreover, seemed to arise from this, that all happiness or unhappiness depends solely on the quality of the object to which we are bound by love. For strife will never arise on account of that which is not loved; there will be no sorrow if it is lost, no envy if it is possessed by another, no fear, no hatred. . . .'[301] Such objects, it became clear to Spinoza, were perishable things. This insight is almost identical to the Buddha's (or, for that matter, Blake's), but Spinoza's inflection of it is a little different: 'But love towards a thing eternal and infinite feeds the mind with joy alone, unmixed with any sadness.' That thing 'eternal and infinite' had only one authorised term in the Western tradition: 'God'. (Plotinus offered 'the One'; Hobbes preferred the Judaic formulation 'I am'; Hume suggested 'MIND'; and Feuerbach, like Descartes, tended to 'infinity'.) One might even suspect that Spinoza's instinct would turn to the devotional, but we shall see that he remains rigorously *jnani*, even if he speaks of love. The Buddha's turning point in his quest for enlightenment was the acceptance that extreme asceticism would damage the fine balance of his body and make enlightenment impossible.[302] Spinoza seems to have also worked out a balance (missing in the extreme *via negativa* or Gnostic traditions) between the life of the spirit and the life of the body. He saw that:

> The acquisition of money, sensual pleasure, and esteem is a hindrance only as long as they are sought on their own account, and not as means to other things. If they are sought as means, they will then be under some restriction, and far from being hindrances, they will do much to further the end for which they are sought, as I shall demonstrate in its proper place.[303]

Spinoza is adamant that the 'end' which would be furthered by moderation in money, sensual pleasure and esteem, is spiritual. In his own life his moderation was closer to voluntary poverty than to what we would understand by 'sensual pleasure' today: rather he can be seen as rejecting only the extreme of asceticism. In this next extract we have a further indication of the *via positiva*; it also includes a terse *jnani* statement of transcendence:

> Thus he is urged to seek the means that will bring him to such a perfection, and all that can be the means that will bring him to such perfection is called a true good, while the supreme good is

> to arrive at the enjoyment of such a nature, together with other individuals, if possible. What that nature is we shall show in its proper place; namely, the knowledge of the union which the mind has with the whole of Nature.[304]

It is of interest that he would like this enjoyment to be in the company of 'other individuals, if possible'. In fact he continues with the idea that he wants to bring as many as possible 'to think as I do', and proposes a social order favourable to that end, including rules for living together. His prospects for achieving a social goal like this were remote in the extreme, as he was treated with suspicion on account of his ideas, and was excommunicated from his Jewish community early in his life. His philosophy was later described by Hume as 'the most monstrous hypothesis imaginable, the most absurd', or as a 'hideous hypothesis',[305] before Goethe and Coleridge later rescued Spinoza's reputation. Spinoza's confidence and his *intent* at least match that of the Buddha, Pythagoras or Ficino in creating a religious community or *sangha*. And in his assertion that the supreme good is 'knowledge of the union which the *mind* has with the whole of Nature' (our italics) we have a *jnani* statement of transcendence (one which caused him to be called a pantheist).

If we now turn to *The Ethics* we find that it is suffused with references to the infinite and the eternal, and to a God that is those things, and also a God to be loved and who 'loves himself with an infinite intellectual love' (Proposition 35). This phrase alone alerts us to an utterly original expression of the spiritual-intellectual (i.e. *jnani*) life.

The text of the *Ethics* is laid out as a pseudo-Euclidean argument consisting of an intertwined sequence of Propositions (the central ideas), Corollaries (related ideas), Scholia (commentaries), Postulates, Definitions and Axioms, but there is no maths and precious little philosophical logic to it. Even Russell, committed to the entire project of Western philosophy, admits that the 'proofs' in Spinoza's Euclidean style of exposition are mainly spurious. In the *TEI* Spinoza makes it clear that it was however Euclidean mathematics that *inspired* him to seek the truth, and in this sense Spinoza is the natural heir to Pythagoras, Plato and the entire Neoplatonist tradition, one that is uniquely Western. Presumably he hoped that the pseudo-mathematical style would hide his ideas from Inquisitorially-minded authorities. Here are a few extracts from Part I of *The Ethics* called 'Concerning God':

> By God I mean an absolutely infinite being; that is, substance consisting of infinite attributes, each one of which expresses eternal and infinite essence. *(Definition 6, I)*
>
> *There can be, or be conceived, no other substance than God. (Proposition 14, I)*
>
> *Whatever is, is in God, and nothing can be or be conceived without God. (Proposition 15, I)*
>
> Some imagine God in the likeness of man, consisting of mind and body, and subject to passions. But it is clear from what has already been proved how far they stray from the true knowledge of God. These I dismiss, for all who have given any consideration to the divine nature deny that God is corporeal. *(Scholium to Proposition 15,)*

These extracts clearly show that the accusations of atheism levelled at Spinoza are absurd, but at the same time that his 'God' is not the anthropomorphic 'God' of mainstream devotional religion: here instead is the 'God of the philosophers' that Pascal complained about, and which is natural to the *jnani* impulse. Neither does Spinoza agree with Hobbes's conception of 'God' as corporeal. Spinoza also puts forward a simple and direct argument for the perfection of existence; that everything we perceive as imperfect is only from our own narrow perspective and not from the perspective of the object in question. This larger way of seeing things is called *sub specie aeternitatis*, that is 'under the form of eternity', a phrase that has entered the English language through him.

Spinoza shows a *via positiva* outlook when he says: 'For laughter, and likewise merriment, are pure pleasure, and so, provided that they are not excessive, they are good in themselves (Pr.41,IV).' Spinoza goes on to list good food and drink, blossoming plants, dress, music, sporting activities, theatres 'and the like' as 'nourishment' to the various parts of the body. The idea that greater pleasure leads us to greater perfection and to a greater participation in the divine nature is a rare sentiment in the religious life, and shows again Spinoza's originality. Yet something of this must also come from the great Jewish tradition of the love of life, and the enjoyment of its pleasures. Overall, the *Ethics* is almost a poem of divine love, representing one of the greatest statements of love from a Western *jnani* in the cultural context of a theistic language. Spinoza's case suggests that *jnani* and *bhakti* ultimately reach the same goal, where the heart of the *bhakti* is illuminated by spiritual intelligence, and the mind of the *jnani* illuminated by spiritual love.

Spinoza, Science and Damasio

Spinoza made a living making telescopes and microscopes, grinding the lenses for them, but was ultimately an amateur scientist, making no significant contribution to the field. In *TEI* he made it clear that for him science was merely the method to improve society as the precondition for the pursuit of 'God' in favourable circumstances, but in itself would provide no answers. It might have been this which made him much less keen than Descartes to separate mind and matter, and he devoted the opening paragraphs of Part V of the *Ethics* to dismissing the pineal gland as the meeting point of the two, going so far as to call it an occult theory. But Spinoza devoted his early career to teaching the system of Descartes, and found, like all subsequent philosophers, that the Cartesian split presented a challenge worth taking up. His own solution was to assert, with little supporting argument, that there was only one substance: that mind and matter were simply attributes of 'God' seen under different aspects.

Antonio Damasio much prefers Spinoza to Descartes, devoting one of his books on neuroscience to the memory of Spinoza. Damasio claims that there are three widely recognised 'Spinozas': first as radical religious scholar, second as a political architect, third as philosopher using a geometric method. Damasio is interested in a fourth however, Spinoza as protobiologist,[306] and outside of which Damasio largely discounts his thinking. Damasio's belief is that philosophy has prefigured science – which we suggest is a problematic notion given the struggle of the emerging scientists against Aristotle, and the subsequent total disengagement from science of philosophers like Hume. However, given Damasio's claim: what was it in Spinoza that apparently makes him a protobiologist relevant to Damasio's work on neurobiology and emotion? Damasio lauds Spinoza for proposing an ideal state marked by freedom of speech, separation of church and state, and many other ideas that were ahead of later thinkers like Voltaire. Damasio usefully points out that Spinoza's tomb is inscribed with the Latin word *Caute!* – be careful! – in a telling parallel to what we know Descartes intended for his own grave.[307] But it is Spinoza's opposition to Cartesian dualism that attracted Damasio, the idea that thought and extension are attributes of the same substance. Damasio has found in the *Ethics* the idea of 'correspondences' between bodily and mental events, which is at the heart of modern neuroscience: the neural correlates between brain function and subjective experience. This is enough to makes Spinoza a protobiologist it seems. But Damasio's *Spinoza* book founders at exactly the same point as his *Descartes* book, as do all theories of consciousness depending on neural correlates: it avoids Chalmer's 'hard problem' of lived experience. Damasio also ignores Spinoza's insistence that mind and matter are differently modified attributes of 'God'. They are not differently modified attributes of neurons. But this does not put him off in his quest to wrench feeling from the domain of mind-stuff – *res cogitans* – to the domain of extended stuff – *res extensans*. To be precise, Damasio, having moved mind or spirit from its 'nowhere pedestal' into the brain, is now doing the same with feeling. Damasio's error is compounded here: has a scalpel or CAT-scan has ever found in the extended stuff of the brain anything like mind, spirit or feeling? All of which remain unmeasurable, not even down to angstroms or the sub-atomic level?

Damasio's works on Descartes and Spinoza are illuminating however: they show how thinkers of the pre-modern period, when stripped of the religious and spiritual core of their thinking, become almost a blank slate to write upon. Damasio responds to the stronger instinct of Spinoza to the *via positiva* (compared to Descartes), and to the genuinely joyous element in his writing, but to suggest that there is any relationship between Spinoza's sacred unitive vision, and reductionist brain science is absurd.

Leibniz

Leibniz published little of his thought in his lifetime, and then only the most populist of his works. He operated under the same pan-European constraints as Descartes and Spinoza, but was much more visible, and more politically minded. His thinking was influenced, like all the Enlightenment philosophers, by the emerging science, and he participated in its development and promotion. Where Hume wanted to be a 'Newton of the mind' Leibniz believed it possible to apply the scientific method to the areas of politics and religious reconciliation. Leibniz invented the doctrine that this is the best of all possible worlds, and was caricatured for it by Voltaire as Dr Pangloss in *Candide*. There is evidence that Leibniz studied the Kabbalah, put forward by scholar Allison Coudert.[308] She points out that two of Europe's leading scholars of the Kabbalah, Christian Knorr von Rosenroth and Francis Mercury von Helmont, had completed a translation called the *Cabala Denudata* in the seventeenth century. Both Leibniz and John Locke knew von Helmont well and included his Kabbalist works in their libraries. Furthermore, the very term 'monad' used so crucially by Leibniz is widely deployed in the Kabbalah, as well as present in the writings of Dionysius and Giordano Bruno.

We will just consider Leibniz's key work here, the *Monadology*. It can be understood as a profound restatement of the Renaissance insight of 'man as microcosm'. Christian de Quincey (editor of the *Institute of Noetic Science Review*) considers that this idea goes back to Pythagoras and beyond, to the Egyptians, and to shamanic practitioners in much earlier times.[309] He also sees Leibniz's *Monadology* as a key work going beyond monism or dualism, one which restates the Renaissance ideal, and which underpins de Quincey's concept of 'panpsychism' – a world everywhere imbued with consciousness.[310] The *Monadology* is an extraordinary text of the *jnani, via positiva* temperament, and is a fitting summary of the Western esoteric *jnani* traditions since Pythagoras. The *Monadology* was one of the last writings of Leibniz (written in 1714, two years before his death), so it is plausible that he placed in this short text a summary of his own intellectual and spiritual journey.

Leibniz was writing partly as a response to the challenge of Descartes, rejecting both his dualism, and also the solution put forward by Spinoza, that matter and mind are two aspects of 'God'. It is also a response to the challenge posed by the new materialistic science emanating from Newton and Descartes: Leibniz rejected the view of matter as dead, inert, extended stuff that interacted mechanically, leaving 'soul' shut out from the physical universe. He says instead: 'Thus there is nothing fallow, nothing sterile, nothing dead in the universe, no chaos, no confusion, save in appearance. . . .' (69) [311]

His key idea is of a hierarchy of entities known as 'monads' which are simple and unextended: the 'real atoms of nature' (3). Extension is possible by the aggregation of monads into compounds, where the compounds themselves become monads at a higher level. The radical departure from modern materialism is apparent from the start: monads are nothing like the atoms of materialist science, but exist at all levels of hierarchy, are 'windowless', (7) and no two are alike (9). By windowless, Leibniz means that they neither act on each other, nor are changed by another monad, but are synchronised by the pre-existing harmony. This entirely solves the problem that Descartes raised with his division, and which cannot be solved by the pineal gland: in modern terms the problem of upward and downward causation. Of course, no modern thinker would accept such a solution, because it starts by denying that monads can act upon each other. But this shouldn't be such an obstacle: Hume denies causation – thus undermining the very foundation of science – and modern philosophers don't seem to balk at this, but Leibniz's monads cannot be understood in scientific terms: the *Monadology* is more of a sacred poetics.

While monads cannot affect each other because they are 'windowless', what monads do undergo is Perception (14), each one viewing the world according to its Appetition (15) or desires. Leibniz uses an Aristotelian word, *entelechy*, to indicate the purpose or goal-directed nature of the monads, each actualising its own potential. More than that, monads are souls, though the simplest of aggregations, such as a rock, has very primitive perceptions, and the term 'soul' is better used for 'those in which perception is more distinct, and is accompanied by memory' (19). To put it another way, consciousness is intrinsic to matter, but gradated according to the complexity or sophistication of the monad.

This concept of 'panpsychism' stands as an utter refutation of Cartesian dualism; it is at the heart of the Hermetic-Pythagorean-Neoplatonist conception of the universe. It is also close to the ancient shamanic insight that all of Nature is imbued with spirit. It is also utterly incompatible with mechanistic science, though de Quincey and other New Age thinkers believe that quantum theory renews the possibility of panpsychism. Mechanistic evolutionary theory believes that consciousness emerged 'at some point' in the evolutionary chain, being a property of higher animals, or even just of humans, or even only late in human history. Panpsychism finds this claim absurd: how can a category shift take place on the basis of gradual increasing complexity alone? It requires the arbitrary selection of a point in evolution, upon which no agreement can be found, or even the conditions for that agreement. Descartes was forced through his theory to deny animals souls, but Leibniz not only insists that animals have souls, but so do all simpler aggregations of elements.

Leibniz also anticipates some of the key arguments in Consciousness Studies, including Searle's Chinese Room Paradox.[312] Leibniz imagines if 'there were a machine, so constructed as to think, feel, and have perception, it might be conceived as increased in size, while keeping the same proportions, so that one might go into it as into a mill.' (17) If one should then examine its interior one would find nothing to explain perception. Leibniz says that this is because the mill, as a compound object, is not a monad, and only monads have perception. In other words it is the unity of the collective, rather than in the working of the parts, that perception and consciousness is found. This is one of the simplest arguments against the 'neurologising' of self that scientists like Dennett, Damasio and Pert are pursuing: that nothing in the parts will account for subjective conscious experience; instead it is the whole that must be approached for any account of consciousness. At the heart of Leibniz's vision in the *Monadology* is what the New Age call 'holism' – the attempt to think, not in terms of the working of the parts, but of wholes.

In Leibniz's hierarchy of monads he imagines that the human body is made up of the monads of its cells (he was responding here to the advances in microscopy and the findings of Dutch biologist Leeuwenhoek). The cells in turn are made up of lower level monads, each level capable of its own kind of desires and perceptions, unaware of those of the levels below or above. The highest level of monad, capable of the sum total of all desire and perception is of course 'God'. More than this, all lower substances or monads emanate from the highest monad (47): this is an emanationist philosophy such as that of Erigena, and which finds different forms throughout the Neoplatonist tradition. Monads cannot influence each other as they are windowless, only through the mediation of 'God' the highest monad. (51) Leibniz then states the key Renaissance idea: that each monad is 'a perpetual living mirror of the universe.' (56) This is a reframing of the idea of 'man as microcosm', or 'as above so below', and allows each monad, at whatever level of complexity, to participate fully in the universe. More than that, the world is a plenum in which each monad perceives distinctly what is close to it, but is also 'mediately affected by bodies adjoining those with which it itself is in immediate contact. Wherefore it follows that this inter-communication of things extends to any distance, however great. And consequently every body feels the effect of all that takes place in the universe. . . .' (61) In this Leibniz anticipates some of the findings of quantum theory (quantum entanglement), making the *Monadology* again remarkably prescient.

But Leibniz is cautious about the prospects for material science. We saw earlier that Holbach was confident, as scientists are today, that wherever science now fails to explain a thing it is merely a matter of time before it will. The 'particle zoo' of quantum theory shows that such optimism is unfounded, at least where it comes to ultimate explanations. Leibniz didn't need quantum science to reach that conclusion; for him it was obvious from the scientific method itself:

> And as all this detail involves other prior or more detailed contingent things, each of which still needs a similar analysis to yield its reasons, we are no further forward: and the sufficient or final reason must be outside of the sequence or series of particular contingent things, however infinite this series may be.[313]

Leibniz is saying 'sufficient or final reason' – by which is meant the ultimate source or cause of things – cannot be found through the ever-increasing details of materialist science, as the sequence would be infinite. Instead the ultimate source – and he uses the only term for it available to him, 'God' – has to be approached unmediated.

There is a hymnic or rhapsodical quality to the work: Leibniz is certain that the universe is benign, the best of all possible worlds, or a 'City of God'. (85) Whether it is ultimately reconcilable with modern science through quantum mechanics is a question that may preoccupy New Age thinkers, but, more simply, we are interested in demonstrating its *jnani, via positiva* esoteric credentials. These guarantee that it will be dismissed by Christian theology, modern philosophy, and reductionist science. While Russell may be right that within traditions of magic and occultism there are elements properly to be considered as 'antique nonsense', the *Monadology* is worthy of serious study as having unbroken roots with the wisdom of antiquity, while framed in a radically new way. It is also worth pointing out that while Plato (so revered by Russell and other modern philosophers), held to the 'antique nonsense' of reincarnation, Leibniz happens to deny it in the *Monadology* (72). On the other hand there is good empirical evidence for past life recall, made in many academic studies: who can judge so quickly what is 'nonsense'?

The Religious Rationalists

Descartes, Spinoza and Leibniz, as the so-called Continental Rationalists, are clearly deeply rational thinkers, and they applied themselves to almost all the topics of the day, except perhaps the arts. Russell considers Leibniz to be 'one of the supreme intellects of all time'.[314] Yet clearly all three men devoted their exemplary intelligence to profoundly spiritual questions, holding radical religious views as a result, with no conception that religion itself would later be held by intelligent thinkers as *only* superstition and ignorance.

The *Ethics* of Spinoza, the *Monadology* of Leibniz, and to a lesser degree the *Meditations* of Descartes, must rank with the works of Dionysius, Erigena and Eckhart: these six men, that we characterise as deeply *jnani* spiritual thinkers, may disagree at many points, but their engagement with questions of the spirit is profound. Rationality and spirituality find no contradiction in pre-modern Western history, even, it seems, at the point when rationality burst free from its suppressed historical containment.

5.1.3 The Heyday of the Enlightenment

But Descartes and Spinoza are thinkers of the early Enlightenment (as we have chosen to delineate it): what about the Enlightenment in full swing? As we saw earlier, the first atheists and true precursors of the secular mind appear within the eighteenth century. We can list some thirty-nine key figures of this period, in order of date of birth: Locke, Malebranche, Newton, Leibniz, Pierre Bayle, Shaftesbury, Dumarsais, Anthony Collins, Christian Wolff, Swedenborg, Montesquieu, Voltaire, Benjamin Franklin, Buffon, La Mettrie, Hume, Rousseau, Diderot, d'Alembert, Adam Smith, Holbach, Kant, Edmund Burke, George Washington, Joseph Priestley, John Adams, Edward Gibbon, Paine, Jefferson, Lavoisier, Condorcet, Jeremy Bentham, Laplace, Goethe, James Madison, William Blake, Hegel, Faraday and Auguste Comte.

But only four of these are truly atheists, in that they made arguments *against* religion. These are La Mettrie, Diderot, Holbach, and Bentham, of which Bentham fully represents modern atheism, while La Mettrie is more agnostic, and Diderot is borderline, having pursued Deist thinking in many of his works. But even if we assume that Diderot, in assisting his friend Holbach in writing his *System of Nature*, was as atheist as his friend (though Diderot denied it to Voltaire), only about ten percent of leading thinkers of the main period of the Enlightenment wanted the end of religion. How could this small minority have become the intellectual forbears of the secular mind? How could the ninety percent or more, the majority, fail to prevent the death of 'God'? Many commentators today criticise the structures and operation of democracy, but only a tiny minority seriously argue for its end. Similarly many Enlightenment thinkers criticised the structure and operation of religion, but only a tiny minority argued for its abandonment.

If the Enlightenment comprises a spectrum of views, focussed around the concerns of toleration, freethought, rationalism, science, and the separation of Church and state, then these four atheists represent an extreme. But we first need to show that a progressive counter-Enlightenment sprang up alongside the rise of rationalism, one we can broadly call Romantic. Of the Enlightenment figures listed above, Rousseau, Goethe, Blake and Hegel contributed in some way to the Romantic worldview, one that was as anti-clerical as anything one could find in the Enlightenment, but which opposed the growing *materialism* of the period. The Romantics as a larger group (if we include further figures such as Wordsworth and Coleridge) make La Mettrie, Diderot, Holbach, and Bentham look even lonelier as atheists.

5.1.4 Romantics and Idealists

Romantic thinkers and writers of the eighteenth century were anticlerical and individualistic, and spiritually they had more in common with the Renaissance and Ficino than with the emerging materialist science of their time. J. A. Stewart sums up the Platonism that infused the Romantic poets:

> Platonism I would describe, in the most general terms, as the mood of one who has a curious eye for the endless variety of this visible and temporal world and a fine sense of its beauties, yet is haunted by the presence of an invisible and eternal world behind, or, when the mood is most pressing, within the visible and temporal world, and sustaining both it and himself – a world not perceived as external to himself, but inwardly lived by him, as that which, at moments of ecstasy, or even habitually, he is become one.[315]

The individualism of the Romantics was a natural expression of a period during which the dominant role of the Church waned (and was fragmented in fringe movements and schism) and the fascination with science grew. It was its *aesthetic* impulse which drew Romanticism close to the spiritual, quite unlike the rational impulse that drove Descartes, Spinoza and Leibniz. Romanticism also turned to Nature as the source of its aesthetics, its values and its spiritualities.

In the *Marriage of Sense and Soul* Ken Wilber makes an interesting analysis of the Romantics. He suggests that their spirituality was regressive, because of their focus on Nature and their desire to invoke a better, mythical past.[316] Although he criticises the New Age, he effectively supports its progressive stance, one he sharply differentiates from the regressive stance of Romanticism, out of which movement the New Age stems. But Wilber's assessment is predicated on his preference for one side of a massive shift in cultural values that he highlights so clearly: the shift from a past-oriented thinking to a future-oriented thinking. As he says: 'One of the most astonishing and radical differences between premodern and modern cultures is the *direction* in which the universe is said to be unfolding.'[317] In older cultures there was a universal belief in a Golden Age, from which to the present time there was a spiritual and physical decline. He goes on: 'But sometime in the modern era – it is impossible to pinpoint exactly – the idea of history as devolution (or a fall from God) was slowly replaced by the idea of history as evolution (or growth toward God).' (Actually one might point to the Enlightenment writings of Condorcet as marking that turn.[318]) Wilber believes that the Romantics held to the devolution idea, and hence sought their spiritual growth in a return to the primitive, the mythical, and the 'natural'. Wilber misses two things here: firstly that the shift from past- to future-orientation is *arbitrary*, and secondly that future-orientation has strong roots in both Hellenic Aristotelian teleology and in Abrahamic messianic teleology. But Wilber has a limpet-like adherence to evolution and developmental psychology that makes him a quintessentially New Age thinker: the belief that there is not just historical progress in the spiritual life, but that it is culminating – in the near future – in a collective transformation of consciousness. This is an arbitrary position, because, in the Popperian sense, it is not falsifiable. A more sober look at the Abrahamic millenarianism of at least the last three thousand years, shows that many groups throughout history have believed in the imminent 'kingdom of heaven', including the first Christians as an intensely apocalyptical group. An examination of the Jehovah's Witnesses for example reveals the psychology of a group seemingly never disillusioned by the failure of each prophesied date of the new dawn. Their belief is not falsifiable because they are endlessly inventive in justifying the postponement of that date, just as New Age prophets are.

However Wilber does very clearly articulate the key flaw in the Romantic spirit, one that ultimately discredits it: narcissism. It is of course a grave charge to accuse an entire movement of narcissism, and risky, because this accusation can be levelled at any kind of spirituality that is not immediately concerned with social improvement. But Wilber insists that all the ills of Romanticism, including its narcissism, are due ultimately to its regressive nature. Our analysis differs, because there is no special reason to privilege later spiritualities over earlier ones. The dynamic in a sense is a simpler one: once the Romantic thinkers had freed themselves from Christian church tradition, the core corrective of that tradition – which we have called the Kierkegaard Corrective – was lost. Instead of the ethical pull that continually forces a return to love for the neighbour, Romanticism put in its place the *aesthetic* imperative. We can illustrate this in two of Goethe's works that were massively influential on Europe of the late eighteenth and early nineteenth centuries: *The Sorrows of Young Werther*, and *Faust*. Werther is a semi-autobiographical character who falls hopelessly in love and finally shoots himself when he fails to win his beloved Charlotte. It was the *aesthetic* of unrequited love and suicide, so poetically described by Goethe, that allegedly led many young men in Europe to dress in the same 'English-style' blue suit and yellow waistcoat of Werther and blow their brains out. (Censors in Leipzig and Denmark banned the novel to stem this tide.) Goethe's magnificent prose and poetry raised him to the status of cultural hero, with almost no question raised then or now as to the *ethics* of his literary works. It is not that the Christian tradition simply says that suicide is a sin in the eyes of 'God', it is that Werther, as an archetypal Romantic, was so self-absorbed that his 'love' for Charlotte did not extend to joy at her happiness, whoever she chose as a husband. It did not extend to all those others who – he readily admits – loved him, including Charlotte's much younger siblings who adored Werther and his kindnesses. For the Romantics, the blood-spattered remains of Werther represent their highest values, oblivious to the unnecessary pain caused to all around him. Far worse even than this is Goethe's Faust, whose behaviour towards Gretchen, the 14-year old peasant girl that he 'loves', is so monstrously self-absorbed that he thinks nothing of making her pregnant, killing her brother, and subsequently abandoning her in order to wallow in his narcissist misery. *Faust* represents a radical denial of Christianity because its protagonist is rewarded for his vile deeds, not with hell, but with heaven. The work is received as an aesthetic masterpiece, as are operatic versions complete with thundering orchestral accompaniment, but to the serious religionist of any tradition, or to the serious humanist, no defence of irony or artistic merit can mitigate the ethical ugliness of *Faust*. Any spirituality that puts such acts of self-absorption forward as representative of its values has not discovered the essential feature of spirituality: the connectedness that makes another person's happiness and suffering as important and as poignant as one's own – the Kierkegaard Corrective.

There is no reason however why the aesthetic impulse and sexual love in themselves cannot be part of or lead to a higher spirituality, such as expressed by Stuart's description above of the Platonist vision within Romanticism. But from the time of Calvin, the rejection of the Catholic tradition was marked by extremism, and Romanticism at its worst was as extreme as the Calvinist ethos it opposed. It is no surprise that the dour iconoclasm of much Protestant thought should have its mirror image in the hedonist aesthetics of the Romantics. In their haste to reject the elaboration of medieval Catholicism, Protestants returned to the iconoclastic instincts of the fourth-century Christian mobs, and churches were stripped of art and ornamentation – and even sometimes of their bells. If even a church bell was a frivolity and an aesthetic titillation too far for the Calvinists, what common ground could the Romantic find with them? Stuart is right that within Romanticism there is a deeply Neoplatonist spirituality and aesthetic. Though William Blake is not a typical Romantic figure, he does epitomise its rejection of science, and exemplifies the marriage of a profound spirituality with a prodigious aesthetic impulse, and celebration of sexuality. But his work seems to be free of the taint of narcissism (of the Goethean type), and it is tempting to wonder if this is because he retained such powerful biblical roots.

The German Idealists and Historicism

Wilber regards the narcissism of the Romantics as entirely due to the regressive nature of their spirituality: we disagree and say that earlier spiritualities (and the modern re-adoption of them) are no more likely to engender narcissism than later ones. He believes that German Idealism – 'This glorious spiritual flower, perhaps the finest the modern West has ever known'[319] – failed just as much in its spirituality as Romanticism, but for different reasons. Wilber is deeply drawn to the Idealists because they are progressive: the Idealists believe that we are evolving towards 'God' instead of away from him. In Descartes, Spinoza and Leibniz we saw no trace of the modern belief in progress, or at least not in the world of spirit. Theirs was a spirituality of neither evolution nor devolution, and neither did their great successor, Kant, hold to either of these extremes. In a discussion typical of his instinct to find the common ground between opposites, Kant concedes that it is hard to find evidence for either position, though he detected a tendency to a belief in progress amongst contemporary philosophers and educators.[320] But Kant's successors, Fichte, Hegel and Schelling, abandoned his balanced view, and quite naturally Wilber seizes on their ideas. He says: 'Friedrich Schelling took that initial developmental insight and worked it into a profound philosophy of spiritual unfolding, and Georg Hegel hammered out its details in a series of brilliantly difficult treatises.'[321]

It is through Hegel (1770-1831) that we can examine the ideas of evolutionary spirituality and its secular counterpart: *historicism*. Historicism is the more general belief that humanity makes progress, and that history shows that certain social and cultural destinies are *inevitable*. We look first at how this plays out as a *jnani* spirituality in Hegel, one which later influenced spiritual thinkers like Sri Aurobindo (1872-1950) and Teilhard de Chardin (1881-1955). Hegel, though he lived a century after Spinoza, wrote in a period that was not much easier for radical thinkers. He took his first teaching post in the renowned University of Jena, where Schelling was a professor, in the year 1799, which saw Fichte, one of the principal early influences on Hegel, dismissed from the university for alleged atheism. Yet

Hegel was quite radical in his religious thinking, introducing into philosophy as a central idea the term 'Spirit'. He intended this abstraction to stand for the ultimate principle which manifests the world, forgets its origins in the play of world, and then seeks return to itself – remarkably close to the Hindu conception. The evolution of the world, as Spirit alienated from itself, is a manifestation and expression of that return to Spirit. Hegel's interpretation of history is constructed to show this inevitable progress towards self-realisation and Absolute Knowledge. He also recapitulates in himself the entire involution and evolution of Spirit in his life as a philosopher, going through his own 'dark night of the soul':

> Probably everyone has such a turning point in his life, the nocturnal point of the contraction of his essence in which he is forced through a narrow passage by which his confidence in himself and everyday life grows in strength and assurance. . . . It is science which has led you into this labyrinth of the soul, and science alone is capable of leading you out again and healing you.[322]

Hegel's use of the word 'science' here is foreign to any modern usage, and appears throughout the remarkable preface to his *Phenomenology of Spirit* (perhaps his most important book).[323] He makes clear early on that 'To help bring philosophy closer to the form of Science, to the goal where it can lay aside the title '*love* of knowing' and be actual knowing – that is what I have set myself to do.'[324] In other words Hegel wants to restore *gnosis* to philosophy, though he does not use that word – neither is he acquainted with the term *jnani*. He is in favour of the historical movement away from the inwardness of the medieval period:

> The eye of Spirit had to be forcibly turned and held fast to the things of this world; and it has taken a long time before the lucidity which only heavenly things used to have could penetrate the dullness and confusion in which the sense of worldly things was enveloped, and so make attention to the here and now as such, attention to what has been called 'experience,' an interesting and valid enterprise.[325]

That enterprise is of course science, but science in the form that had now triumphed in the materialism of the Newtonian tradition, and which in turn needed overcoming. For Hegel, a new 'science' of a revitalised philosophy was required, one that would lead back to Spirit away from dead materialism:

> Now we seem to need just the opposite: sense is so fast rooted in earthly things that it requires just as much force to raise it. The Spirit shows itself as so impoverished that, like a wanderer in the desert craving for a mere mouthful of water, it seems to crave for its refreshment only the bare feeling of the divine in general. By the little which now satisfies Spirit, we can measure the extent of its loss.[326]

Hegel is writing at a time when the march to secularism was already a worrying problem, and it is clear that he does not believe that traditional religion is the cure. His historicism places the spiritual modalities and practices of the past as obsolete. He clearly considers the mystics of Christianity to be seeking mere 'edification' in denial of the manifest world, or in our terminology he is against the *via negativa* as inadequate. 'Whosoever wants to shroud in a mist the manifold variety of his earthly existence and of thought, in order to pursue the indeterminate enjoyment of this indeterminate divinity, may look where he likes to find all this.'[327] This may be a direct reference to the *Cloud of Unknowing*, or to the Divine Darkness of Dionysius, and reminds us of the rejection that Rudolf Steiner makes of mysticism. Another passage might be a direct attack on the devotional mystics: 'Such minds when they give themselves up to the uncontrolled ferment of the divine substance, imagine that, by drawing a veil over self-consciousness and surrendering understanding they become the beloved of God to who He gives wisdom in sleep; and hence what they in fact receive, and bring to birth in their sleep, is nothing but dreams.'[328] Only a *jnani* would say this. Neither did Hegel have sympathy for the 'agnosia' of Dionysius, or the 'learned ignorance' of Nicholas of Cusa (or would he have had for the 'no-mind' of Zen). His is a *jnani via positiva* route, and his historicism insists that the *via negativa,* whether *jnani* or *bhakti,* is consigned to the dustbin of history.

So we return to why Wilber considers Hegel and the German Idealists to have failed; after all, Wilber is utterly wedded to Hegel's spiritual historicism. Having devoted some space in the *Marriage of Sense and Soul* to the 'glory' of their vision Wilber provides for us an insight that is relevant not only to the Idealists, but to the whole Western *jnani* tradition once it shifted its allegiance from Christianity to philosophy. Wilber says quite simply that it 'possessed no yoga – that is, no tried and tested practice for reliably reproducing the transpersonal and superconscious insights that formed the very core of the great Idealist vision.'[329] Wilber is right, that the Idealists along with all the *jnani* thinkers of the Enlightenment lacked a shared spiritual practice to underpin their thought: their philosophy had great insights at times, but was often ungrounded. The traditional practice of Christianity was prayer, suited best to those of a *bhakti* orientation, but even so serving the great *jnanis* such as Eckhart to some extent. Both the lay and monastic traditions of Christianity had prayer at the heart of their ritual and practice, but the philosophers of the Enlightenment, moving away from that tradition, had nothing to replace it with. Wilber, as a Buddhist, is well aware of the great *jnani* traditions of meditation which ground and underpin Eastern non-devotional modalities of the spirit. He says: 'Idealism in this regard degenerated into abstract speculations without the means of experiential confirmation or rejection.' He is placing emphasis on transpersonal and superconscious states in meditation, but there is another dimension to any spiritual practice: its role as a corrective. The sheer discipline of a daily practice, conducted in a meditation hall with others and under the guidance of a spiritual teacher, requires an act of humility, the outward sign of the inward discipline of ego-transcendence.

Whether Wilber really shares the political naivety of the New Age, or chose to ignore it because he so admires their historicism, he omits any mention of the charge that the German Idealists probably provided the intellectual ground for Nazism. At the very least it is clear that Fichte and Hegel promoted a German nationalism that proclaims other races and nations as inferior. Hegel's historicism claims that Germany is the inevitable carrier of human destiny, giving the Chinese for example the role of merely representing the ancient historic stage of humanity where only the ruler is free. Nietzsche contributed with his unfortunate vision of the 'superman', and Martin Heidegger, the German philosopher who inherited much of the Idealist belief system, actively encouraged the German people to vote for Hitler, and refused to apologise for Nazism after the war. Postmodernists attribute much of the evils of fascism to the fascination with 'grand narratives' and German Idealism certainly provided a grand if not grandiose myth that the German people found irresistible. Perhaps Wilber as a *modernist* and one who has constructed a philosophy that is in itself a grand narrative – a theory of everything – isn't convinced of the easy slide from Idealism to nationalism.

Wilber's spiritual historicism is deeply entrenched in the New Age, having made its way via Hegel through Sri Aurobindo and Teilhard de Chardin. Aurobindo's case is interesting: he studied at Cambridge and took back to India Western teleology and historicism, and inserted it into Indian spiritual thought. India is a good example of Wilber's assertion that premodern societies believed in devolution, so Aurobindo's spiritual teachings were nothing less than a revolution for India. At the same time Gandhi brought with him Western ideas of political progress, and it is no coincidence that many Westerners drawn to India gravitate first to Aurobindo and Gandhi as a natural reflection of recent Western thought.

German Idealists as Spiritual Teachers

We have suggested that broadly speaking the monotheist traditions of the West can be understood as traditions of texts, while the Eastern traditions can be understood as traditions of teachers. This is not to say that the Western traditions do not have a role for the spiritual teacher, or that Eastern traditions do not have key or semi-canonical texts. However, the East developed cultural checks and balances on the teacher that provided correctives to the kind of abuses that have arisen in the West, as we saw earlier. We largely saw spiritual teachers in twentieth-century Western society as a New Age phenomenon, but we may suggest that it began in the Renaissance, with Ficino's Academy, at the point of some initial independence from the Church. In a small way this re-emerged in Descartes, Spinoza and Leibniz: their original spiritual teachings attracted small groups of followers. Prince Eugene of Savoy so treasured Leibniz's teachings that he was reputed to have kept a summary of them – possibly the *Monadology* – like a 'jewel in a case', rather as his cousins the Dukes of Savoy kept the Turin Shroud.[330]

If we understand the German Idealist philosophers, Fichte, Hegel and Schelling as effectively setting themselves up as spiritual teachers, then we can better understand how their ideas ran aground. Wilber describes how Fichte gave his students an exercise in introspective inquiry that he hoped would help them contact the 'absolute Self'.[331] But philosophy as a discipline in the late eighteenth century was not like philosophy in the context of Pythagoras: its basis was no longer a religious one. For Fichte's philosophy students there was no context of piety, no rituals or practice grounded in the religious impulse, and no tradition that could be transposed from the monastery into the philosophy class. Wilber is keenly aware of this: 'In Zen, for example, if the koan (or meditation theme) is 'Who am I?' or 'Who chants the name of the Buddha?,' it takes an average six years, according to Yasutani Roshi, for the successful student to have the first profound satori, or genuine breakthrough to the True

Self (which is the True World as well).'[332] It is not just a question of the length of time for a daily spiritual practice to be effective: the entire context is crucial. Fichte's students were effectively three times removed from a spiritual practice like Zen: their environment was not explicitly religious; their practice was sporadic and without grounding in an appropriate spiritual tradition; and their instructor was at best making a shot in the dark as spiritual teacher. Above all the intentionality was coloured quite differently: they pursued knowledge at a period where knowledge had become, not a question of interiority and divinity, but of materialist science or speculative logic. Because the West was a tradition of texts, not teachers, the Idealists were also hamstrung, by the bewitching glamour of the 'Word', and by a lack of ground-rules and correctives as spiritual teachers. The habit of philosophy had also become obscurantist and prolix in its writings, for the reasons we have explored earlier on. Marx's prose is extraordinarily persuasive, and much of it is due to his literary skill: the use of rhythm, repetition and inversion. This is also visible in Hegel, from whom he drew so much. In the writings of the Idealists generally, rooted in a Romantic aesthetic of the word, the reader is easily hypnotised. A person in the thrall of beautiful words is easily led.

Hegel leaves us with two points, eloquently made by him. Firstly that, after the Dark Ages, 'the eye of Spirit had to be forcibly turned and held fast to the things of this world', a process begun at the time of Aquinas, and becoming a lyrical materialism in the Enlightenment. Secondly, when he says: 'By the little which now satisfies Spirit, we can measure the extent of its loss', he signals for us how the Enlightenment was producing, in Holbach and others, an extreme of materialism now demanding equally that the 'eye of Spirit' be turned back to the spiritual again, because: 'sense is so fast rooted in earthly things that it requires just as much force to raise it.'

Feuerbach

We suggested earlier that Ludwig Feuerbach, whose writings on religion are sometimes assumed to be atheist, and to be the source of Marx's atheism, is better understood to stand with those making an argument within religion, not against it. Engels thought so when he wrote: 'The real idealism of Feuerbach becomes evident as soon as we come to his philosophy of religion and ethics. He by no means wishes to abolish religion; he wants to perfect it. Philosophy itself must be absorbed in religion.'[333] Marx, in his famous *Theses on Feuerbach*, also finds him religious rather than atheist, and dismisses him thus: 'Feuerbach, consequently, does not see that the 'religious sentiment' is itself a social product, and that the abstract individual whom he analyses belongs to a particular form of society.'[334] Marx and Engels, living closer to the period, understood Feuerbach better than contemporary secular scholarship. In fact, Feuerbach perfectly epitomises the Renaissance and Enlightenment turn to a *jnani, via positiva,* engaged spirituality, a turn towards the manifest world that is now understood by socialists merely as an imperfect form of materialism, a materialism that would find its dialectical perfection in Marx. But Feuerbach, despite his rejection of many Idealist principles, is not only lyrically spiritual, but occasionally also has the tone of the spiritual teacher or guru, as Wilber found in the Idealists. Feuerbach's best known work, translated into English by George Eliot, is *The Essence of Christianity,* in which he says:

> And so far as thy nature reaches, so far thy unlimited self-consciousness, so far art thou God.[335]
>
> It follows that if thou thinkest the infinite, thou perceivest and affirmest the infinitude of the power of thought; if thou feelest the infinite thou feelest and affirmest the infinitude of the power of feeling.[336]

His work was a profound challenge to orthodox Christianity, as these extracts show, but we can spot immediately the resonance with Descartes' mysticism in his frequent use of the term 'infinite', and, in the work taken as a whole, there is a typical *jnani,* even Neoplatonist feel. Right at the start of the book, Feuerbach defines religion as 'consciousness of the infinite', and proposes, in a similar way to Descartes: 'Consciousness, in the strict or proper sense, is identical with consciousness of the infinite; a limited consciousness is no consciousness; consciousness is essentially infinite in nature.'[337]

The Political Damage to Spirituality

Marx inherited Hegel's historicism as well as his literary style, but applied it in a radically different way. In addition he made it politically unacceptable and intellectually incredible to believe in spirituality. Where Marx's legacy is a total disengagement with spirituality or religion, the Idealist legacy is to taint it with the fascisms of the Right. In fact the Nazi experiment not only discredited the spirituality of the German Idealists, which we can characterise as *jnani, via positiva* and largely non-esoteric, but also discredited all the esoteric traditions that fed the fevered imaginings of the Third Reich. These were precisely the regressive, mythic, Romantic occultisms that Wilber criticises, and which took a populist and nationalistic form in Nazism. While Blavatsky and the Theosophists were largely genuine and well-meaning in their global search for esoteric truths, they did also indulge in the widespread habit of the late nineteenth and early twentieth century of speculation on race, in their case esoteric speculations. Blavatsky's theories of 'root races' easily fed into the Nazi imagination, and the experience of WW2 left, for the modern mind, esotericism just as much as Idealism tainted by fascism. It is a characteristic of most New Age thinkers to ignore this problem, after all it is not the fault of the spirituality itself that it was distorted to meet fascist ends. But any attempt to understand the complex secular geography of mind that has shut the door on religion must take into account the landmark of the *perceived* fascist appropriation of spirituality.

But the extremes of German Idealism, where it became intertwined with nationalism, should not stand for its main body of thought, which was deeply spiritual. Marx and Lenin, in rightly arguing that Feuerbach did not want the end of religion, but its perfection, were making a point that applied to all the Idealists. And neither is it true that Romantics wanted an end to religion, even if their anticlericalism was as vigorous as in the mainstream of Enlightenment rationalism. Neither Romanticism nor Idealism contributed to the emergence of a reductionist materialism that would spell the end of religion as a cultural force, so the question remains: how is that their defence of 'God' in his dying failed? Why did Marx and Bentham succeed?

5.1.5 Bentham and Fundamentalist Materialism

The year 1822 saw the publication of Bentham's *The Influence of Natural Religion on the Temporal Happiness of Mankind*. Prior to this work there were only two atheist texts in circulation: La Mettrie's *Man a Machine* and Holbach's *System of Nature* (or perhaps three if we include Diderot's 'Letter on the Blind'). Bentham is best known as the founder of Utilitarianism, a political philosophy that defines good government as that which produces the greatest good for the greatest number. There is nothing in this that is opposed to spirituality, or even religion when understood more broadly, but it is, like most Enlightenment thought, anticlerical. Where Bentham is astonishing for his time is that he dismisses religion in its totality, in a manner prescient of twentieth-century atheism, and creating the standard to which it would conform. In his influential *Principals of Morals and Legislation* he champions the principle of utility, juxtaposing it with that of *asceticism*, defined by him as follows:

> By the principle of asceticism I mean that principle, which, like the principle of utility, approves or disapproves of any action, according to the tendency which it appears to have to augment or diminish the happiness of the party in question; but in an inverse manner: approving of actions in as far as they tend to diminish his happiness; disapproving of them in as far as they tend to augment it.[338]

Where asceticism (if not pursued to extremes) can be understood as a natural expression of the *via negativa* and solitary spiritual impulses, leading to a profound religious happiness, Bentham presents it as an *inversion* of normal values, and therefore pathological. He goes on to say that asceticism is the positive hope of moralists (those who accept pain in the present for pleasure in the future) and the negative hope of religionists (those who accept pain in the present in fear of worse pain in the future). In his *The Influence of Natural Religion on the Temporal Happiness of Mankind* he then elaborates on how the religionist acts in fear of the future such as to inevitably decrease his 'temporal happiness'. Although this book was published under the pseudonym Philip Beauchamp, and was never widely known, it comprises perhaps the most systematic and devastating attack on religion ever mounted up to that time. It is extreme in its materialism, not because it uses intemperate language, but rather, as in modern secularism since Marx, it denies any interiority to the human being. Bentham restricts his definition of religion to the narrowest of terms: 'By the term religion is meant the belief in the existence of an almighty Being, by whom pains and pleasures will be dispensed to mankind, during an infinite and future state of existence.'[339] This is now the modern habit, as in Dawkins, to reduce the immensity of religion to the most simplistic and superficial terms, and then attack that, effectively a straw man. Worse, Bentham then assumes that the Deity is like a 'mad, capricious and incomprehensible Being exalted to this overwhelming sway [i.e. to omnipotence]',[340] and insists that he is a despot with an 'insatiate love of dominion and thirst for its increase.'[341] But Bentham then follows this with an argument of such clarity, and so well structured, that it becomes easy to overlook the audacity of the original premises, and be persuaded that atheism is the only rational course.

Even if this extraordinary atheist work is not widely known, Bentham makes a useful icon of contemporary atheism. Indeed his 'auto-icon' stands as a shrine to secularism in University College London, an institution inspired by all his positive pronouncements on progress, equality of opportunity, and the promotion of reason, science and freethought. His preserved body and wax head, mythically present at meetings of the College Council, effectively gazes approvingly on all Universities as defenders of secular values. What is not apparent to the secular mind is that the materialism underpinning his philosophy, and almost all of higher education in the West, is a *fundamentalist* materialism. In Marx's hands this materialism reduces all of human experience to the economic interaction of people within certain types of society. When Sartre reduced Jean Genet and Teresa of Avila to equivalent non-producers in a bourgeois 'society of consumption', he was following Marx's reductionism, and which produces the absurd inversion that Genet is superior to Teresa because his acts of theft did not have cultural sanction, while her dependency on charity did. Marx only allowed a single source of happiness: economic, and considered religious happiness an illusion, incompatible with his dialectical materialism.

But there is another materialism which has its roots in a *jnani via positiva* spiritual orientation; and is the more natural milieu of the Renaissance and the majority of Enlightenment thinkers, including Bacon, Hobbes, Newton, Descartes, Spinoza, Leibniz, Voltaire, the American Deists and Transcendentalists. It is lyrically present in Feuerbach, and found in many twentieth-century physicists, including Einstein and Erwin Schroedinger. It is a materialism which has, with the Neoplatonists, 'the mood of one who has a curious eye for the endless variety of this visible and temporal world and a fine sense of its beauties, yet is haunted by the presence of an invisible and eternal world behind.' This materialism is enchanted, lyrical, rhapsodical, oceanic and sacred; it is based on the inner sense of a profound connectedness to all things, leading to the divine, and characterises the majority of Enlightenment thought.

In contrast the materialism of Bentham and Marx is disenchanted, atheistic, utilitarian, and reductionist, leading to a profound disconnection with an atomised, mechanistic universe, and to a debilitating sense of alienation, and is found in a tiny minority of Enlightenment thinkers. Rousseau thought that such a materialism was like the deaf obstinately denying the existence of sound. Why then should an obscure text promoting this materialism, pseudonymously published by Bentham, stand as the hallmark of the secular mind? We saw how Descartes, Spinoza and Leibniz, amongst the thinkers of the early Enlightenment, were deeply religious men, almost 'lost buddhas of the West', so how could Bentham, as part of the Enlightenment legacy be so casually atheist? His book represents not just the 'death of God' but the death of all interiority, all spirituality, all possibility for a profound connectedness with cosmos; it represents a reduction of the human being to an economic entity. So why didn't the vast majority of Enlightenment, Romantic and Idealist thinkers mount a concerted credible defence of 'God', of human interiority, of spirituality?

5.2 *Deism*

5.2.1 A Broader Conception of Deism

We have seen that the majority of Enlightenment, Romantic and Idealist thinkers were creative spiritual innovators whose detailed approach may have differed, but in all of which there was a coherent religiosity, according to our analysis. This broadly *jnani via positiva* movement can, with only the smallest of elasticity in definition, be termed *Deist*. Deism as more narrowly understood appeared in the middle seventeenth century and was a proper movement in Christianity for only about a century. We have used the term loosely up to now, and note also that Gore Vidal comments: 'Jefferson said he was a deist, which could mean anything or nothing. . . .'[342] We have to disagree with Vidal, and say that Deism means something very precise, but only when one has to hand a detailed articulation of the spiritual life. However for the modern world Deism as a term has become a vague anachronism, for two reasons: firstly the secular world fails to recognise it as more than a passing fad within a larger discredited religion, and secondly the Christian world fails to recognise it as anything more than a heretical near-atheism. We can define its character quite precisely however: it was a Christian inflection of the *jnani via positiva* impulse, with an anti-anthropomorphism shared with the apophatic or negative theology. The *Catholic Encyclopedia* pinpoints one of its features: 'deism is no more than an application of critical principles to religion' – which as usual betrays both Catholic prejudice ('deism is no more than . . .') and keen insight ('application of critical principles').

The early British advocates of Deism were Lord Herbert of Cherbury, Hobbes and Locke. Hume's thought is much better characterised as Deist than it is atheist, whatever his contemporaries thought. Deism denied two aspects of the Bible: the revealed nature of religion, and the miracles, but retained 'God' as the ultimate principle and creator of the universe – closer to the 'God' of the philosophers than the 'God' of a devotional religion. To that extent it was always distrusted by the Christian mainstream, though the educated – whose intellectual milieu was deeply stimulated by the emerging science – were more inclined to it, in private at least. Lord Herbert of Cherbury laid down five principles of Deism which barely distinguish it from conventional Christianity, but both secular histories and the *Catholic Encyclopedia* quote these as the defining principles of Deism. However, it is more useful to consider Deism in broader terms, as we have indicated. While Cherbury's Deist principles were not disputed by most Enlightenment figures (apart from the very few atheists), their freethought context gave reign to a veritable explosion of religious creative insight that went far beyond these principles. Freethought itself was advocated by the leading Deist Anthony Collins (1676-1729) as the pre-requesite for 'knowing that a perfectly Good, Just, Wise and Powerful Being made and governs the World.'[343] In Deism – in the broadest sense – the suppressed *jnani* impulse of the West broke free to express itself in a multitude of ways while attempting to remain within the Christian envelope.

The national character in Britain, France, Germany and America ensured that Deism looked different in each country: the British version was mostly quiet-spoken (as it was the most tolerant country); the French was the most outspoken (as its protests against Roman Catholic oppression led up to the Revolution); the German was the most Romantic (flowering in Goethe, Hegel and Feuerbach); and in American it was the most directly political (being the religion of America's founding fathers). Benjamin Franklin best epitomises in one person the Enlightenment Deist project as a totality, including entrepreneurialism, and the uniquely American spirit. His generous sanguine temperament did not make him a critical thinker of great distinction, but his contributions to public life were enormous and make him, for many historians, the pre-eminent Enlightenment figure. Franklin, in a letter to Ezra Stiles, confirmed his deep religiosity but didn't mind admitting that he had some doubts over the divinity of Jesus, as did 'dissenters in England'.[344] What would have utterly shocked him would have been Bentham's mode of dissent. We have in these two men the contrast between two avowedly materialist individuals: Franklin who saw nothing in the science, technology and politics of his day that contradicted his religion, and Bentham whose materialism utterly shut out the sacred. Bentham's *The Influence of Natural Religion on the Temporal Happiness of Mankind* was explicitly an attack on 'natural religion' – a term that is often synonymous with Deism.

But Deism was under attack from both sides: the accelerating number of atheists in the nineteenth century attacked it from the perspective of secularism, while, since its beginnings in the seventeenth century, Deism was perceived as such a threat to establishment religion that laws were passed against it, and Deist books regularly burned in public. 'Deist' was a dirty word in the eighteenth century, often used in the same way as 'atheist' (and its favour is not much greater today amongst many American Christians). Deists were not so much troubled by what other Christians thought of them or impugned to them, as they were committed freethinkers, but were considerably exercised against atheists. Voltaire, for example, attacked Holbach's *System of Nature* for its atheism, calling it a 'pernicious book',[345] while also defending Descartes against the accusation of atheism.[346] But collectively, Deists (and we need to include Hume here) indulged in a defence of religion through the Argument from Design – to them, the beauty and efficacy of the created world was proof of the existence of 'God'. For the less educated the Deist conception of 'God' was perhaps too abstract, but the Argument from Design held great popular appeal.

5.2.2 Deism and Freethought in Thomas Paine

While Benjamin Franklin was a 'respectable' Deist in America, Thomas Paine (1737-1809), initially hailed as a leading advocate of independence, gradually became suspect over his Deist principles. Paine was born in England of humble origins, rising to become an excise officer, but largely failing to make a living or be successful in England. In 1772 he met Benjamin Franklin in London, who advised him to go to America and gave him letters of introduction. Paine arrived in Philadelphia in 1774 at the height of tension between America and its colonial ruler, England. He worked as a newspaper editor and wrote polemical works against slavery. His arguments for independence were published in 1776 in the pamphlet 'Common Sense', which had enormous circulation and influence: he became the leading voice of the American Revolution as his ideas underpinned the Declaration of Independence. He was greatly influenced by the French Revolution of 1789 and published *The Rights of Man* as a direct answer to Burke's attack on the Revolution. This was to be Paine's most influential work, drawing on Voltaire, and promoting ideas of the rights of ordinary people; it circulated widely in the US and UK, putting forward arguments for political change underpinned by Deist ideas on religion. His status as cultural hero seemed assured, but it ran aground after the publication of his second major work in

1794, *The Age of Reason*. This made clear the radical freethought of his Deist position, and he was quickly branded an atheist in America, Theodore Roosevelt later going so far as to call him 'a filthy little atheist'. Here for example is an extract in which Paine states his position on 'revelation', a view that was at the heart of Deism, but is a radical development of Hobbes' position on the subject:

> No one will deny or dispute the power of the Almighty to make such a communication, if he pleases. But admitting, for the sake of a case, that something has been revealed to a certain person, and not revealed to any other person, it is revelation to that person only. When he tells it to a second person, a second to a third, ... and so on, it ceases to be a revelation to all those persons. It is revelation to the first person only, and hearsay to every other, and consequently they are not obliged to believe it. It is a contradiction in terms and ideas to call anything a revelation that comes to us at second-hand, either verbally or in writing.[347]

That Paine insists on the right to freethought is beyond question, but a secular world only knows him for this. It is therefore instructive to look at the opening passage of *The Age of Reason*, in which he makes it utterly clear that he is religious, and regards with dismay the abolition in France of all religion since the Revolution:

> The circumstance that has now taken place in France of the total abolition of the whole national order of priesthood, and of everything appertaining to compulsive systems of religion, and compulsive articles of faith, has not only precipitated my intention, but rendered a work of this kind exceedingly necessary, lest in the general wreck of superstition, of false systems of government, and false theology, we lose sight of morality, of humanity, and of the theology that is true.[348]

Paine wants the 'wreck of superstition' like every other Enlightenment thinker, but, like over ninety percent of the rest of them, is utterly opposed to the idea of an end to religion itself. Paine wants a 'theology that is true'. He would not have agreed with the trend of thought in La Mettrie, Diderot, Holbach and Bentham, that made freethought the wreck of superstition *and* of all religion. He was a materialist in the sense of his complete engagement with the world of politics, science, and even engineering, receiving a patent in Europe for the single-span iron bridge, and working on steam engines. But his materialism was truer to the Enlightenment than that of La Mettrie, Diderot, Holbach and Bentham, because he saw the world as the work of the divine. In his own words: 'The word of God is the Creation we behold'[349] – almost a defining Deist principle. Aquinas began the revolution that we call Deism by declaring, against Christian tradition up to that time, that revelation came in two books, the Bible and Nature. Paine completes the transition from Bible to Nature by ruling out the first book as 'hearsay'.

But his freethought had far-reaching consequences for secularism beyond the influence of the few Enlightenment atheists, because it reached the mass of ordinary people. What Paine represents is the movement of freethought from the domain of the privileged and educated, who often were Deists themselves, to the domain of the ordinary person. Many educated Deists may have doubted the wisdom of this, but freethought as the right of every citizen was an unstoppable tide, and Paine's works did much to promote it. It just so happened that along the way religion got lost. Conventional Christians recoiled from his advanced Deist views, and secularists simply ignored his plea for a 'theology that was true'.

5.3 Failure of the Western Jnani *Religion*

5.3.1 William Paley and the Argument from Design

Energy and intelligence was haemorrhaging from religion, and 'God' was under a death sentence: first Hegel quietly pronounced his death in the *Phenomenology of Spirit*,[350] and then Nietzsche shouted it from the rooftops in *Zarathustra*.

That 'God' was under a death sentence was true only for a few atheists of the eighteenth and nineteenth centuries, but gradually the shapers of culture were moving the mainstream away from its Christian base. The old weapons of Christianity to stem dissent had been lost in the humanist advances since the French Revolution, and the quieter moves to toleration in the UK. Hence it fell to what were largely Deist principles to defend Christian beliefs in what had been, going back to Augustine and Aquinas, the gentler way of countering the infidels: 'proofs' of the existence of 'God'. But none of the proofs from Aristotle to Leibniz counted for anything next to the Argument from Design. This was the proof forced on Christianity by the science of Newton and the rationalism of Descartes, and the early failure of which created the atheism of La Mettrie, Diderot, and Holbach. In La Mettrie's *Man a Machine* that failure was not yet complete, as he sits on the fence in regard to the existence of 'God'. Diderot's first atheist text, the *Letter on the Blind,* is perhaps the first true body-blow to the Argument from Design, but his personal position is ambiguous, as he responded to Voltaire's complaint about the work that he himself believed in God.[351] Holbach's *System of Nature* was the second significant attack on the Argument from Design. But Voltaire's condemnation of it was typical of the age, and Holbach's work failed to slow down the widespread publication of texts promoting the Argument from Design, which continued into the next century.

An influential and inspired work in this genre was published in 1802, an exposition of the wonders of the natural world: *Natural Theology; or, Evidences of the Existence and Attributes of the Deity* by the Rev. William Paley (1743-1805). Richard Dawkins says of it: 'It is a book that I greatly admire, for in his own time its author succeeded in doing what I am struggling to do now. . . . He had a proper reverence for the complexity of the living world, and he saw that it demands a very special kind of explanation. The only thing he got wrong – admittedly quite a big thing! – was the explanation itself.'[352] That explanation was the core Deist principle, that 'God' was evidenced by his creation. But, as we suggested earlier, *explanation* is not the point of religion.

Paley describes various natural phenomena, including the human eye, as conclusive evidence of intelligent design. He is clearly filled with both the reverence for Nature that Dawkins so passionately displays, and his firm conviction as a theologian that the explanation for so much beautiful design is 'God'. He made a now-famous analogy with a watch: anyone finding anything so intricate as a watch, so clearly expressing a purpose, would assume that it had a designer. Similarly all the creations of Nature, including the marvellous optical properties of the eye, whether in animals or humans, must be due to a Designer. His concluding chapters are titled: 'Personality of the Deity', 'Of the Natural Attributes of the Deity', 'Of the Unity of the Deity', and 'The Goodness of the Deity', having satisfied himself and his thoughtful Christian audience of the validity of his 'proof'. The book was well received, but in effect it was the culmination of the Argument from Design tradition that was a time-bomb waiting to go off.

In 1859 its powder was lit by Darwin's *On the Origin of Species by Means of Natural Selection*. The first edition sold out immediately and in the next thirteen years ran into six editions. Scientists almost universally accepted it, because it had precursors in the works of earlier researchers, and also in the writings of Darwin's contemporaries. Clergy rushed to denounce it however, but their attempts to argue against it on rational grounds were easily defeated.

5.3.2 Darwinism and the Discrediting of Deism

The furore created by Darwin's thesis marks the point where the incongruence between science and the Christian religion finally became starkly obvious. Newton may have discovered the laws behind the movement of inanimate objects, but living creatures, especially humans, had retained their special status up to this time, despite the best efforts of Holbach, Diderot and La Mettrie. Even if the workings of the body could be described in mechanical terms, the whole human being had retained in religion the status of 'God's' favoured creature, custodian of the natural world, and placed within it to pursue a moral and holy purpose. Paley's work was in keeping with the feelings of both scientists and religionists up to the time of Darwin that an increasingly intimate knowledge of the workings of Nature only added to a sense of awe and wonder, properly understood as religious. But the concept of evolution by natural selection, a blind process which could produce everything from the simplest organism right up to the human being, drove a wedge of hollow despair between science and religion. Science, far from inspiring awe and wonder, now removed at a stroke the meaning that had been provided by religion, a sense of meaning that was for celebration during times of plenty, and the rock on which one stood during crisis and loss.

Dawkins brilliantly borrowed from Paley in the title of one of his books: *The Blind Watchmaker*. Dawkins relishes in the telling that there is no creator 'God'; rather natural selection through random mutation had slowly evolved every last living thing. No intellectual concept in the history of man so spectacularly imploded under modern science as the Argument from Design, and the undermining of no other concept ever had such an impact on the world. Although it is highly unlikely that anyone was ever converted to Christianity by *any* of the proofs of the existence of 'God', the spectacular undermining of the Argument from Design called into question the whole intellectual edifice of Deist Christianity. In particular it undermined the concept of the 'creator God', one which is at the heart of monotheism. As we saw, Buddhism, Taoism and much of Hinduism manage very well without a 'creator' and so would not have been damaged by Darwinism; indeed these religions were neither theistic in the first place, or had any recourse to 'proofs' of any kind. But Christianity since Aquinas had constructed an intellectual edifice around its devotional basis. It was unable to shrug off Darwinism by extolling its devotional core, because the language of devotion had lost its currency (mostly due to the Protestant Reformation and the work of the Scholastics).

As Dawkins believes, it was now possible to be an intellectually fulfilled atheist, or more to the point *it was now impossible to be an intellectually credible Christian.* If Deism was the intellectual boat that Christianity launched to save its credibility, then its running aground on the reef of Darwinism left Christianity with no other vessel to reach the secular mind. Atheists like Dawkins still dance around the wreck of that vessel, while Christianity on a distant shore is condemned to be an intellectual third world country. Deism had run its *jnani via positiva* course from Aquinas to Paine, and even its name became equally lost to Christianity and secularism.

5.3.3 The Floodgates of Secularism Open

With a social phenomenon as complex as secularism it is impossible to set an exact date for its beginning, but our analysis has shown that Enlightenment thinkers had created a project which turned the religious impulse from its earlier inward emphasis to an outward contemplation of Nature, a project that was encapsulated in the Argument from Design. A thousand variations of it existed, but its central religious idea was that 'God' was the more joyously visible in Nature than in the revelation of earlier Christianity. It was a rhapsodical materialism, but which was shattered by Darwin, leaving ultimately only a fundamentalist reductionist materialism epitomised by Bentham and Marx. Hence the date of 1859, the date of the publication of *On the Origin of Species,* becomes the symbolic choice for the start of secularism.

Darwin's work was the dagger into the heart of Deism, the discrediting of which opened the floodgates of secularism. It mattered little that Diderot had already recapitulated in advance the journey from theism to Deism to atheism (if this really was the case): the great Encyclopedia was read only by the literate, and was anyway written in Diderot's Deist heyday. It was the propagation of secular ideas to the ordinary person that made it the cultural norm in the twentieth century, and that propagation was enormously accelerated by Paine's Deist refutation of revelation, and later by Thomas Huxley's tireless exposition of Darwinism. He taught it directly in working men's clubs, and indirectly through the great secularists of the age: Charles Bradlaugh in Britain and Robert Green Ingersoll in America. The scientific promotion of atheism took this route, while the social promotion of atheism journeyed through Bentham and Marx, and the artistic promotion of atheism came about through the collapse of the Romantic and Idealist movements, tainted by association with fascism. But it was the scientific promotion of atheism, through Darwinism, that was the greatest of these, and is virulently alive into the twenty-first century in the work of Dawkins et al.

5.3.4 The Hidden Origins of the Secular Mind

We saw that a list of forty-odd Enlightenment thinkers living in the eighteenth century only four were atheists, and then not all of them in a modern sense. Of these it is Jeremy Bentham who typifies modern atheism, adhering to a casual fundamentalist materialism that would become the hallmark of Marxist thought and what we have also termed Monty Python atheism. No conventional account of the rise of atheism, relying only on the twin factors of the emergence of science and the horror of religious cruelty, could account for this. How come the remaining ninety percent of Enlightenment thinkers did not defend religion in a broad sense against the fundamentalist materialist atheism of the few? In other words, we posed the question: what are the hidden origins of the secular mind?

Our answer, now spelled out, is that the Enlightenment was far from the incubator of atheism: it was in fact a melting pot of religious creativity giving rise to a new religious form that we broadly call Deism. *And it was the collapse of Deism which opened the floodgates of secularism*. The huge range of thinkers whose religiosity could have contributed to Deism had one thing in common (in our terms): their 'God' was the objectless object of the *jnani* spiritual impulse. In a thousand different ways this 'God' was defended against a reductionist atheist materialism, but those who did so were defending the wrong 'God'. It was not, ultimately, a 'God' that Christianity could really own, with Christianity's origins in a jealous Judaic 'God' and its reframing as the object of the *bhakti* devotional impulse. The most visible defence of 'God' mounted by the Enlightenment was the Argument from Design – which itself was merely a reflex of the greater *jnani* religiosity of the time – and it most visibly failed. Effectively the Western 'God' was not defended in terms that either worked for Christians, or worked for the emerging materialists. Ultimately, the *jnani* impulse moves towards an objectless object that does not need the name 'God', but in the proliferation of other, quite suitable, terms, it seemed that 'God' had been put to death. In a sense, it is inevitable: if the anthropomorphism of 'God' continues for millennia, despite the best efforts of great religious thinkers to resist it, then 'God' becomes human, and, like all humans, has to die.

Secularism certainly drew on emerging scientism and positivism: an exaggerated reductionist rationality. Vituperative atheism is most probably emotionally fuelled by the horror felt at the history of religious cruelty. But the *hidden* origins of the secular mind lie in the collective *religious* creativity of the Enlightenment, one that had a *jnani via positiva* basis, and produced what we broadly call Deism: its collapse ensured that the religious impulse mostly returned to its historical containment within Christianity, while intelligence and creativity burst forth into the whole spectrum of secular activities: science, arts and politics. The struggle of the Enlightenment for a new form of religion was over, and the West was polarised into two communities with little common ground or communication. The secular-religious détente of the twentieth century had arrived, and Deism as a common ground disappeared as both sides heaped scorn on it. The failure of Deism as a Western *jnani* religion – *the true project of the Enlightenment* – is what really gave rise to secularism.

But the death of the 'God' of devotion was only because the great minds of the time were largely non-devotional, and could not anticipate that if the devotional 'God' were to die, so would its counterpart. Hence the 'God' of the philosophers also disappeared with the collapse of Deism, as Christianity retreated bruised to one corner of the ring, and secularism appeared to stand triumphant. This was a defeat, not just for mainstream Christianity, but also for its great rival and potential partner, the *jnani* confluences of spiritual insight that we have, for convenience, lumped together as 'Neoplatonism'. What survived of it then became the marginal cultural force of the New Age. In this arena little survived of the grandeur of spiritual thought that we have discovered in thinkers like Descartes, Spinoza, Leibniz and Hegel, let alone an impressive list of other spiritual heavyweights that could include Berkeley, Kierkegaard, other German Idealists, Feuerbach, and the American Transcendentalists. The profoundest of religious intelligence had left no mark on any of the three domains: old religion, secularism, or the New Age. Intelligence and religion had parted company, leaving secularism condemned to an arid reductionist playground, and Christianity condemned to an intellectual and cultural ghetto.

Atheists say that religion is nothing more than ignorance and superstition which was swept aside by the rise of science and rationalism. More than that, they are unconcerned as to the details of the origins of the secular mind because they hold that secularism is the *inevitable* outcome of the maturing of the human mind. But we have shown that this simplistic account of the rise of secularism is nothing more than another unfounded historicism. We have shown that there was nothing inevitable about secularism, rather that a number of complex forces came into play to bring it, perhaps temporarily, into prominence. We can briefly recapitulate the history of those forces. Firstly, Judaism evolved a single creator 'God' that excluded other gods and all previous modalities of the spirit. Secondly, Judaic monotheism was spread throughout the Roman Empire as Constantine adopted Christianity: there arose the unfortunate combination of an absolutist religion with an absolutist militarism. Thirdly, as early Christianity coalesced from its nebulous sources, it set itself against the Greek philosophical, scientific and *jnani* mindset, forging a deeply *bhakti, via negativa,* anti-esoteric monotheism, and subsequently created various Inquisitions to root out other clusterings of the religious impulse as heresy. Fourthly, the first revival of Greek spirituality in the Renaissance was quashed, building up a reservoir of suppressed *jnani* spiritual energy. Fifthly, that repressed *jnani* energy burst forth in the Enlightenment as a 'revenge of rationalism', triggered by the rise of science. Sixthly, the emerging spiritual revolution of the Enlightenment foundered on the Argument from Design: a possible revitalised *jnani via positiva* religion of the West then drained away through the sands of reductionist materialism.

Secularism then emerged by default.

Conclusions

The argument in this book suggests that the secular mind came into being not because the Enlightenment was a successful secular revolution, but because the Enlightenment was a failed spiritual revolution. Here lies the hidden origins of the secular mind, only comprehensible however when the religiosity of the Enlightenment, and of the West leading up to that time, is understood through the concept of *spiritual difference*. Key to this idea of spiritual difference is that of a non-devotional orientation, termed *jnani* in this book, but which could equally have been called gnosis, or many other terms, and which was unfamiliar to the Christian mainstream and rejected by it.

Once the hidden origins of the secular mind are laid bare, many of its shibboleths in respect of religion are shown to be wrong. But the importance of this is not just academic: the détente between the secular world and faith traditions is now breaking down, or under close scrutiny. This détente came into being after the Enlightenment as a mutual ignorance pact, allowing the secular world and the world of faith traditions to move in parallel tracks, virtually unknown to each other, and failing to mutually enliven and enrich each other. This gulf of comprehension lay mostly unremarked upon until 9/11 and its aftermath. Since that event the secular world has begun to inquire into spirituality and religion with a new urgency, but with little awareness of its own origins. This book should make a contribution to secularism's own perception of itself, and also open up its history to sympathetic understanding from faith traditions.

The post 9/11 era opens up two possibilities: a retrenchment on both sides such that secularism hardens its stance against religion, succumbing even further to its shibboleths, its unexamined prejudices, while faith traditions move towards fundamentalist extremism. America is a crucial indicator of this tendency as its constitutional wall of separation between state and religion comes under fire. The other possibility is a more benign one: that 9/11 and its aftermath forces a new debate leading to mutual understanding. The analysis in this book suggests that the division between secularism and religion is a false one, an accident of history, a default position due to the failure of a new religion to emerge in the Enlightenment period. That religion, given the umbrella term Deism, both rejects the negative legacy of old religion (intolerance and cruelty), and embraces much that is central to the modern world, such as science, democracy, and the non-devotional (*jnani*) spiritual impulse. It offers a lyrical materialism as opposed to a reductionist materialism. We offered an image of this Deist impulse as vividly present in the writings of most Enlightenment thinkers, but ultimately failing to carry popular opinion in its time, and draining away through the sands of an emerging reductionist materialism. To pursue this image: perhaps, as a liquid, it did drain through the sands of materialism, but has since built up a vast spiritual reservoir, a water-table, the level of which is now rising and becoming visible again.

An examination of many apparently secular fields of enquiry, including physics, consciousness studies, transpersonal psychology, the arts, postmodernism, feminism and ecology – not to mention the New Age and new religious movements – suggest that a new spiritual impulse *is* on the rise. Rather than give this collective movement an old and rather discredited name – Deism – a new term has suggested itself in the writings of contemporary thinkers, including that of theologian John D. Caputo: postsecularism. If the hidden origins of the secular mind are truly understood; if the shibboleths of secularism are abandoned while holding on to all that is vital in secularism and hard-won in the Enlightenment (democracy, freedom of expression, and critical thinking applied to religion); then could not a postsecular sensibility emerge, one that reclaims the spiritual impulse which was at the core of the Enlightenment? Does not a proper understanding of the origins of the secular mind, based on the idea of spiritual difference, naturally lead to postsecularism? Would this not provide the essential bridge of understanding between the secular world and faith traditions, so desperately needed in the early twenty-first century?

References

1. Tucker, Mike (Ed.), *Alan Davie – The Quest for the Miraculous*, Brighton: University of Brighton Gallery (Lund Humphries), 1993, p. 10
2. Marx, K. and Engels, F., *Collected Works*, Vol. 5 London: Laurence and Wishart, 1976, (*Theses On Feuerbach*, 6), p. 4
3. *The Guardian*, 23 December 2006
4. Government statistics for 2001 census, available for example via Wikipedia
5. *The Guardian*, 17 July 2004
6. Atkins, Peter, 'Letter Inciting Religious Hatred', *The Times Higher Education Supplement*, 19 October 2001
7. www.secularism.org.uk
8. Vidal, Gore, 'Monotheism and its Discontents', The Lowell Lecture, Harvard University, April 20, 1992, cited in McGrath, Alister, *The Twilight of Atheism – the Rise and Fall of Disbelief in the Modern World*, New York, London, Toronto, Sidney, Auckland: Doubleday, 2004, p.229
9. Toynbee, Polly, 'Last chance to speak out', 2001-10-05, *Guardian Unlimited*
10. Rummel, R. J., *Death by Government*, Somerset, NJ: Transaction Publishers, 1997
11. Dawkins, Richard, 'What Use is Religion?' *Free Inquiry*, Volume 24, Number 5, 2004
12. Lyotard Jean-Francois, *The Postmodern Condition – a Report on Knowledge*, Manchester: Manchester University Press 1984.
13. Jung, C.G., *Memories, Dreams, Reflections*, London: Fontana 1993, p. 276
14. Goleman, Daniel, *Emotional Intelligence – Why it Can Matter More Than IQ*, London: Bloomsbury, 1996
15. Berry, Thomas, *The Dream of the Earth*, San Francisco: Sierra Club Books, 1988, pp. 16-17
16. Beauvoir, Simone de, *The Prime of Life*, London: Andre Deutsch and Weidenfeld and Nicholson, 1962, p. 169
17. Sartre, Jean-Paul, *Nausea*, London: Penguin, 1965, p. 139
18. King, Thomas, M., *Sartre and the Sacred*, Chicago and London: University of Chicago Press, 1974, p. 43
19. Freud, Sigmund, *Civilization and Its Discontents*, New York, London: W.W.Norton and Co., 1961, p. 72

20. Lodge, David, *Thinks. . .*, London: Penguin, 2001, p. 62
21. Lodge, David, *Thinks. . .*, London: Penguin, 2001, p. 61
22. Lodge, David, *Thinks. . .*, London: Penguin, 2001, p. 28
23. Lodge, David, *Thinks. . .*, London: Penguin, 2001, p. 35
24. Huxley, A. *The Perennial Philosophy*, London: Chatto and Windus, 1950, p. 1
25. Katz, Steven *Mysticism and Language*, Oxford University Press 1992, p. 5
26. Rorty, Richard, *Philosophy and the Mirror of Nature*, Princeton, New Jersey: Princeton University Press, 1979, p. 10
27. Wittgenstein, Ludwig, *Philosophical Investigations*, Oxford: Blackwell, 1967, Aphorism 1
28. Abram, David, *The Spell of the Sensuous*, New York: Vintage Books, 1997
29. Damasio, Antonio R., *Descartes' Error – Emotion, Reason and the Human Brain*, London: MacMillan, 1996, p. 165
30. Huxley, A. *The Perennial Philosophy*, Chatto and Windus, London, 1950, ch. 21.
31. Watts, Alan, *The Way of Zen*, New York: Random House, 1989, p. 11
32. Popper, Karl, The Open Society and Its Enemies, Vol. 2: Hegel and Marx, London: Routledge, 1973, p. 271
33. Gandhi, M., *The Collected Works of Mahatma Gandhi*, Vol. XLI, The Publications Division, Ministry of Information and Broadcasting, Government of India, Ahmedabad, 1970, p. 125
34. Blond, Phillip (Ed.), *Postsecular Philosophy – Between Philosophy and Theology*, London and New York: Routledge, 1988
35. Abram, David, *The Spell of the Sensuous*, New York: Vintage Books, 1997, p.102
36. Underhill, E. *Mysticism – The Nature and Development of Spiritual Consciousness*, Oneworld Publications, Oxford, UK, 1993, p. 99
37. Ward, Keith, *God: A Guide for the Perplexed*, Oxford: Oneworld, 2003, p. 159
38. *The Guardian* August 30 2000
39. Reproduced by kind permission of Ivan Ward, Director of Education, the Freud Museum, www.freud.org.uk
40. Austin, James, *Zen and the Brain – Towards an Understanding of Meditation and Consciousness*, Cambridge, Massachusetts, London: The MIT Press, 1999, p. xix
41. Damasio, Antonio R., *Looking for Spinoza: Joy, Sorrow, and the Feeling Brain*, Orlando, Austin, New York, San Diego, Toronto, London: Harcourt, Inc, 2003, p. 53
42. Pert, Candace B., *Molecules of Emotion – The Science Behind Mind-Body Medicine*, New York, London, Toronto, Sydney: Scribner, 2003
43. Austin, James, *Zen and the Brain – Towards an Understanding of Meditation and Consciousness*, Cambridge, Massachusetts, London: The MIT Press, 1999
44. Toynbee, Polly, 'God is a Brain Disorder', Radio Times, 24-30 May 2004
45. 'What the Bleep do We Know' is directed by William Arntz, Betsy Chasse and Mark Vicente (2004). It was largely conceived by students from the 'Ramtha School of Enlightenment', Ramtha allegedly being a 35,000 year old warrior spirit channelled by spiritual teacher JZ Knight.
46. Graves, Robert, *The Greek Myths*, London: Penguin, 1992, p. 476

47. Brody, Hugh, The Other Side of Eden – Hunters, Farmers, and the Shaping of the World, New York: North Point Press, 2000, p. 237
48. Longchenpa, *Kindly bent to Ease Us – Part 1: Mind*, (translated and annotated by Guenther, H.V.) Dharma Publishing, Berkely, USA, 1975
49. Wilber, Ken, *The Atman Project, A Transpersonal View of Human Development*, Wheaton, Madras, London: Theosophical Publishing House, 1980, p. 5
50. Rawlinson, Andrew, The Book of Enlightened Masters, Western Teachers in Eastern Traditions, Chicago and La Salle, Illinois: Open Court, 1997, p. 98
51. Richardson, Peter Tufts, *Four Spiritualities – Expressions of Self, Expressions of Spirit*, Palo Alto, California: Davies-Black Publishing, 1996
52. Richardson, Peter Tufts, *Four Spiritualities – Expressions of Self, Expressions of Spirit*, Palo Alto, California: Davies-Black Publishing, 1996, p. 80
53. Harner, Michael, *The Way of the Shaman*, New York: HarperSanFrancisco, 1990, p. xi
54. Stone, Merlin, *When God was a Woman*, Harcourt, 1976
55. Paper, Jordan, *The Spirits are Drunk – Comparative Approaches to Chinese Religion*, Albany: State University of New York Press, 1995, p. 5
56. Yogananda, Paramahansa, *Autobiography of a Yogi*, London, Sydney, Auckland, Johannesburg: Rider, 1969
57. Hemleben, J. *Rudolf Steiner – A Documentary Biography*, East Grinstead: Henry Goulden, 1975, p. 24
58. See for example: Castaneda, Carlos, *Tales of Power*, Harmondsworth: Penguin, 1976
59. Eliade, Mircea, *A History of Religious Ideas, Vol.2*, Chicago and London: University of Chicago Press, 1982, p. 24
60. Osborne, Arthur, *Ramana Maharshi and the Path of Self-Knowledge*, Maine: Samuel Weiser, 1970, p. 153
61. Happold, F.C. *Mysticism – a Study and an Anthology*, London: Penguin Books, 1970, Chapter 8
62. Smith, Huston, *The Religions of Man*, New York: Harper and Row, 1986, p. 53
63. Blake, William, *The Complete Poetry and Prose*, Edited by David V. Erdman, New York, London, Toronto, Sydney, Auckland: Doubleday, 1988, 'The Marriage of Heaven and Hell: Proverbs of Hell', p. 36
64. Buber, Martin, *I and Thou*, Edinburgh: T&T Clark, 1999
65. Waddell, Norman (Ed.), *Wild Ivy – The Spiritual Autobiography of Zen Master Hakuin*, Boston and London: Shambhala, 2001, p. 66
66. Howlin, Patricia, 'What Wittgenstein and Warhol had in common', *Times Higher Educational Supplement*, 28 March 2003
67. Einstein, Albert, *The World as I See It*, Citadel Press, 1998, p. 91
68. Dawkins, Richard, 'God's just too small', *Times Higher Educational Supplement* Oct 29 2004
69. Wilber, Ken, *Sex, Ecology, Spirituality*, Boston: Shambhala, 1995, p. 123
70. Nanamoli, Bhikku, and Bhodi, Bhikku, *The Middle Length Discourses of the Buddha – a New Translation of the Majjhima Nikaya*, Boston: Wisdom Publications, 1995, 67.4, p. 560
71. Armstrong, Karen, *A History of God*, London: Mandarin, 1994, p. 64
72. Nietzsche, F. *Thus Spoke Zarathustra*, London: Penguin Books, 1969, p. 42

73. Hume, David, *The Natural History of Religion*, London: Penguin Classics, 1990, section VIII
74. Wilber, Ken, *The Marriage of Sense and Soul*, Dublin: Gateway, 1998, p. 110
75. Shlain, Leonard, *The Alphabet Versus the Goddess – The Conflict Between Words and Images*, Penguin / Compass, 1998
76. Kierkegaard, *The Essential Kierkegaard*, Princeton, New Jersey: Princeton University Press, 1990, p. 279
77. Lawson, Neal, 'If they preach the cause of the poor, they're my people', *The Guardian*, January 3 2007
78. Marx, Karl, *Capital – A Critical Analysis of Capitalist Production*, London: William Glashier, 1912, p. 45
79. Paper, Jordan, *The Spirits are Drunk – Comparative Approaches to Chinese Religion*, Albany: State University of New York Press, 1995, p. 3
80. Pascal, Blaise, *Pensées*, Harmondsworth: Penguin, 1980, p. 309
81. Hume, David, *The Natural History of Religion*, London: Penguin Classics, 1990, section XV
82. Ahmad, Shaykh Zulfiqar, *Love for Allah*, Chicago: Faqir Publications, 2001 p. 54
83. M., (Trans. Swami Nikhilnanda), *The Gospel of Sri Ramakrishna*, New York: Ramakrishna-Vivekananda Centre, 1984
84. Rolland, Romain, *The Life of Ramakrishna*, Calcutta: Advaita Ashrama (Publication Department), 1992, p. 1
85. Rolland, Romain, *The Life of Ramakrishna*, Calcutta: Advaita Ashrama (Publication Department), 1992, p. 236
86. Tillich, Paul, *The Courage to Be*, Glasgow: William Collins and Sons, 1980, p.44
87. Kierkegaard, *The Essential Kierkegaard*, Princeton, New Jersey: Princeton University Press, 1990, p. x
88. Magee, Brian, *Confessions of a Philosopher*, London: Weidenfeld and Nicolson, 1997, p. 116
89. Gould, Stephen Jay, *Rocks of Ages*, London: Jonathon Cape, 2001, pp. 27-45
90. www.imshealth.com
91. *The Guardian*, 20 September 2003
92. www.healthyplace.com
93. Elliott, Carl and Tod Chambers (Eds.), *Prozac as a Way of Life (Studies in Social Medicine)*, Chapel Hill and London: University of North Carolina Press, 2004, p. 13
94. Elliott, Carl and Tod Chambers (Eds.), *Prozac as a Way of Life (Studies in Social Medicine)*, Chapel Hill and London: University of North Carolina Press, 2004, p. 54
95. Elliott, Carl and Tod Chambers (Eds.), *Prozac as a Way of Life (Studies in Social Medicine)*, Chapel Hill and London: University of North Carolina Press, 2004, p. 52
96. Elliott, Carl and Tod Chambers (Eds.), *Prozac as a Way of Life (Studies in Social Medicine)*, Chapel Hill and London: University of North Carolina Press, 2004, p. 159
97. Hume, David, *Dialogues Concerning Natural Religion*, 1854 Works, Kemp

Smith edition, Pt. II, quoted in Dennett, Daniel C., *Darwin's Dangerous Idea*, London: Penguin Books, 1995, p. 29

98. Dennett, Daniel C., *Darwin's Dangerous Idea*, London: Penguin Books, 1995, p. 32
99. Hume, David, *The Natural History of Religion*, London: Penguin Classics, 1990, Introduction
100. Hume, David, *Dialogues Concerning Natural Religion*, 1854 Works, Kemp Smith edition, *Dialogues*, p. 216
101. Hume, David, *Dialogues Concerning Natural Religion*, 1854 Works, Kemp Smith edition, *Dialogues*, p. 141
102. Crick, Francis, *The Astonishing Hypothesis – The Scientific Search for the Soul*, Simon and Schuster, 1994, p. 3
103. Crick, Francis, *What Mad Pursuit: A Personal View of Science*, New York: Basic Books, 1988
104. Reported in the London *Daily Telegraph*, March 24, 2003
105. Crick, Francis, *The Astonishing Hypothesis – The Scientific Search for the Soul*, Simon and Schuster, 1994, p. 258
106. *Observer*, 25 July 2004
107. Dawkins, R. *The Selfish Gene*, Oxford University Press, Oxford, New York, 1992, p. 198
108. Dawkins, R. *The Selfish Gene*, Oxford University Press, Oxford, New York, 1992, p. 330
109. Dawkins, Richard, *The Blind Watchmaker*, Penguin 1991, p. 4
110. Dawkins, Richard, *The Blind Watchmaker*, Penguin 1991, p. 6
111. Dawkins, Richard, 'Special report: terrorism in the US', *Guardian*, September 15, 2001
112. Dawkins, Richard, 'When Religion Steps on Science's Turf: The Alleged Separation Between the Two Is Not So Tidy', *Free Inquiry* magazine, Volume 18, Number 2, Council for Secular Humanism
113. Kramnick, Isaac, (Ed.) *The Portable Enlightenment Reader*, London: Penguin, 1995, p. 138
114. Blake, William, The Complete Poetry and Prose, Edited by David V. Erdman, New York, London, Toronto, Sydney, Auckland: Doubleday, 1988, p. 2
115. www.gc.cuny.edu/faculty/research_briefs/aris/aris_index.htm
116. Martin, David, *On Secularization – Towards a Revised General Theory*, Aldershot: Ashgate, 2005
117. Martin, David, *On Secularization – Towards a Revised General Theory*, Aldershot: Ashgate, 2005, p. 173
118. Heelas, Paul, *The New Age*, Oxford, Malden: Blackwell Publishers, 1996
119. Heelas, Paul and Woodhead, Linda, *The Spiritual Revolution – why religion is giving way to spirituality*, Blackwell Publishing, 2005
120. Armstrong, Karen, *A History of God*, London: Mandarin, 1994, p. 38.
121. Pontifical Council for Culture, *Jesus Christ The Bearer Of The Water Of Life – A Christian reflection on the 'New Age'*, Vatican: Holy See, codice 7424-x, 2003
122. Feuerstein, Georg, *Holy Madness*, London: Arkana, 1990
123. Storr, Anthony, *Feet of Clay – A Study of Gurus*, London HarperCollins

Publishers, 1997
124. Caplan, Mariana, *Halfway Up the Mountain – the Error of Premature Claims of Enlightenment*, Prescott, Arizona: Hohm Press, 1999
125. Cohen, Andrew, *Autobiography of an Awakening*, Corte Madera, CA: Moksha Foundation, 1992
126. Rawlinson, Andrew, *The Book of Enlightened Masters, Western Teachers in Eastern Traditions*, Chicago and La Salle, Illinois: Open Court, 1997, p. 614
127. Tarlo, Luna, *The Mother of God*, New York: Plover Press, 1997
128. Cohen, Andrew, *Autobiography of an Awakening*, Corte Madera, CA: Moksha Foundation, 1992, p. 84
129. Heelas, Paul, *The New Age*, Oxford, Malden, Blackwell Publishers, 1996, p. 137
130. Hemleben, J. *Rudolf Steiner – A Documentary Biography*, East Grinstead: Henry Goulden, 1975
131. Lissau, R. *Rudolf Steiner: Life, work, inner path and social initiatives*, Hawthorn Press, Stroud, 1987
132. Steiner, R. *The Course of My Life*, New York: Anthroposophic Press, 1986
133. Steiner, R. *The Course of My Life*, New York: Anthroposophic Press, 1986, p. 124
134. Jung, C.G., *Memories, Dreams, Reflections*, London: Fontana 1993, p. 306
135. Rawlinson, Andrew, *The Book of Enlightened Masters, Western Teachers in Eastern Traditions*, Chicago and La Salle, Illinois: Open Court, 1997, p. 614
136. Chandmal, Asit, *One Thousand Moons – Krishnamurti at Eighty-Five*, New York: Abrams 1985, p. 9
137. Krishanmurti, J. *The Only Revolution*, New York and Evanston: Harper and Row, 1970, p. 114
138. Heelas, Paul, *The New Age*, Oxford, Malden, Blackwell Publishers, 1996, p. 137
139. Dawkins, Richard, *Unweaving the Rainbow*, London: Penguin 1998, p. 191
140. Gell-Mann, Murray, *The Quark and the Jaguar*, London: Abacus, 2002
141. White, Michael, *Isaac Newton, the Last Sorcerer*, London: Fourth Estate, 1998, p. 188
142. White, Michael, *Isaac Newton, the Last Sorcerer*, London: Fourth Estate, 1998, p. 182
143. Kramnick, Isaac, (Ed.) *The Portable Enlightenment Reader*, London: Penguin, 1995, p. 127
144. Plato, *The Republic*, London: Penguin Books, 1987, (V.531), p. 279
145. Plato, *Phaedrus and Letters VII and VIII*, Trans.: Walter Hamilton, Harmondsworth: Penguin, 1973, p. 26
146. White, Michael, *Leonardo: The First Scientist*, London: St. Martin's Press, 2000
147. Tarnas, Richard, *The Passion of the Western Mind*, London: Pimlico (Random House) 1996, p. 48
148. Augustine (Saint), *Confessions*, (trans. Pine-Coffin, R.S.) Penguin Books, Middlesex, England 1986, p. 96
149. Brooke, John Hedley, *Science and Religion: Some Historical Perspectives*, Cambridge University Press, 1991, p. 157
150. Brooke, John Hedley, *Science and Religion: Some Historical Perspectives*,

Cambridge University Press, 1991, p.14
151. Damasio, Antonio R., *Descartes' Error – Emotion, Reason and the Human Brain*, London: MacMillan, 1996, p. 249
152. Russell, Paul 'Atheism and the Title-Page of Hume's *Treatise*', *Hume Studies* Volume XIV, Number 2 (November, 1988) pp. 408-423.
153. Bacon, Francis, *The Essays of Sir Francis Bacon*, George Macy, 1944, chapter 16
154. Chappell, Vere, (Ed.), *Hobbes and Bramhall on Liberty and Necessity*, Cambridge: Cambridge University Press, 1999
155. Ball, B.W., *Mortality, Resurrection and Immortality*, Cambridge: James Clarke, 2007
156. Hobbes, Thomas, *Leviathan*, London: Everyman, 1994, Ch. XII
157. Hume, David, *The Natural History of Religion*, London: Penguin Classics, 1990, section VI
158. Hume, David, *A Treatise Of Human Nature*, Mineola, NY: Dover Publications, 2004, section V
159. Smith, T.V. and Grene, M. (Eds.) *Berkely, Hume and Kant*, Chicago: University of Chicago Press, 1957, p. 129
160. Hume, David, *A Treatise Of Human Nature*, Mineola, NY: Dover Publications, 2004, Sect. XIV.
161. White, Michael, *Isaac Newton – The Last Sorcerer*, London: Fourth Estate, 1998, p. 233
162. Clarke, J. J. , *Oriental Enlightenment: The Encounter Between Asian and Western Thought*, London and New York: Routledge, 1997
163. Kramnick, Isaac, (Ed.) *The Portable Enlightenment Reader*, London: Penguin, 1995, p. 208
164. Dennett, Daniel C., *Consciousness Explained*, Allen Lane, The Penguin Press, 1991, p. 33
165. Dennett, Daniel C., *Consciousness Explained*, Allen Lane, The Penguin Press, 1991, p. 37
166. Kramnick, Isaac, (Ed.) *The Portable Enlightenment Reader*, London: Penguin, 1995, p. 208
167. Buckley, Michael J., *At the Origins of Modern Atheism*, New Haven and London: Yale University Press, 1987, p. 222
168. Holbach, Paul-Henri Dietrich, (transl. H.D. Robinson) *The System of Nature*, Kitchener: Batoche Books, 2001
169. Holbach, Paul-Henri Dietrich, (transl. H.D. Robinson) *The System of Nature*, Kitchener: Batoche Books, 2001, p.29
170. Gell-Mann, Murray, *The Quark and the Jaguar*, London: Abacus, 2002, p. 129
171. Marx, K. and Engels, F., Collected Works, Vol. 3, London: Laurence and Wishart, 1976Marx, Karl, p. 175
172. Marx, Karl, *Capital – A Critical Analysis of Capitalist Production*, London: William Glashier, 1912, p. 45
173. Marx, K. and Engels, F., *On Religion*, Moscow: Progress Publishers, 1972, p. 212
174. Marx, K. and Engels, F., *On Religion*, Moscow: Progress Publishers, 1972, p. 13

175. Marx, K. and Engels, F., *On Religion*, Moscow: Progress Publishers, 1972, p.10
176. Gould, Stephen Jay, *Rocks of Ages*, London: Jonathon Cape, 2001, p. 35
177. Nietzsche, F. *Thus Spoke Zarathustra*, Penguin Books, London, 1969
178. Marx, K. and Engels, F., *On Religion*, Moscow: Progress Publishers, 1972, p. 126
179. Peters, Edward, *Inquisition*, Berkeley and Los Angeles: University of California Press, 1989
180. Blotzer, Joseph, 'Inquisition', *Catholic Encyclopedia*, Robert Appleton Company, 1910
181. Hilgers, Joseph, 'Censorship of Books', *Catholic Encyclopedia*, Robert Appleton Company, 1908
182. Figure cited by the American Association of Librarians
183. Thomas, Keith, *Religion and the Decline of Magic – Studies in Popular Beliefs in Sixteenth- and Seventeenth-century England*, London: Penguin Books 1971, p. 210
184. Starhawk, *The Spiral Dance – A Rebirth of the Ancient Religion of the Great Goddess*, HarperSanFrancisco, 1999, p. 30
185. *Witches of Eastwick*, Dir.: George Miller, 1987
186. Starhawk, *The Spiral Dance – A Rebirth of the Ancient Religion of the Great Goddess*, HarperSanFrancisco, 1999, p. 32
187. Chadwick, Henry, *The Early Church*, Middlesex: Penguin, 1967, p. 29
188. Hume, David, *The Natural History of Religion*, London: Penguin Classics, 1990, section IX
189. Chadwick, Henry, *The Early Church*, Middlesex: Penguin, 1967, p. 157
190. Chadwick, Henry, *The Early Church*, Middlesex: Penguin, 1967, p. 100
191. Rudolf, Kurt, *Gnosis – The Nature and History of Gnosticism*, New York: HarperSanFrancisco, 1980, p. 15
192. Seneviratna, Anuradha, *King Asoka and Buddhism – Historical and Literary Studies*, Kandy, Sri Lanka: Buddhist Publication Society, 1994, p. 12
193. Aiken, Charles F., 'Buddhism', *Catholic Encyclopedia*, Robert Appleton Company, 1910
194. Gandhi, M. *The Collected Works of Mahatma Gandhi*, Vol. XLI, The Publications Division, Ministry of Information and Broadcasting, Government of India, Ahmedabad, 1970, p. 93
195. Jacoby, Susan, *Freethinkers – A History of American Secularism*, New York: Metropolitan Books, 2004, p. 44
196. Armstrong, Karen, *A History of God*, London: Mandarin, 1994, p. 226
197. Scholem, Gershom, *Major Trends in Jewish Mysticism*, New York: Schocken Books, 1995, p. 55
198. Scholem, Gershom, *Major Trends in Jewish Mysticism*, New York: Schocken Books, 1995, p. 94
199. Buber, Martin, *I and Thou*, Edinburgh: T&T Clark, 1999, p. 15
200. Buber, Martin, *I and Thou*, Edinburgh: T&T Clark, 1999, p. 120
201. Scholem, Gershom, *Major Trends in Jewish Mysticism*, New York: Schocken Books, 1995, p. 7
202. Scholem, Gershom, *Major Trends in Jewish Mysticism*, New York: Schocken Books, 1995, p. 8

203. Sheldrake, R., *The Rebirth of Nature – New Science and the Revival of Animism*, Rider London, Sidney, Aukland, Johannesburg 1993, p. 16
204. Freke, Timothy, and Gandy, Peter, *The Jesus Mysteries – Was the 'Original Jesus' a Pagan God?*, New York: Three Rivers Press, 1999
205. Kramnick, Isaac, (Ed.) *The Portable Enlightenment Reader*, London: Penguin, 1995, p. 108
206. Hutchins, Robert M., (Ed. in Chief) *Locke, Berkeley, Hume*, Chicago, London, Toronto: Encyclopedia Britannica, 1952, p. 14
207. Peters, Edward, *Inquisition*, Berkeley and Los Angeles: University of California Press, 1989, p. 310
208. Paper, Jordan, *The Spirits are Drunk – Comparative Approaches to Chinese Religion*, Albany: State University of New York Press, 1995, p. 3
209. Cupitt, Don, *Mysticism After Modernity*, Oxford: Blackwell Publishers, 1998, p. 45
210. Cupitt, Don, *The Sea of Faith – Christianity in Change*, London: British Broadcasting Corporation, 1984, p. 178
211. Hobbes, Thomas, Leviathan, London: Everyman, 1994, chapter XXXI
212. Rolle, Richard, *The Fire of Love*, London: Penguin 1972, p. 14
213. Rolle, Richard, *The Fire of Love*, London: Penguin 1972, p. 121
214. Julian of Norwich, *Revelations of Divine Love*, Harmondsworth: Penguin, 1986, p. 67
215. Thorold, Algar, (Ed.), *The Dialogue of the Seraphic Virgin Catherine of Siena*, London: Kegan Paul, Trench, Trubner & Co., 1907, p. 9
216. Thorold, Algar, (Ed.), *The Dialogue of the Seraphic Virgin Catherine of Siena*, London: Kegan Paul, Trench, Trubner & Co., 1907, p. 14
217. St John of the Cross, *The Dark Night of the Soul*, London: Hodder and Staughton, 1988, p. 2
218. Sartre, Jean-Paul, *Saint Genet*, London: Heinemann, 1988, p. 11
219. Sartre, Jean-Paul, *Saint Genet*, London: Heinemann, 1988, p. 195
220. Sartre, Jean-Paul, *Saint Genet*, London: Heinemann, 1988, p. 218
221. Russell, Bertrand, *A History of Western Philosophy*, London, Sidney, Wellington: Unwin Paperbacks, 1989, p. 351.
222. Augustine (Saint), *Confessions*, (trans. Pine-Coffin, R.S.), Middlesex, England: Penguin Books 1986, XI.22
223. Augustine (Saint), *Confessions*, (trans. Pine-Coffin, R.S.), Middlesex, England: Penguin Books 1986, V:3 (p. 94)
224. Augustine (Saint), *Confessions*, (trans. Pine-Coffin, R.S.), Middlesex, England: Penguin Books 1986, VII:9 (p. 146)
225. Augustine (Saint), *Confessions*, (trans. Pine-Coffin, R.S.), Middlesex, England: Penguin Books 1986, VII:20 (p. 154)
226. Augustine (Saint), *Confessions*, (trans. Pine-Coffin, R.S.), Middlesex, England: Penguin Books 1986, VII:20 VII: 20 (p. 154)
227. Russell, Bertrand, *A History of Western Philosophy*, London, Sidney, Wellington: Unwin Paperbacks, 1989, p. 444
228. Fox, Matthew, *Sheer Joy – Conversations with Thomas Aquinas on Creation Spirituality*, HarperSanFrancisco, 1992, p. 5
229. Fox, Matthew, *Sheer Joy – Conversations with Thomas Aquinas on Creation Spirituality*, HarperSanFrancisco, 1992, p. 32

230. Nanamoli, Bhikku, and Bhodi, Bhikku, *The Middle Length Discourses of the Buddha – a New Translation of the Majjhima Nikaya*, Boston: Wisdom Publications, 1995, 87, p. 718
231. Fox, Matthew, *Sheer Joy – Conversations with Thomas Aquinas on Creation Spirituality*, HarperSanFrancisco, 1992, p. 4
232. Fox, Matthew, *Sheer Joy – Conversations with Thomas Aquinas on Creation Spirituality*, HarperSanFrancisco, 1992, p. 31
233. Augustine (Saint), *Confessions*, (trans. Pine-Coffin, R.S.) Penguin Books, Middlesex, England 1986, p. 96
234. Fortune, Dion, *The Mystical Qabalah*, London and Tonbridge: Ernest Benn Limited, 1976, p. 5
235. Armstrong, Karen, *A History of God*, London: Mandarin, 1994, p. 102
236. Eliade, Mircea, *A History of Religious Ideas, Vol.2*, Chicago and London: University of Chicago Press, 1982, p. 350
237. Smith, Huston, *The Religions of Man*, New York: Harper and Row, 1986, p. 53
238. Cupitt, Don, *The Sea of Faith – Christianity in Change*, London: British Broadcasting Corporation, 1984, p. 178
239. Augustine (Saint), *Confessions*, (trans. Pine-Coffin, R.S.) Penguin Books, Middlesex, England 1986, book VII, pp. 155-6
240. Feuerstein, Georg, *The Yoga Sutras – A New Translation and Commentary*, Rochester, Vermont: Inner Traditions International, 1989, 1:23-26
241. Bailey, Alice *The Light of The Soul*, London: Lucis Press 1950
242. Nanamoli, Bhikku, and Bhodi, Bhikku, *The Middle Length Discourses of the Buddha – a New Translation of the Majjhima Nikaya*, Boston: Wisdom Publications, 1995, 70.21, p. 582
243. Nanamoli, Bhikku, and Bhodi, Bhikku, *The Middle Length Discourses of the Buddha – a New Translation of the Majjhima Nikaya*, Boston: Wisdom Publications, 1995, 35.27, p. 330
244. Nanamoli, Bhikku, and Bhodi, Bhikku, *The Middle Length Discourses of the Buddha – a New Translation of the Majjhima Nikaya*, Boston: Wisdom Publications, 1995, 72.15, p. 592
245. Buber, Martin, *I and Thou*, Edinburgh: T&T Clark, 1999, p. 120
246. Wilber, Ken, *Eye of Spirit – An Integral Vision for a World Gone Slightly Mad*, Boston and London: Shambhala, 1998, p. 279
247. Prabhavananda, Swami, and Christopher Isherwood (Eds.) *Shankara's Crest-Jewel of Discrimination*, Hollywood, California: Vedanta Press, 1975, p. 33
248. Osborne, Arthur, *Ramana Maharshi and the Path of Self-Knowledge*, Maine: Samuel Weiser, 1970, p. 153
249. Palmer, Martin and Elizabeth Breuilly (Eds.) *The Book of the Chuang Tzu*, London: Penguin Arkana, 1996, pp. 60-64
250. Carpenter, Edward, *Days with Walt Whitman*, London: George Allen, 1906, p. 49
251. Hourihan, Paul, *Mysticism in American Literature – Thoreau's Quest and Whitman's Self*, Redding, CA: Vedantic Shores Press, 2004
252. Jung, C.G., *Memories, Dreams, Reflections*, London: Fontana 1993, p. 306
253. Steiner. R., *The Course of My Life*, New York: Anthroposophic Press,

1986, p. 124
254. Nanamoli, Bhikku, and Bhodi, Bhikku, *The Middle Length Discourses of the Buddha – a New Translation of the Majjhima Nikaya*, Boston: Wisdom Publications, 1995, 72, p. 591
255. Murti, T.R.V., *The Central Philosophy of Buddhism – A Study of Madhyamika System*, New Delhi: Munshiram Manoharlal, 1998
256. Guthrie, Kenneth Sylvan, *The Pythagorean Sourcebook and Library*, Grand Rapids: Phanes Press, 1987
257. Guthrie, Kenneth Sylvan, *The Pythagorean Sourcebook and Library*, Grand Rapids: Phanes Press, 1987, p. 30
258. Guthrie, Kenneth Sylvan, *The Pythagorean Sourcebook and Library*, Grand Rapids: Phanes Press, 1987, p. 71
259. Nanamoli, Bhikku, and Bhodi, Bhikku, *The Middle Length Discourses of the Buddha* – a New Translation of the Majjhima Nikaya, Boston: Wisdom Publications, 1995, 72, p. 676 and p. 696
260. All Heraclitus extracts from: *The Fragments of Heraclitus*, Bray: Guild Press, 1976
261. Kramnick, Isaac, (Ed.) *The Portable Enlightenment Reader*, London: Penguin, 1995, p. 165
262. Plato, *The Last Days of Socrates*, (Trans.: Hugh Tredennick), Harmondsworth: Penguin, 1969, p. 136
263. O'Brien, Elmer (Trans.), *The Essential Plotinus*, Indiana: Hackett Publishing Company, 1964
264. Plotinus, *The Enneads*, (Trans. Stephen MacKenna), London: Penguin Books, 1991
265. Turner, William, 'John Scotus Eriugena', *Catholic Encyclopedia*, Robert Appleton Company, 1910
266. Caputo, John D., *The Mystical Element in Heidegger's Thought*, New York: Fordham University Press, 1986, p. 14
267. Caputo, John D., *The Mystical Element in Heidegger's Thought*, New York: Fordham University Press, 1986, p. 10
268. Cupitt, Don, *Mysticism After Modernity*, Oxford: Blackwell Publishers, 1998, p. 95
269. Eckhart *Selected Treatises and Sermons*, The Fontana Library, Collins, 1963, p. 156
270. Eckhart *Selected Treatises and Sermons*, The Fontana Library, Collins, 1963, p. 164
271. White, Michael, *Leonardo: The First Scientist*, London: St. Martin's Press, 2000
272. *Observer*, 6 April 1997
273. Russell, Bertrand, *A History of Western Philosophy*, London, Sidney, Wellington: Unwin Paperbacks, 1989, p. 483
274. Russell, Bertrand, *A History of Western Philosophy*, London, Sidney, Wellington: Unwin Paperbacks, 1989, p. 489
275. Tarnas, Richard, *Cosmos and Psyche : Intimations of a New World View*, New York, N.Y. ; London : Viking, 2006
276. Pagels, Elaine, *The Gnostic Gospels*, New York: Vintage Books, 1979, (18), p. 27

277. Ross, J.B. and McLaughlin, M.M. (Eds. and translators) *The Portable Renaissance Reader*, New York: Penguin, 1977
278. Mill, John Stuart, *On Liberty and Other Essays*, Oxford: Oxford University Press, 1998, p. 12
279. *National Catholic Reporter*, Vol. 35, No. 17, Feb. 27, 1999
280. *Letter to the Bishops of the Catholic Church on Some Aspects of Christian Meditation*, Vatican: Sacred Congregation for the Doctrine of the Faith, Oct. 15, 1989
281. Kramnick, Isaac, (Ed.) *The Portable Enlightenment Reader*, London: Penguin, 1995Kramnick, p.xii
282. Kramnick, Isaac, (Ed.) *The Portable Enlightenment Reader*, London: Penguin, 1995, p.x
283. Russell, Bertrand, *A History of Western Philosophy*, London, Sidney, Wellington: Unwin Paperbacks, 1989, p. 566
284. Russell, Bertrand, *A History of Western Philosophy*, London, Sidney, Wellington: Unwin Paperbacks, 1989, p. 566
285. Descartes, Rene, *Discourse on Method Etc*, London: Dent and Sons, 1949, p. 66
286. Descartes, Rene, *Discourse on Method Etc*, London: Dent and Sons, 1949, p. 89
287. Descartes, Rene, *Discourse on Method Etc*, London: Dent and Sons, 1949, p. 167
288. Descartes, Rene, *The Passions of the Soul*, Indianapolis, Cambridge: Hackett Publishing Co., 1989, p. 37
289. Descartes, Rene, *Discourse on Method Etc*, London: Dent and Sons, 1949, p. 95
290. Descartes, Rene, *Discourse on Method Etc*, London: Dent and Sons, 1949, p. 76
291. Descartes, Rene, *Discourse on Method Etc*, London: Dent and Sons, 1949, p. 104
292. Gandhi, M. *The Collected Works of Mahatma Gandhi*, Vol. XLI, The Publications Division, Ministry of Information and Broadcasting, Government of India, Ahmedabad, 1970, p. 125
293. Dennet, Daniel C., *Consciousness Explained*, Allen Lane, The Penguin Press, 1991, p. 34
294. Holbach, Paul-Henri Dietrich, (transl. H.D. Robinson) *The System of Nature*, Kitchener: Batoche Books, 2001, p. 51
295. Losee, John, *A Historical Introduction to the Philosophy of Science*, Oxford, New York: Oxford University Press, 1993, p. 78
296. Damasio, Antonio R., *Descartes' Error – Emotion, Reason and the Human Brain*, London: MacMillan, 1996, p. 250
297. Damasio, Antonio R., *Descartes' Error – Emotion, Reason and the Human Brain*, London: MacMillan, 1996, p. 252
298. Chalmers, David, 'Facing up to the Problem of Consciousness', Journal of Consciousness Studies, Vol. 2, No. 3, 1995, p. 201
299. Spinoza, Baruch, *Ethics – TEI – Selected Letters*, Indianapolis and Cambridge: Hackett Publishing Company, 1992, p. 10
300. Spinoza, Baruch, *Ethics – TEI – Selected Letters*, Indianapolis and

Cambridge: Hackett Publishing Company, 1992, (TEI 7), p. 234
301. Spinoza, Baruch, *Ethics – TEI – Selected Letters,* Indianapolis and Cambridge: Hackett Publishing Company, 1992, (TEI 9), p. 235
302. Nanamoli, Bhikku, and Bhodi, Bhikku, *The Middle Length Discourses of the Buddha* – a New Translation of the Majjhima Nikaya, Boston: Wisdom Publications, 1995, (36) p. 340
303. Spinoza, Baruch, *Ethics – TEI – Selected Letters,* Indianapolis and Cambridge: Hackett Publishing Company, 1992, (TEI 11), p. 235
304. Spinoza, Baruch, *Ethics – TEI – Selected Letters,* Indianapolis and Cambridge: Hackett Publishing Company, 1992, (TEI 13), p. 235
305. Hume, David, *A Treatise Of Human Nature,* Mineola, NY: Dover Publications, 2004, Sect. V
306. Damasio, Antonio R., *Looking for Spinoza: Joy, Sorrow, and the Feeling Brain,* Orlando, Austin, New York, San Diego, Toronto, London: Harcourt, Inc, 2003, p. 14.
307. Damasio, Antonio R., *Looking for Spinoza: Joy, Sorrow, and the Feeling Brain,* Orlando, Austin, New York, San Diego, Toronto, London: Harcourt, Inc, 2003, p. 19
308. Coudert, Alison, 'Leibniz, Mysticism and Religion', Eds. Alison Coudert, Richard H. Popkin, Gordon M. Weiner, *Archives Internationales D'Histoire des Idees,* p. 158
309. Quincey, Christian de, *Radical Nature – Rediscovering the Soul of Matter,* Montpelier, Vermont: Invisible Cities Press, 2002, p. 107
310. Quincey, Christian de, *Radical Nature – Rediscovering the Soul of Matter,* Montpelier, Vermont: Invisible Cities Press, 2002, p. 128
311. Latta, Robert, (Trans. and Ed.) *Leibniz: The Monadology and other Philosophical Writings,* London: Oxford University Press, 1968. The quotations here are numbered following the paragraphs of the *Monadology* as they are usually reproduced. Translation, Robert Latta.
312. Searle, John, 'Minds, Brains, and Programs', *Behavioral and Brain Sciences,* 3, 417-424, 1980
313. Latta, Robert, (Trans. and Ed.) *Leibniz: The Monadology and other Philosophical Writings,* London: Oxford University Press, 1968, (37) p. 238
314. Russell, Bertrand, *A History of Western Philosophy,* London, Sidney, Wellington: Unwin Paperbacks, 1989, p. 563
315. Gregory, John, *The Neoplatonists – A Reader,* London and New York: Routledge, 1999, p. 180
316. Wilber, Ken, *The Marriage of Sense and Soul,* Dublin: Gateway, 1998, ch. 7
317. Wilber, Ken, *The Marriage of Sense and Soul,* Dublin: Gateway, 1998, p. 102
318. Kramnick, Isaac, (Ed.) *The Portable Enlightenment Reader,* London: Penguin, 1995, p. 26
319. Wilber, Ken, *The Marriage of Sense and Soul,* Dublin: Gateway, 1998, p. 113
320. Kant, Immanuel, *Religion Within the Boundaries of Mere Reason,* Cambridge, New York: Cambridge University Press, 1998, p. 45
321. Wilber, Ken, *The Marriage of Sense and Soul,* Dublin: Gateway, 1998, p. 106
322. Spencer, Lloyd, and Krauze, Andrzej, *Introducing Hegel,* Cambridge: Icon Books, 1997, p. 66
323. Hegel, G.W.F., *Phenomenology of Spirit,* Oxford University Press, 1977,

Preface
324. Hegel, G.W.F., *Phenomenology of Spirit*, Oxford University Press, 1977, (Preface 5), p. 3
325. Hegel, G.W.F., *Phenomenology of Spirit*, Oxford University Press, 1977, (Preface 8), p. 5
326. Hegel, G.W.F., *Phenomenology of Spirit*, Oxford University Press, 1977, (Preface 8), p. 5
327. Hegel, G.W.F., *Phenomenology of Spirit*, Oxford University Press, 1977, (Preface 9), p. 5
328. Hegel, G.W.F., *Phenomenology of Spirit*, Oxford University Press, 1977, (Preface 10), p. 6
329. Wilber, Ken, *The Marriage of Sense and Soul*, Dublin: Gateway, 1998, p. 112
330. Latta, Robert, (Trans. and Ed.) *Leibniz: The Monadology and other Philosophical Writings*, London: Oxford University Press, 1968, p. 215
331. Wilber, Ken, *The Marriage of Sense and Soul*, Dublin: Gateway, 1998, p. 112
332. Wilber, Ken, *The Marriage of Sense and Soul*, Dublin: Gateway, 1998, p. 112
333. Engels, *Ludwig Feuerbach and the End of Classical German Philosophy*, part 3, Marx Engels Internet Archive 1994 (www.marxists.org)
334. Marx, K. and Engels, F., *Collected Works*, Vol. 5 London: Laurence and Wishart, 1976, (Theses On Feuerbach, 7), p. 5
335. Feuerbach, Ludwig, *The Essence of Christianity*, New York: Prometheus Books, 1989, p. 8
336. Feuerbach, Ludwig, *The Essence of Christianity*, New York: Prometheus Books, 1989, p. 9
337. Feuerbach, Ludwig, *The Essence of Christianity*, New York: Prometheus Books, 1989, p. 2
338. Kramnick, Isaac, (Ed.) *The Portable Enlightenment Reader*, London: Penguin, 1995, p. 311
339. Bentham, Jeremy, *The Influence of Natural Religion on the Temporal Happiness of Mankind*, New York: Prometheus Books, 2003, p. 30
340. Bentham, Jeremy, *The Influence of Natural Religion on the Temporal Happiness of Mankind*, New York: Prometheus Books, 2003, p. 45
341. Bentham, Jeremy, *The Influence of Natural Religion on the Temporal Happiness of Mankind*, New York: Prometheus Books, 2003, p. 55
342. Vidal, Gore, 'Monotheism and its Discontents', The Lowell Lecture, Harvard University, April 20, 1992, cited in McGrath, Alister, *The Twilight of Atheism – the Rise and Fall of Disbelief in the Modern World*, New York, London, Toronto, Sidney, Auckland: Doubleday, 2004, p.229
343. Kramnick, Isaac, (Ed.) The Portable Enlightenment Reader, London: Penguin, 1995, p. 103
344. Kramnick, Isaac, (Ed.) The Portable Enlightenment Reader, London: Penguin, 1995, p. 166
345. Kramnick, Isaac, (Ed.) The Portable Enlightenment Reader, London: Penguin, 1995, p. 131
346. Kramnick, Isaac, (Ed.) The Portable Enlightenment Reader, London: Penguin, 1995, p. 57
347. Kramnick, Isaac, (Ed.) The Portable Enlightenment Reader, London: Penguin, 1995, p. 176

348. Paine, Thomas, *The Age of Reason*, Mineola, NY: Dover Publications, 2004, p. 1
349. Kramnick, Isaac, (Ed.) The Portable Enlightenment Reader, London: Penguin, 1995, p. 178
350. Hegel, G.W.F., *Phenomenology of Spirit*, Oxford University Press, 1977, p. 476
351. Buckley, Michael J., *At the Origins of Modern Atheism*, New Haven and London: Yale University Press, 1987, p. 224
352.. Dawkins, Richard, *The Blind Watchmaker*, London: Penguin 1991, p. 4

Bibliography

Abram, David, *The Spell of the Sensuous*, New York: Vintage Books, 1997

Ahmad, Shaykh Zulfiqar, *Love for Allah*, Chicago: Faqir Publications, 2001

Aiken, Charles F., 'Buddhism', *Catholic Encyclopedia*, Robert Appleton Company, 1910

Armstrong, Karen, *A History of God*, London: Mandarin, 1994

Augustine (Saint), *Confessions*, (trans. Pine-Coffin, R.S.) Penguin Books, Middlesex, England 1986

Austin, James, *Zen and the Brain – Towards an Understanding of Meditation and Consciousness*, Cambridge, Massachusetts, London: The MIT Press, 1999

Bacon, Francis, *The Essays of Sir Francis Bacon*, George Macy, 1944

Bailey, Alice *The Light of The Soul*, London: Lucis Press 1950

Ball, B.W., *Mortality, Resurrection and Immortality*, Cambridge: James Clarke, 2007

Beauvoir, Simone de, *The Prime of Life*, London: Andre Deutsch and Weidenfeld and Nicholson, 1962

Bentham, Jeremy, *The Influence of Natural Religion on the Temporal Happiness of Mankind*, New York: Prometheus Books, 2003

Berry, Thomas, *The Dream of the Earth*, San Francisco: Sierra Club Books, 1988

Blake, William, *The Complete Poetry and Prose*, Edited by David V. Erdman, New York, London, Toronto, Sydney, Auckland: Doubleday, 1988

Blond, Phillip (Ed.), *Postsecular Philosophy – Between Philosophy and Theology*, London and New York: Routledge, 1988

Blotzer, Joseph, 'Inquisition', *Catholic Encyclopedia*, Robert Appleton Company, 1910

Brody, Hugh, *The Other Side of Eden – Hunters, Farmers, and the Shaping of the World*, New York: North Point Press, 2000

Brooke, John Hedley, *Science and Religion: Some Historical Perspectives*, Cambridge University Press, 1991

Buber, Martin, *I and Thou*, Edinburgh: T&T Clark, 1999

Buckley, Michael J., *At the Origins of Modern Atheism*, New Haven and London: Yale University Press, 1987

Caplan, Mariana, *Halfway Up the Mountain – the Error of Premature Claims of Enlightenment,* Prescott, Arizona: Hohm Press, 1999

Caputo, John D., *The Mystical Element in Heidegger's Thought,* New York: Fordham University Press, 1986

Carpenter, Edward, *Days with Walt Whitman,* London: George Allen, 1906

Castaneda, Carlos, *Tales of Power,* Harmondsworth: Penguin, 1976

Chadwick, Henry, *The Early Church,* Middlesex: Penguin, 1967

Chandmal, Asit, *One Thousand Moons – Krishnamurti at Eighty-Five,* New York: Abrams 1985

Chappell, Vere, (Ed.), *Hobbes and Bramhall on Liberty and Necessity,* Cambridge: Cambridge University Press, 1999

Clarke, J. J. , *Oriental Enlightenment: The Encounter Between Asian and Western Thought,* London and New York: Routledge, 1997

Cohen, Andrew, *Autobiography of an Awakening,* Corte Madera, CA: Moksha Foundation, 1992

Coudert, Alison, 'Leibniz, Mysticism and Religion', Eds. Alison Coudert, Richard H. Popkin, Gordon M. Weiner, *Archives Internationales D'Histoire des Idees*

Crick, Francis, *The Astonishing Hypothesis – The Scientific Search for the Soul,* Simon and Schuster, 1994

Crick, Francis, *What Mad Pursuit: A Personal View of Science,* New York: Basic Books, 1988

Cupitt, Don, *Mysticism After Modernity,* Oxford: Blackwell Publishers, 1998

Cupitt, Don, *The Sea of Faith – Christianity in Change,* London: British Broadcasting Corporation, 1984

Damasio, Antonio R., *Descartes' Error – Emotion, Reason and the Human Brain,* London: MacMillan, 1996

Damasio, Antonio R., *Looking for Spinoza: Joy, Sorrow, and the Feeling Brain,* Orlando, Austin, New York, San Diego, Toronto, London: Harcourt, Inc, 2003

Dawkins, R. *The Selfish Gene,* Oxford University Press, Oxford, New York, 1992

Dawkins, Richard, *The Blind Watchmaker,* London: Penguin, 1991

Dawkins, Richard, *Unweaving the Rainbow,* London: Penguin, 1998

Dennett, Daniel C., *Consciousness Explained,* Allen Lane, The Penguin Press, 1991

Dennett, Daniel C., *Darwin's Dangerous Idea,* London: Penguin Books, 1995

Descartes, Rene, *Discourse on Method Etc,* London: Dent and Sons, 1949

Descartes, Rene, *The Passions of the Soul,* Indianapolis, Cambridge: Hackett Publishing Co., 1989

Eckhart *Selected Treatises and Sermons,* The Fontana Library, Collins, 1963

Einstein, Albert, *The World as I See It,* Citadel Press, 1998

Eliade, Mircea, *A History of Religious Ideas, Vol.2,* Chicago and London: University of Chicago Press, 1982

Elliott, Carl and Tod Chambers (Eds.), *Prozac as a Way of Life (Studies in Social*

Medicine), Chapel Hill and London: University of North Carolina Press, 2004
Feuerbach, Ludwig, *The Essence of Christianity*, New York: Prometheus Books
Feuerstein, Georg, *Holy Madness*, London: Arkana, 1990
Feuerstein, Georg, *The Yoga Sutras – A New Translation and Commentary*, Rochester, Vermont: Inner Traditions International, 1989
Fortune, Dion, *The Mystical Qabalah*, London and Tonbridge: Ernest Benn Limited, 1976
Fox, Matthew, *Sheer Joy – Conversations with Thomas Aquinas on Creation Spirituality*, HarperSanFrancisco, 1992
Freke, Timothy, and Gandy, Peter, *The Jesus Mysteries – Was the 'Original Jesus' a Pagan God?*, New York: Three Rivers Press, 1999
Freud, Sigmund, *Civilization and Its Discontents*, New York, London: W.W.Norton and Co., 1961
Gandhi, M. *The Collected Works of Mahatma Gandhi*, Vol. XLI, The Publications Division, Ministry of Information and Broadcasting, Government of India, Ahmedabad, 1970
Gell-Mann, Murray, *The Quark and the Jaguar*, London: Abacus, 2002
Goleman, Daniel, *Emotional Intelligence – Why it Can Matter More Than IQ*, London: Bloomsbury, 1996
Gould, Stephen Jay, *Rocks of Ages*, London: Jonathon Cape, 2001
Graves, Robert, *The Greek Myths*, London: Penguin, 1992
Gregory, John, *The Neoplatonists – A Reader*, London and New York: Routledge, 1999
Guthrie, Kenneth Sylvan, *The Pythagorean Sourcebook and Library*, Grand Rapids: Phanes Press, 1987
Happold, F.C. *Mysticism – a Study and an Anthology*, London: Penguin Books, 1970
Harner, Michael, *The Way of the Shaman*, New York: HarperSanFrancisco, 1990
Heelas, Paul and Woodhead, Linda, *The Spiritual Revolution – why religion is giving way to spirituality*, Blackwell Publishing, 2005
Heelas, Paul, *The New Age*, Oxford, Malden, Blackwell Publishers, 1996
Hegel, G.W.F., *Phenomenology of Spirit*, Oxford University Press, 1977
Hemleben, J. *Rudolf Steiner – A Documentary Biography*, East Grinstead: Henry Goulden, 1975
Heraclitus, *The Fragments of Heraclitus*, Bray: Guild Press, 1976
Hilgers, Joseph, 'Censorship of Books', *Catholic Encyclopedia*, Robert Appleton Company, 1908
Hobbes, Thomas, *Leviathan*, London: Everyman, 1994
Holbach, Paul-Henri Dietrich, (transl. H.D. Robinson) *The System of Nature*, Kitchener: Batoche Books, 2001
Hourihan, Paul, *Mysticism in American Literature – Thoreau's Quest and Whitman's Self*, Redding, CA: Vedantic Shores Press, 2004
Hume, David, *A Treatise Of Human Nature*, Mineola, NY: Dover Publications, 2004

Hume, David, *Dialogues Concerning Natural Religion*, 1854 Works, Kemp Smith edition

Hume, David, *The Natural History of Religion*, London: Penguin Classics, 1990

Huxley, A. *The Perennial Philosophy*, London: Chatto and Windus, 1950

Jacoby, Susan, *Freethinkers – A History of American Secularism*, New York: Metropolitan Books, 2004

John of the Cross, St, *The Dark Night of the Soul*, London: Hodder and Stoughton, 1988

Julian of Norwich, *Revelations of Divine Love*, Harmondsworth: Penguin, 1986

Jung, C.G., *Memories, Dreams, Reflections*, London: Fontana 1993

Kant, Immanuel, *Religion Within the Boundaries of Mere Reason*, Cambridge, New York: Cambridge University Press, 1998

Katz, Steven *Mysticism and Language*, Oxford University Press 1992

Kierkegaard, *The Essential Kierkegaard*, Princeton, New Jersey: Princeton University Press, 1990

King, Thomas, M., *Sartre and the Sacred*, Chicago and London: University of Chicago Press, 1974

Kramnick, Isaac, (Ed.) *The Portable Enlightenment Reader*, London: Penguin, 1995

Krishanmurti, J. *The Only Revolution*, New York and Evanston: Harper and Row, 1970

Latta, Robert, (Trans. and Ed.) *Leibniz: The Monadology and other Philosophical Writings*, London: Oxford University Press, 1968

Lissau, R. *Rudolf Steiner: Life, work, inner path and social initiatives*, Hawthorn Press, Stroud, 1987

Lodge, David, *Thinks. . .* , London: Penguin, 2001

Longchenpa, *Kindly bent to Ease Us – Part 1: Mind*, (translated and annotated by Guenther, H.V.) Dharma Publishing, Berkely, USA, 1975

Losee, John, *A Historical Introduction to the Philosophy of Science*, Oxford, New York: Oxford University Press, 1993

Lyotard Jean-Francois, *The Postmodern Condition – a Report on Knowledge*, Manchester: Manchester University Press, 1984.

M., (Trans. Swami Nikhilnanda), *The Gospel of Sri Ramakrishna*, New York: Ramakrishna-Vivekananda Centre, 1984

Magee, Brian, *Confessions of a Philosopher*, London: Weidenfeld and Nicolson, 1997

Martin, David, *On Secularization – Towards a Revised General Theory*, Aldershot: Ashgate, 2005

Marx, K. and Engels, F., *Collected Works*, London: Laurence and Wishart, 1976

Marx, K. and Engels, F., *On Religion*, Moscow: Progress Publishers, 1972

Marx, Karl, *Capital – A Critical Analysis of Capitalist Production*, London: William Glashier, 1912

McGrath, Alister, *The Twilight of Atheism – the Rise and Fall of Disbelief in the Modern World*, New York, London, Toronto, Sidney, Auckland: Doubleday, 2004

Mill, John Stuart, *On Liberty and Other Essays*, Oxford: Oxford University Press, 1998

Murti, T.R.V., *The Central Philosophy of Buddhism – A Study of Madhyamika System*, New Delhi: Munshiram Manoharlal, 1998

Nanamoli, Bhikku, and Bhodi, Bhikku, *The Middle Length Discourses of the Buddha – a New Translation of the Majjhima Nikaya*, Boston: Wisdom Publications, 1995

Nietzsche, F. *Thus Spoke Zarathustra*, London: Penguin Books, 1969

O'Brien, Elmer (Trans.), *The Essential Plotinus*, Indiana: Hackett Publishing Company, 1964

Osborne, Arthur, *Ramana Maharshi and the Path of Self-Knowledge*, Maine: Samuel Weiser, 1970

Pagels, Elaine, *The Gnostic Gospels*, New York: Vintage Books, 1979

Paine, Thomas, *The Age of Reason*, Mineola, NY: Dover Publications, 2004

Palmer, Martin and Elizabeth Breuilly (Eds.) *The Book of the Chuang Tzu*, London: Penguin Arkana, 1996

Paper, Jordan, *The Spirits are Drunk – Comparative Approaches to Chinese Religion*, Albany: State University of New York Press, 1995

Pascal, Blaise, *Pensées*, Harmondsworth: Penguin, 1980

Pert, Candace B., *Molecules of Emotion – The Science Behind Mind-Body Medicine*, New York, London, Toronto, Sydney: Scribner, 2003

Peters, Edward, *Inquisition*, Berkeley and Los Angeles: University of California Press, 1989

Plato, *Phaedrus and Letters VII and VIII*, Trans.: Walter Hamilton, Harmondsworth: Penguin, 1973

Plato, *The Last Days of Socrates*, (Trans.: Hugh Tredennick), Harmondsworth: Penguin, 1969

Plato, *The Republic*, London: Penguin Books, 1987

Plotinus, *The Enneads*, (Trans. Stephen MacKenna), London: Penguin Books, 1991

Popper, Karl, *The Open Society and Its Enemies*, Vol. 2: Hegel and Marx, London: Routledge, 1973

Prabhavananda, Swami, and Christopher Isherwood (Eds.) *Shankara's Crest-Jewel of Discrimination*, Hollywood, California: Vedanta Press, 1975

Quincey, Christian de, *Radical Nature – Rediscovering the Soul of Matter*, Montpelier, Vermont: Invisible Cities Press, 2002

Rawlinson, Andrew, The Book of Enlightened Masters, Western Teachers in Eastern Traditions, Chicago and La Salle, Illinois: Open Court, 1997

Richardson, Peter Tufts, *Four Spiritualities – Expressions of Self, Expressions of Spirit*, Palo Alto, California: Davies-Black Publishing, 1996

Rolland, Romain, *The Life of Ramakrishna*, Calcutta: Advaita Ashrama (Publication Department), 1992

Rolle, Richard, *The Fire of Love*, London: Penguin 1972

Rorty, Richard, *Philosophy and the Mirror of Nature*, Princeton, New Jersey: Princeton University Press, 1979

Ross, J.B. and McLaughlin, M.M. (Eds. and translators) *The Portable Renaissance Reader*, New York: Penguin, 1977

Rudolf, Kurt, *Gnosis – The Nature and History of Gnosticism*, New York: HarperSanFrancisco, 1980

Rummel, R. J., *Death by Government*, Somerset, NJ: Transaction Publishers, 1997

Russell, Bertrand, *A History of Western Philosophy*, London, Sidney, Wellington: Unwin Paperbacks, 1989

Sartre, Jean-Paul, *Nausea*, London: Penguin, 1965

Sartre, Jean-Paul, *Saint Genet*, London: Heinemann, 1988

Scholem, Gershom, *Major Trends in Jewish Mysticism*, New York: Schocken Books, 1995

Seneviratna, Anuradha, *King Asoka and Buddhism – Historical and Literary Studies*, Kandy, Sri Lanka: Buddhist Publication Society, 1994

Sheldrake, R., *The Rebirth of Nature – New Science and the Revival of Animism*, Rider London, Sidney, Aukland, Johannesburg, 1993

Shlain, Leonard, *The Alphabet Versus the Goddess – The Conflict Between Words and Images*, Penguin / Compass, 1998

Smith, Huston, *The Religions of Man*, New York: Harper and Row, 1986

Smith, T.V. and Grene, M. (Eds.) *Berkely, Hume and Kant*, Chicago: University of Chicago Press, 1957

Spencer, Lloyd, and Krauze, Andrzej, *Introducing Hegel*, Cambridge: Icon Books, 1997

Spinoza, Baruch, *Ethics – TEI – Selected Letters*, Indianapolis and Cambridge: Hackett Publishing Company, 1992

Starhawk, *The Spiral Dance – A Rebirth of the Ancient Religion of the Great Goddess*, HarperSanFrancisco, 1999

Steiner, R. *The Course of My Life*, New York: Anthroposophic Press, 1986

Stone, Merlin, *When God was a Woman*, Harcourt, 1976

Storr, Anthony, *Feet of Clay – A Study of Gurus*, London HarperCollins Publishers, 1997

Tarlo, Luna, *The Mother of God*, New York: Plover Press, 1997

Tarnas, Richard, *Cosmos and Psyche : Intimations of a New World View*, New York, N.Y. ; London : Viking, 2006

Tarnas, Richard, *The Passion of the Western Mind*, London: Pimlico (Random House) 1996

Thomas, Keith, *Religion and the Decline of Magic – Studies in Popular Beliefs in Sixteenth- and Seventeenth-century England*, London: Penguin Books, 1971

Thorold, Algar, (Ed.), *The Dialogue of the Seraphic Virgin Catherine of Siena*, London: Kegan Paul, Trench, Trubner & Co., 1907

Tillich, Paul, *The Courage to Be*, Glasgow: William Collins and Sons, 1980

Tucker, Mike (Ed.), *Alan Davie – The Quest for the Miraculous*, Brighton: University of Brighton Gallery (Lund Humphries), 1993

Turner, William, 'John Scotus Eriugena', *Catholic Encyclopedia*, Robert Appleton Company, 1910

Underhill, E. *Mysticism – The Nature and Development of Spiritual Consciousness*, Oneworld Publications, Oxford, UK, 1993
Waddell, Norman (Ed.), *Wild Ivy – The Spiritual Autobiography of Zen Master Hakuin*, Boston and London: Shambhala, 2001
Ward, Keith, *God: A Guide for the Perplexed*, Oxford: Oneworld, 2003
Watts, Alan, *The Way of Zen*, New York: Random House, 1989
White, Michael, *Isaac Newton – The Last Sorcerer*, London: Fourth Estate, 1998
White, Michael, *Leonardo: The First Scientist*, London: St. Martin's Press, 2000
Wilber, Ken, *Eye of Spirit – An Integral Vision for a World Gone Slightly Mad*, Boston and London: Shambhala, 1998
Wilber, Ken, *Sex, Ecology, Spirituality*, Boston: Shambhala, 1995
Wilber, Ken, *The Atman Project, A Transpersonal View of Human Development*, Wheaton, Madras, London: Theosophical Publishing House, 1980
Wilber, Ken, *The Marriage of Sense and Soul*, Dublin: Gateway, 1998
Wittgenstein, Ludwig, *Philosophical Investigations*, Oxford: Blackwell, 1967
Yogananda, Paramahansa, *Autobiography of a Yogi*, London, Sydney, Auckland, Johannesburg: Rider, 1969

Index

Printed in the United Kingdom
by Lightning Source UK Ltd.
123936UK00001B/79-288/A

9 780227 172452